# An Introduction to Criminal Law

# An Introduction to Criminal Law

**Philip E. Carlan, PhD**
*Associate Professor, School of Criminal Justice*
*The University of Southern Mississippi*
*Hattiesburg, Mississippi*

**Lisa S. Nored, JD, PhD**
*Associate Professor and Director, School of Criminal Justice*
*The University of Southern Mississippi*
*Hattiesburg, Mississippi*

**Ragan A. Downey, MS, PhD Candidate**
*(The University of Southern Mississippi)*
*Grant Evaluator, Pine Belt Mental Healthcare Resources*
*Hattiesburg, Mississippi*

**JONES AND BARTLETT PUBLISHERS**
*Sudbury, Massachusetts*
BOSTON    TORONTO    LONDON    SINGAPORE

*World Headquarters*
Jones and Bartlett Publishers
40 Tall Pine Drive
Sudbury, MA 01776
978-443-5000
info@jbpub.com
www.jbpub.com

Jones and Bartlett Publishers Canada
6339 Ormindale Way
Mississauga, Ontario L5V 1J2
Canada

Jones and Bartlett Publishers International
Barb House, Barb Mews
London W6 7PA
United Kingdom

Jones and Bartlett's books and products are available through most bookstores and online booksellers. To contact Jones and Bartlett Publishers directly, call 800-832-0034, fax 978-443-8000, or visit our website, www.jbpub.com.

Substantial discounts on bulk quantities of Jones and Bartlett's publications are available to corporations, professional associations, and other qualified organizations. For details and specific discount information, contact the special sales department at Jones and Bartlett via the above contact information or send an email to specialsales@jbpub.com.

Copyright © 2011 by Jones and Bartlett Publishers, LLC

All rights reserved. No part of the material protected by this copyright may be reproduced or utilized in any form, electronic or mechanical, including photocopying, recording, or by any information storage and retrieval system, without written permission from the copyright owner.

This publication is designed to provide accurate and authoritative information in regard to the Subject Matter covered. It is sold with the understanding that the publisher is not engaged in rendering legal, accounting, or other professional service. If legal advice or other expert assistance is required, the service of a competent professional person should be sought.

**Production Credits**
Publisher, Higher Education: Cathleen Sether
Acquisitions Editor: Sean Connelly
Associate Editor: Megan R. Turner
Production Director: Amy Rose
Associate Production Editor: Julia Waugaman
Associate Marketing Manager: Jessica Cormier
Manufacturing and Inventory Control Supervisor: Amy Bacus
Composition: Shepherd Incorporated
Cover and Title Page Design: Scott Moden
Cover Image: © scoutingstock/ShutterStock, Inc.
Printing and Binding: Courier Westford
Cover Printing: Courier Westford

**Library of Congress Cataloging-in-Publication Data**
Carlan, Philip E.
 An introduction to criminal law / by Philip E. Carlan, Lisa S. Nored, and Ragan A. Downey.
   p. cm.
 Includes bibliographical references and index.
 ISBN 978-0-7637-5525-6 (casebound)
 1. Criminal law—United States. I. Nored, Lisa S. II. Downey, Ragan A. III. Title.
 KF9219.C37 2011
 345.73—dc22
                              2009042373
6048

Printed in the United States of America
14 13 12 11 10   10 9 8 7 6 5 4 3 2 1

# BRIEF CONTENTS

| | | |
|---|---|---|
| Foreword | | xvii |
| Preface | | xix |
| Acknowledgments | | xxi |
| **CHAPTER 1** | Substantive Criminal Law: Principles and Working Vocabulary | 1 |
| **CHAPTER 2** | Crime and Punishment: Constitutional Limitations and Protections | 17 |
| **CHAPTER 3** | Theft Offenses and Fraudulent Practices | 35 |
| **CHAPTER 4** | Crimes Against Habitation, Robbery, and Assault | 51 |
| **CHAPTER 5** | Criminal Homicide | 71 |
| **CHAPTER 6** | Sex Offenses | 91 |
| **CHAPTER 7** | Crimes Against Moral Values | 107 |
| **CHAPTER 8** | Crimes Against the Administration of Justice and Public Order | 121 |
| **CHAPTER 9** | Inchoate Offenses and Party Liability | 133 |
| **CHAPTER 10** | Defenses to Criminal Responsibility | 145 |
| **CHAPTER 11** | Organized Crime and Terrorism | 167 |
| **CHAPTER 12** | White Collar Crime | 181 |
| **APPENDIX A** | Practice Test Solutions | 195 |
| **APPENDIX B** | Declaration of Independence | 197 |
| **APPENDIX C** | Constitution for the United States of America | 199 |
| **APPENDIX D** | Bill of Rights | 205 |
| Glossary | | 207 |
| Index | | 217 |

# CONTENTS

Foreword .................................................. xvii
Preface ................................................... xix
Acknowledgments ........................................... xxi

**CHAPTER 1** Substantive Criminal Law: Principles and Working Vocabulary ........... 1

**Key Terms** ................................................ 1
**Introduction** ............................................. 1
**The Republic for Which It Stands** ......................... 2
**Social Construction of Law** ............................... 2
**Origins of Law** ........................................... 2
    **Ancient** .......................................... 3
    **Natural** .......................................... 3
    **Common** ........................................... 3
**Primary Sources of Criminal Law** .......................... 4
    **Common** ........................................... 4
    **Statutory** ........................................ 4
    **Case** ............................................. 6
    **Constitutional** ................................... 6
    **Administrative** ................................... 6
**Types of Legal Wrongs** .................................... 6
    **Private** .......................................... 6
    **Public** ........................................... 8
**Crime Defined** ............................................ 8
    **Commissions and Omissions** ........................ 8
    **Without Legal Defense** ............................ 9
    **Codified** ......................................... 9
**Crime Classifications** .................................... 9
    **Felonies, Misdemeanors, and Violations** ........... 9
    *Mala in Se* and *Mala Prohibita* ................... 9
**Crime and Deviance Distinguished** ......................... 10

**Essential Elements of Crime and Liability** .................................................. 10
    Phase I of *Actus Reus: Corpus Delicti* .................................................. 11
    Phase II of *Actus Reus: Proximate Cause* .................................................. 11
    Role of *Mens Rea* .................................................. 12
    Degrees of *Mens Rea* .................................................. 13
    Attendant Circumstances .................................................. 13

**Crime in America** .................................................. 14

**Summary** .................................................. 14
    Practice Test .................................................. 15
    References .................................................. 16

**CHAPTER 2** Crime and Punishment: Constitutional Limitations and Protections .................................................. 17

**Key Terms** .................................................. 17

**Introduction** .................................................. 17

**United States Constitution and Criminal Law** .................................................. 17

**Constitutional Principles and Limitations** .................................................. 18
    Fourteenth Amendment .................................................. 19
    Fourth Amendment .................................................. 19
    Fifth Amendment .................................................. 20
    Sixth Amendment .................................................. 23
    Eighth Amendment .................................................. 25

**Criminal Punishment** .................................................. 26
    Retribution .................................................. 26
    Deterrence .................................................. 26
    Rehabilitation or Reformation .................................................. 27
    Incapacitation .................................................. 27
    Restorative Justice .................................................. 27

**Types of Criminal Punishment** .................................................. 27
    Fines .................................................. 27
    Forfeiture .................................................. 28
    Incarceration .................................................. 28
    Alternatives to Incarceration .................................................. 30

**Summary** .................................................. 31
    Practice Test .................................................. 31
    References .................................................. 33

**CHAPTER 3** Theft Offenses and Fraudulent Practices .................................................. 35

**Key Terms** .................................................. 35

**Introduction** .................................................. 35

**Theft** .................................................. 35
    Theft in General .................................................. 36
    Differentiating Custody, Possession, and Ownership .................................................. 36

### Larceny at Common Law ... 37
#### Taking and Carrying Away ... 37
#### Personal Property of Another ... 38
#### Intent to Permanently Deprive ... 38
#### Lost, Mislaid, and Abandoned Property ... 39

### Other Custodial Theft Statutes ... 40
#### Receiving Stolen Property ... 40
#### Unauthorized Use ... 41
#### Theft of Services ... 41

### Embezzlement ... 42
#### Conversion or Misappropriation ... 42
#### Entrusted with Property of Another ... 42
#### Intent to Deprive Possession ... 43

### False Pretenses ... 43

### Modern Consolidation of Theft Statutes ... 44

### Federal Theft Law ... 44

### Forgery and Uttering in General ... 45
#### Forgery Defined ... 45
#### Uttering ... 46
#### Modern Forgery and Uttering Examined ... 47

### Summary ... 48
#### Practice Test ... 48
#### References ... 49

## CHAPTER 4  Crimes Against Habitation, Robbery, and Assault ... 51

### Key Terms ... 51

### Introduction ... 51

### Arson ... 51
#### Arson at Common Law ... 52
#### Modern Arson Examined ... 53
#### Federal Arson Law ... 54

### Burglary ... 54
#### Burglary at Common Law ... 54
#### Modern Burglary Examined ... 56
#### Burglary and Criminal Trespass Distinguished ... 57
#### Possession of Burglar Tools ... 58

### Robbery ... 58
#### Robbery at Common Law ... 58
#### Felonious Taking Through Use or Threat of Force ... 59
#### Money or Property of Another from Person or Presence ... 60

### Extortion ... 61
#### Threat to Cause Harm with Intent to Cause Fear ... 61
#### Future Harm to Gain Compliance from Objects of Threat ... 62

| | Burglary, Robbery, and Extortion at Federal Law | 62 |
|---|---|---|
| | **Assault** | **63** |
| | Assault and Battery at Common Law | 63 |
| | Modern Assault Examined | 65 |
| | Aggravated and Simple Assault Distinguished | 65 |
| | Modern Assault-Related Crimes | 66 |
| | Kidnapping | 67 |
| | **Summary** | **68** |
| | Practice Test | 68 |
| | References | 69 |

## CHAPTER 5   Criminal Homicide   71

**Key Terms**   71

**Introduction**   71

**Murder in America**   71

**Homicide Defined**   71
- Human Being   72
- Living Victim   73
- Legally Dead   74

**Noncriminal Homicide Defined**   74
- Justifiable Homicide   75
- Excusable Homicide   75

**Criminal Homicide Defined**   75
- *Corpus Delicti*   76
- Proximate Cause   76
- Year-and-a-Day Rule   76

**Murder in General**   77
- Malice Aforethought   77
- Deadly Weapon Doctrine   78

**Murder Defined**   78
- Intent-to-Kill Murder   79
- Intent-to-Cause-Serious-Bodily-Injury Murder   80
- Doctrine of Transferred Intent   80
- Felony Murder   80
- Depraved-Heart Murder   82

**Manslaughter in General**   82

**Voluntary Manslaughter Defined**   82
- Adequate Provocation   84
- Heat of Passion   84
- No Cooling Period   84
- Causal Connection   84
- Imperfect Self-Defense   85

**Involuntary Manslaughter Defined**   85
- Unlawful Act Manslaughter   85
- Culpable Negligence Manslaughter   86

**Homicide and Genocide in Federal Law**   86

| | **Summary** | 87 |
|---|---|---|
| | Practice Test | 88 |
| | References | 89 |

**CHAPTER 6**    Sex Offenses . . . . . . . . . . . . . . . . . . . . . . . . . . . . . . . . . . . . . . . . . . . . . . . . . . . . . . . . . . . . . . . . . . . . . . . . . . . . 91

**Key Terms** . . . . . . . . . . . . . . . . . . . . . . . . . . . . . . . . . . . . . . . . . . . . . . . . . . . . . . . . . . . . . . . . . . . . . . . . . . . . . . . . . . . . . . 91

**Introduction** . . . . . . . . . . . . . . . . . . . . . . . . . . . . . . . . . . . . . . . . . . . . . . . . . . . . . . . . . . . . . . . . . . . . . . . . . . . . . . . . . . . . 91

**Forcible Rape in America** . . . . . . . . . . . . . . . . . . . . . . . . . . . . . . . . . . . . . . . . . . . . . . . . . . . . . . . . . . . . . . . . . . . . . . . 91

**Forcible Rape Defined** . . . . . . . . . . . . . . . . . . . . . . . . . . . . . . . . . . . . . . . . . . . . . . . . . . . . . . . . . . . . . . . . . . . . . . . . . . 91
    **Carnal Knowledge and Gender of Participants** . . . . . . . . . . . . . . . . . . . . . . . . . . . . . . . . . . . . . . . . . . . . . . . . . 92
    **Force** . . . . . . . . . . . . . . . . . . . . . . . . . . . . . . . . . . . . . . . . . . . . . . . . . . . . . . . . . . . . . . . . . . . . . . . . . . . . . . . . . . . . . . 92
    **Consent** . . . . . . . . . . . . . . . . . . . . . . . . . . . . . . . . . . . . . . . . . . . . . . . . . . . . . . . . . . . . . . . . . . . . . . . . . . . . . . . . . . . 93
    **Marital Rape** . . . . . . . . . . . . . . . . . . . . . . . . . . . . . . . . . . . . . . . . . . . . . . . . . . . . . . . . . . . . . . . . . . . . . . . . . . . . . . 93
    **Penalties** . . . . . . . . . . . . . . . . . . . . . . . . . . . . . . . . . . . . . . . . . . . . . . . . . . . . . . . . . . . . . . . . . . . . . . . . . . . . . . . . . . 94

**Statutory Rape Defined** . . . . . . . . . . . . . . . . . . . . . . . . . . . . . . . . . . . . . . . . . . . . . . . . . . . . . . . . . . . . . . . . . . . . . . . . . 94
    **Carnal Knowledge and Gender of Participants** . . . . . . . . . . . . . . . . . . . . . . . . . . . . . . . . . . . . . . . . . . . . . . . . . 95
    **Age** . . . . . . . . . . . . . . . . . . . . . . . . . . . . . . . . . . . . . . . . . . . . . . . . . . . . . . . . . . . . . . . . . . . . . . . . . . . . . . . . . . . . . . 96
    **Chaste Character** . . . . . . . . . . . . . . . . . . . . . . . . . . . . . . . . . . . . . . . . . . . . . . . . . . . . . . . . . . . . . . . . . . . . . . . . . . 96
    **Consent** . . . . . . . . . . . . . . . . . . . . . . . . . . . . . . . . . . . . . . . . . . . . . . . . . . . . . . . . . . . . . . . . . . . . . . . . . . . . . . . . . . . 96
    **Intent** . . . . . . . . . . . . . . . . . . . . . . . . . . . . . . . . . . . . . . . . . . . . . . . . . . . . . . . . . . . . . . . . . . . . . . . . . . . . . . . . . . . . 96
    **Penalties** . . . . . . . . . . . . . . . . . . . . . . . . . . . . . . . . . . . . . . . . . . . . . . . . . . . . . . . . . . . . . . . . . . . . . . . . . . . . . . . . . . 96

**Sexual Battery Defined** . . . . . . . . . . . . . . . . . . . . . . . . . . . . . . . . . . . . . . . . . . . . . . . . . . . . . . . . . . . . . . . . . . . . . . . . . 97
    **Penetration** . . . . . . . . . . . . . . . . . . . . . . . . . . . . . . . . . . . . . . . . . . . . . . . . . . . . . . . . . . . . . . . . . . . . . . . . . . . . . . . 97
    **Of Another** . . . . . . . . . . . . . . . . . . . . . . . . . . . . . . . . . . . . . . . . . . . . . . . . . . . . . . . . . . . . . . . . . . . . . . . . . . . . . . . 97
    **Consent** . . . . . . . . . . . . . . . . . . . . . . . . . . . . . . . . . . . . . . . . . . . . . . . . . . . . . . . . . . . . . . . . . . . . . . . . . . . . . . . . . . . 98
    **Penalties** . . . . . . . . . . . . . . . . . . . . . . . . . . . . . . . . . . . . . . . . . . . . . . . . . . . . . . . . . . . . . . . . . . . . . . . . . . . . . . . . . . 98

**Sodomy Defined** . . . . . . . . . . . . . . . . . . . . . . . . . . . . . . . . . . . . . . . . . . . . . . . . . . . . . . . . . . . . . . . . . . . . . . . . . . . . . . . 98
    **Force and Consent** . . . . . . . . . . . . . . . . . . . . . . . . . . . . . . . . . . . . . . . . . . . . . . . . . . . . . . . . . . . . . . . . . . . . . . . . 98

**Child Molestation Defined** . . . . . . . . . . . . . . . . . . . . . . . . . . . . . . . . . . . . . . . . . . . . . . . . . . . . . . . . . . . . . . . . . . . . . 99

**Other Sexual Offenses** . . . . . . . . . . . . . . . . . . . . . . . . . . . . . . . . . . . . . . . . . . . . . . . . . . . . . . . . . . . . . . . . . . . . . . . . . 101
    **Child Pornography and Exploitation** . . . . . . . . . . . . . . . . . . . . . . . . . . . . . . . . . . . . . . . . . . . . . . . . . . . . . . . . . 101
    **Incest** . . . . . . . . . . . . . . . . . . . . . . . . . . . . . . . . . . . . . . . . . . . . . . . . . . . . . . . . . . . . . . . . . . . . . . . . . . . . . . . . . . . . 101
    **Seduction** . . . . . . . . . . . . . . . . . . . . . . . . . . . . . . . . . . . . . . . . . . . . . . . . . . . . . . . . . . . . . . . . . . . . . . . . . . . . . . . . 102
    **Voyeurism** . . . . . . . . . . . . . . . . . . . . . . . . . . . . . . . . . . . . . . . . . . . . . . . . . . . . . . . . . . . . . . . . . . . . . . . . . . . . . . . 102

**Sexual Offenders and Megan's Law** . . . . . . . . . . . . . . . . . . . . . . . . . . . . . . . . . . . . . . . . . . . . . . . . . . . . . . . . . . . . . 103

**Federal Law** . . . . . . . . . . . . . . . . . . . . . . . . . . . . . . . . . . . . . . . . . . . . . . . . . . . . . . . . . . . . . . . . . . . . . . . . . . . . . . . . . . 103

**Summary** . . . . . . . . . . . . . . . . . . . . . . . . . . . . . . . . . . . . . . . . . . . . . . . . . . . . . . . . . . . . . . . . . . . . . . . . . . . . . . . . . . . . . . 104
    **Practice Test** . . . . . . . . . . . . . . . . . . . . . . . . . . . . . . . . . . . . . . . . . . . . . . . . . . . . . . . . . . . . . . . . . . . . . . . . . . . . . 104
    **References** . . . . . . . . . . . . . . . . . . . . . . . . . . . . . . . . . . . . . . . . . . . . . . . . . . . . . . . . . . . . . . . . . . . . . . . . . . . . . . . 105

**CHAPTER 7**    Crimes Against Moral Values . . . . . . . . . . . . . . . . . . . . . . . . . . . . . . . . . . . . . . . . . . . . . . . . . . . . . . . . . . 107

**Key Terms** . . . . . . . . . . . . . . . . . . . . . . . . . . . . . . . . . . . . . . . . . . . . . . . . . . . . . . . . . . . . . . . . . . . . . . . . . . . . . . . . . . . . . 107

**Introduction** . . . . . . . . . . . . . . . . . . . . . . . . . . . . . . . . . . . . . . . . . . . . . . . . . . . . . . . . . . . . . . . . . . . . . . . . . . . . . . . . . . 107

**Morality Legislation in America** . . . . . . . . . . . . . . . . . . . . . . . . . . . . . . . . . . . . . . . . . . . . . . . . . . . . . . . . . . . . . . . . 107

**Sex Offenses** .................................................................... 107
    **Bigamy and Polygamy** ......................................................... 108
    **Prostitution** ................................................................. 108
    **Fornication and Adultery** ..................................................... 109
    **Sodomy and Homosexuality** .................................................... 109

**Indecent Exposure** ............................................................... 110

**Obscenity** ....................................................................... 110

**Pornography** ..................................................................... 111

**Gambling** ........................................................................ 112

**Alcohol and Drugs** ............................................................... 112
    **Drug Offenses** ............................................................... 113
    **Alcohol Offenses** ............................................................ 115

**Hate Crimes** ..................................................................... 116

**Summary** ......................................................................... 117
    **Practice Test** ............................................................... 117
    **References** .................................................................. 119

**CHAPTER 8** Crimes Against the Administration of Justice and Public Order ............ 121

**Key Terms** ....................................................................... 121

**Introduction** .................................................................... 121

**Crimes Affecting the Integrity of the Judicial Process** ........................... 121
    **Obstruction of Justice** ...................................................... 121
    **Resisting Arrest** ............................................................ 122
    **Perjury and Subornation of Perjury** .......................................... 122
    **Embracery and Witness Tampering** ............................................. 123
    **Contempt of Court** ........................................................... 124
    **Misprision of Felony and Compounding Crime** .................................. 125
    **Escape** ...................................................................... 125

**Corruption of the Judicial Process by Public Officials** ........................... 126
    **Bribery** ..................................................................... 126
    **Extortion and Blackmail** ..................................................... 126
    **Ethical Violations** .......................................................... 127

**Crimes Against Public Order and Safety** .......................................... 127
    **Unlawful Assembly, Rout, and Riot** ........................................... 127
    **Fighting** .................................................................... 128
    **Disturbing the Peace and Disorderly Conduct** ................................. 128
    **Nuisance** .................................................................... 128
    **Trespass** .................................................................... 128
    **Vandalism and Malicious Mischief** ............................................ 129
    **Vagrancy and Loitering** ...................................................... 129
    **Traffic Offenses** ............................................................ 130

**Summary** ......................................................................... 130
    **Practice Test** ............................................................... 131
    **References** .................................................................. 132

**CHAPTER 9** Inchoate Offenses and Party Liability . . . . . . . . . . . . . . . . . . . . . . . . . . . . . . . . . . . . . . . . . . 133

**Key Terms** . . . . . . . . . . . . . . . . . . . . . . . . . . . . . . . . . . . . . . . . . . . . . . . . . . . . . . . . . . . . . . . . . . . . . . . 133

**Introduction** . . . . . . . . . . . . . . . . . . . . . . . . . . . . . . . . . . . . . . . . . . . . . . . . . . . . . . . . . . . . . . . . . . . . . 133

**Inchoate Offenses at Common Law** . . . . . . . . . . . . . . . . . . . . . . . . . . . . . . . . . . . . . . . . . . . . . . . . . 133

**Attempt** . . . . . . . . . . . . . . . . . . . . . . . . . . . . . . . . . . . . . . . . . . . . . . . . . . . . . . . . . . . . . . . . . . . . . . . . 133
    **What Constitutes an Act?** . . . . . . . . . . . . . . . . . . . . . . . . . . . . . . . . . . . . . . . . . . . . . . . . . . . . . . 134
    **Intent** . . . . . . . . . . . . . . . . . . . . . . . . . . . . . . . . . . . . . . . . . . . . . . . . . . . . . . . . . . . . . . . . . . . . . . 134
    **Indispensable Element Test** . . . . . . . . . . . . . . . . . . . . . . . . . . . . . . . . . . . . . . . . . . . . . . . . . . . 135
    **Defenses to Attempt** . . . . . . . . . . . . . . . . . . . . . . . . . . . . . . . . . . . . . . . . . . . . . . . . . . . . . . . . . 135

**Solicitation** . . . . . . . . . . . . . . . . . . . . . . . . . . . . . . . . . . . . . . . . . . . . . . . . . . . . . . . . . . . . . . . . . . . . . 136
    **Intermediaries** . . . . . . . . . . . . . . . . . . . . . . . . . . . . . . . . . . . . . . . . . . . . . . . . . . . . . . . . . . . . . . 137
    **Defenses to Solicitation** . . . . . . . . . . . . . . . . . . . . . . . . . . . . . . . . . . . . . . . . . . . . . . . . . . . . . . 137

**Conspiracy** . . . . . . . . . . . . . . . . . . . . . . . . . . . . . . . . . . . . . . . . . . . . . . . . . . . . . . . . . . . . . . . . . . . . . 137
    **Agreement Between Parties** . . . . . . . . . . . . . . . . . . . . . . . . . . . . . . . . . . . . . . . . . . . . . . . . . . . 137
    **Specific Intent** . . . . . . . . . . . . . . . . . . . . . . . . . . . . . . . . . . . . . . . . . . . . . . . . . . . . . . . . . . . . . . 138
    **Crimes or Lawful Objectives by Unlawful Means** . . . . . . . . . . . . . . . . . . . . . . . . . . . . . . . . . . . 138
    **Overt Act** . . . . . . . . . . . . . . . . . . . . . . . . . . . . . . . . . . . . . . . . . . . . . . . . . . . . . . . . . . . . . . . . . . 139
    **Special Considerations** . . . . . . . . . . . . . . . . . . . . . . . . . . . . . . . . . . . . . . . . . . . . . . . . . . . . . . . 139
    **Defenses to Conspiracy** . . . . . . . . . . . . . . . . . . . . . . . . . . . . . . . . . . . . . . . . . . . . . . . . . . . . . . 140
    **Evidentiary Considerations** . . . . . . . . . . . . . . . . . . . . . . . . . . . . . . . . . . . . . . . . . . . . . . . . . . . 140

**Conspiracy and Solicitation and Federal Law** . . . . . . . . . . . . . . . . . . . . . . . . . . . . . . . . . . . . . . . . 140

**Party Liability** . . . . . . . . . . . . . . . . . . . . . . . . . . . . . . . . . . . . . . . . . . . . . . . . . . . . . . . . . . . . . . . . . . 141
    **Defenses** . . . . . . . . . . . . . . . . . . . . . . . . . . . . . . . . . . . . . . . . . . . . . . . . . . . . . . . . . . . . . . . . . . 142

**Summary** . . . . . . . . . . . . . . . . . . . . . . . . . . . . . . . . . . . . . . . . . . . . . . . . . . . . . . . . . . . . . . . . . . . . . . 142
    **Practice Test** . . . . . . . . . . . . . . . . . . . . . . . . . . . . . . . . . . . . . . . . . . . . . . . . . . . . . . . . . . . . . . . 142
    **References** . . . . . . . . . . . . . . . . . . . . . . . . . . . . . . . . . . . . . . . . . . . . . . . . . . . . . . . . . . . . . . . . 143

**CHAPTER 10** Defenses to Criminal Responsibility . . . . . . . . . . . . . . . . . . . . . . . . . . . . . . . . . . . . . . . . . . . 145

**Key Terms** . . . . . . . . . . . . . . . . . . . . . . . . . . . . . . . . . . . . . . . . . . . . . . . . . . . . . . . . . . . . . . . . . . . . . . . 145

**Introduction** . . . . . . . . . . . . . . . . . . . . . . . . . . . . . . . . . . . . . . . . . . . . . . . . . . . . . . . . . . . . . . . . . . . . . 145

**Legal and Moral Rationale for the Allowance of Defenses** . . . . . . . . . . . . . . . . . . . . . . . . . . . . . . 145

**Affirmative Defenses: Justification and Excuse Defenses Distinguished** . . . . . . . . . . . . . . . . . . 146
    **Justification Defenses** . . . . . . . . . . . . . . . . . . . . . . . . . . . . . . . . . . . . . . . . . . . . . . . . . . . . . . . 146
    **Excuse Defenses** . . . . . . . . . . . . . . . . . . . . . . . . . . . . . . . . . . . . . . . . . . . . . . . . . . . . . . . . . . . 149

**Other Defenses** . . . . . . . . . . . . . . . . . . . . . . . . . . . . . . . . . . . . . . . . . . . . . . . . . . . . . . . . . . . . . . . . . 160
    **Alibi** . . . . . . . . . . . . . . . . . . . . . . . . . . . . . . . . . . . . . . . . . . . . . . . . . . . . . . . . . . . . . . . . . . . . . . 160
    **Constitutional and Statutory** . . . . . . . . . . . . . . . . . . . . . . . . . . . . . . . . . . . . . . . . . . . . . . . . . . 161

**Summary** . . . . . . . . . . . . . . . . . . . . . . . . . . . . . . . . . . . . . . . . . . . . . . . . . . . . . . . . . . . . . . . . . . . . . . 163
    **Practice Test** . . . . . . . . . . . . . . . . . . . . . . . . . . . . . . . . . . . . . . . . . . . . . . . . . . . . . . . . . . . . . . . 165
    **References** . . . . . . . . . . . . . . . . . . . . . . . . . . . . . . . . . . . . . . . . . . . . . . . . . . . . . . . . . . . . . . . . 166

**CHAPTER 11** Organized Crime and Terrorism . . . . . . . . . . . . . . . . . . . . . . . . . . . . . . . . . . . . . . . . . . . . . . . . . . . . 167

    **Key Terms** . . . . . . . . . . . . . . . . . . . . . . . . . . . . . . . . . . . . . . . . . . . . . . . . . . . . . . . . . . . . . . . . . . . . . . . . . . . . 167

    **Introduction** . . . . . . . . . . . . . . . . . . . . . . . . . . . . . . . . . . . . . . . . . . . . . . . . . . . . . . . . . . . . . . . . . . . . . . . . . 167

    **Organized Crime** . . . . . . . . . . . . . . . . . . . . . . . . . . . . . . . . . . . . . . . . . . . . . . . . . . . . . . . . . . . . . . . . . . . 167

        **Misconceptions of Organized Crime in America** . . . . . . . . . . . . . . . . . . . . . . . . . . . . . . . . . . . . . . . . 167

        **History of Organized Crime in America** . . . . . . . . . . . . . . . . . . . . . . . . . . . . . . . . . . . . . . . . . . . . . 168

        **Nature of Organized Crime** . . . . . . . . . . . . . . . . . . . . . . . . . . . . . . . . . . . . . . . . . . . . . . . . . . . . . . . . 168

        **Legal Issues in Organized Crime** . . . . . . . . . . . . . . . . . . . . . . . . . . . . . . . . . . . . . . . . . . . . . . . . . . . 169

        **Offenses Associated with Organized Crime** . . . . . . . . . . . . . . . . . . . . . . . . . . . . . . . . . . . . . . . . . 170

        **Emerging Issues in Organized Crime** . . . . . . . . . . . . . . . . . . . . . . . . . . . . . . . . . . . . . . . . . . . . . . . 172

        **Organized Crime and Gangs** . . . . . . . . . . . . . . . . . . . . . . . . . . . . . . . . . . . . . . . . . . . . . . . . . . . . . . 172

    **Terrorism** . . . . . . . . . . . . . . . . . . . . . . . . . . . . . . . . . . . . . . . . . . . . . . . . . . . . . . . . . . . . . . . . . . . . . . . . . . 172

        **Types of Terrorism** . . . . . . . . . . . . . . . . . . . . . . . . . . . . . . . . . . . . . . . . . . . . . . . . . . . . . . . . . . . . . . . 173

        **Legal Issues in Terrorism** . . . . . . . . . . . . . . . . . . . . . . . . . . . . . . . . . . . . . . . . . . . . . . . . . . . . . . . . . 174

        **Other Terrorism-Related Crimes** . . . . . . . . . . . . . . . . . . . . . . . . . . . . . . . . . . . . . . . . . . . . . . . . . . . 176

    **Summary** . . . . . . . . . . . . . . . . . . . . . . . . . . . . . . . . . . . . . . . . . . . . . . . . . . . . . . . . . . . . . . . . . . . . . . . . . . 178

        **Practice Test** . . . . . . . . . . . . . . . . . . . . . . . . . . . . . . . . . . . . . . . . . . . . . . . . . . . . . . . . . . . . . . . . . . . . 179

        **References** . . . . . . . . . . . . . . . . . . . . . . . . . . . . . . . . . . . . . . . . . . . . . . . . . . . . . . . . . . . . . . . . . . . . . . 180

**CHAPTER 12** White Collar Crime . . . . . . . . . . . . . . . . . . . . . . . . . . . . . . . . . . . . . . . . . . . . . . . . . . . . . . . . . . . . . . . . . . . . . . . 181

    **Key Terms** . . . . . . . . . . . . . . . . . . . . . . . . . . . . . . . . . . . . . . . . . . . . . . . . . . . . . . . . . . . . . . . . . . . . . . . . . . . . 181

    **Introduction** . . . . . . . . . . . . . . . . . . . . . . . . . . . . . . . . . . . . . . . . . . . . . . . . . . . . . . . . . . . . . . . . . . . . . . . . . 181

    **Corporate Crime and Liability in America** . . . . . . . . . . . . . . . . . . . . . . . . . . . . . . . . . . . . . . . . . . . . . . . . 182

    **Tax Evasion** . . . . . . . . . . . . . . . . . . . . . . . . . . . . . . . . . . . . . . . . . . . . . . . . . . . . . . . . . . . . . . . . . . . . . . . . . 182

    **False Advertising** . . . . . . . . . . . . . . . . . . . . . . . . . . . . . . . . . . . . . . . . . . . . . . . . . . . . . . . . . . . . . . . . . . . . 183

    **Harmful Products** . . . . . . . . . . . . . . . . . . . . . . . . . . . . . . . . . . . . . . . . . . . . . . . . . . . . . . . . . . . . . . . . . . . 184

    **Food and Drug Administration** . . . . . . . . . . . . . . . . . . . . . . . . . . . . . . . . . . . . . . . . . . . . . . . . . . . . . . . . 185

    **Environmental Offenses** . . . . . . . . . . . . . . . . . . . . . . . . . . . . . . . . . . . . . . . . . . . . . . . . . . . . . . . . . . . . . . 185

    **Securities Fraud and Insider Trading** . . . . . . . . . . . . . . . . . . . . . . . . . . . . . . . . . . . . . . . . . . . . . . . . . . . 187

    **Mail Fraud** . . . . . . . . . . . . . . . . . . . . . . . . . . . . . . . . . . . . . . . . . . . . . . . . . . . . . . . . . . . . . . . . . . . . . . . . . . 187

    **Wire Fraud** . . . . . . . . . . . . . . . . . . . . . . . . . . . . . . . . . . . . . . . . . . . . . . . . . . . . . . . . . . . . . . . . . . . . . . . . . . 188

    **Bad Checks** . . . . . . . . . . . . . . . . . . . . . . . . . . . . . . . . . . . . . . . . . . . . . . . . . . . . . . . . . . . . . . . . . . . . . . . . . 189

        **Worthless Checks** . . . . . . . . . . . . . . . . . . . . . . . . . . . . . . . . . . . . . . . . . . . . . . . . . . . . . . . . . . . . . . . 189

        **Check Kiting** . . . . . . . . . . . . . . . . . . . . . . . . . . . . . . . . . . . . . . . . . . . . . . . . . . . . . . . . . . . . . . . . . . . . 189

    **Credit Card Theft/Fraud** . . . . . . . . . . . . . . . . . . . . . . . . . . . . . . . . . . . . . . . . . . . . . . . . . . . . . . . . . . . . . . 190

    **Identity Theft** . . . . . . . . . . . . . . . . . . . . . . . . . . . . . . . . . . . . . . . . . . . . . . . . . . . . . . . . . . . . . . . . . . . . . . . 191

    **Summary** . . . . . . . . . . . . . . . . . . . . . . . . . . . . . . . . . . . . . . . . . . . . . . . . . . . . . . . . . . . . . . . . . . . . . . . . . . . 192

        **Practice Test** . . . . . . . . . . . . . . . . . . . . . . . . . . . . . . . . . . . . . . . . . . . . . . . . . . . . . . . . . . . . . . . . . . . . 192

        **References** . . . . . . . . . . . . . . . . . . . . . . . . . . . . . . . . . . . . . . . . . . . . . . . . . . . . . . . . . . . . . . . . . . . . . . 194

| APPENDIX A | Practice Test Solutions | 195 |
|---|---|---|
| APPENDIX B | Declaration of Independence | 197 |
| | **IN CONGRESS, July 4, 1776** | 197 |
| APPENDIX C | Constitution for the United States of America | 199 |

- **Article I** . . . 199
  - **Section 1** . . . 199
  - **Section 2** . . . 199
  - **Section 3** . . . 199
  - **Section 4** . . . 200
  - **Section 5** . . . 200
  - **Section 6** . . . 200
  - **Section 7** . . . 200
  - **Section 8** . . . 201
  - **Section 9** . . . 201
  - **Section 10** . . . 201
- **Article II** . . . 202
  - **Section 1** . . . 202
  - **Section 2** . . . 202
  - **Section 3** . . . 203
  - **Section 4** . . . 203
- **Article III** . . . 203
  - **Section 1** . . . 203
  - **Section 2** . . . 203
  - **Section 3** . . . 203
- **Article IV** . . . 203
  - **Section 1** . . . 203
  - **Section 2** . . . 203
  - **Section 3** . . . 204
  - **Section 4** . . . 204
- **Article V** . . . 204
- **Article VI** . . . 204
- **Article VII** . . . 204

| APPENDIX D | Bill of Rights | 205 |
|---|---|---|

- **Amendment I** . . . 205
- **Amendment II** . . . 205
- **Amendment III** . . . 205
- **Amendment IV** . . . 205
- **Amendment V** . . . 205
- **Amendment VI** . . . 205

**Amendment VII** . . . . . . . . . . . . . . . . . . . . . . . . . . . . . . . . . . . . . . . . . . . . . . . . . . . . . . 205
**Amendment VIII** . . . . . . . . . . . . . . . . . . . . . . . . . . . . . . . . . . . . . . . . . . . . . . . . . . . . 206
**Amendment IX** . . . . . . . . . . . . . . . . . . . . . . . . . . . . . . . . . . . . . . . . . . . . . . . . . . . . . 206
**Amendment X** . . . . . . . . . . . . . . . . . . . . . . . . . . . . . . . . . . . . . . . . . . . . . . . . . . . . . 206

Glossary . . . . . . . . . . . . . . . . . . . . . . . . . . . . . . . . . . . . . . . . . . . . . . . . . . . . . . . . . 207
Index . . . . . . . . . . . . . . . . . . . . . . . . . . . . . . . . . . . . . . . . . . . . . . . . . . . . . . . . . . . 217

# FOREWORD

Mastery of fundamental doctrines of criminal law is important for undergraduate students interested in careers in law, law enforcement, corrections, forensics, and mental health. Expansive in scope yet accessible to all readers, *An Introduction to Criminal Law* provides thorough treatment of all important areas of criminal law. The work presents historical context, noting the common law and social antecedents to modern criminal law. It depicts traditional and modern theories and types of punishment. Constitutional sources of rights of the accused citizen and related limits on government autonomy and law enforcement are likewise presented.

As would be expected in any authoritative text on criminal law, *An Introduction to Criminal Law* offers a complete treatment of the legal elements of crimes and defenses available to the criminally accused. Importantly, however, the work includes chapters on modern commercial, organized, and international crimes. It covers terrorism and organized and white-collar crimes, which unfortunately appear to be the future of criminality in our shrinking, interconnected, digital world. Throughout, this work profits from the diverse perspectives of its authors, who bring to bear their professional backgrounds in criminal justice, mental health, and criminal defense practice.

What is perhaps most remarkable about this work is that it departs from presenting legal doctrine through judicial opinions. Although the case method is so important for the education of law students, students of criminal justice—including prelaw students—will find it easier to comprehend legal doctrine and concepts as presented within *An Introduction to Criminal Law*. A work on criminal law that is both comprehensive and comprehensible is no small thing. As a law professor, former state and federal prosecutor, and drug court judge, I welcome the publication of *An Introduction to Criminal Law*. Students aspiring to various careers in criminal justice will be enriched by its pages.

Patricia Bennett, Professor of Law
Mississippi College School of Law
Jackson, Mississippi

# PREFACE

*An Introduction to Criminal Law* aims to transmit substantive law and its elemental components in a simplistic and practical manner. Criminal justice students often express frustration concerning the general presentations of criminal law textbooks. Primarily written for law school studies, most criminal law textbooks are rich in legalese and far surpass the fundamental underpinnings required of criminal justice professionals. The unfortunate result is that those most responsible for the law's enforcement often become entrenched in a continuous struggle to decipher legalistic presentations.

Because most criminal law textbooks are authored by attorneys, they often fail to simplify the language and approach of criminal law. Although their methods appear quite successful for preparing future lawyers, their pedagogical "learn it on your own" approach tends to confuse and frustrate professionally oriented students attracted to criminal justice programs. Criminal justice students, much like those of other occupations, learn best from practical, hands-on exercises. Through the collaboration of two nonattorneys with an attorney, *An Introduction to Criminal Law* abandons the case approach while retaining all comprehensive principles of substantive law. *An Introduction to Criminal Law* "holds the hand" of students while walking them through a chronological and simplistic (yet detailed) dissection of the legal labyrinth.

*An Introduction to Criminal Law* is a gift to students who aspire to master the complexities of substantive law. Legal jargon is unavoidable, but clarification is added when the meaning of language is evasive. Offering students the opportunity to test emerging knowledge of the law, each chapter presents opportunities for critical thought and practice test scenarios. With *An Introduction to Criminal Law*, current and future employment duties related to substantive law are made simple.

## Ancillary Materials

A comprehensive set of instructor's materials, including PowerPoint Presentations and a TestBank are available online.

# ACKNOWLEDGMENTS

We offer sincere thanks to the Criminal Justice editorial staff at Jones & Bartlett Publishers. We are deeply grateful for the commitment and guidance received from all Acquisitions Editors. First, we thank Jeremy Spiegel for the publishing opportunity. Without his vision, *An Introduction to Criminal Law* would be little more than an idea. We also appreciate Cathleen Sether for assuming the editorial reigns in the midst of the process. Without her intervention and commitment, the production schedule for *An Introduction to Criminal Law* would have been delayed substantially. Gratitude also is extended to Sean Connelly for providing the editorial direction necessary to bring *An Introduction to Criminal Law* to successful completion. We reserve deepest thanks for Megan Turner (Associate Editor), who was the singular constant from beginning to end. Without her hard work and diligence, *An Introduction to Criminal Law* would not have evolved into a quality product. Finally, we are indebted to Julia Waugaman (Associate Production Editor), for her contributions regarding the many tasks associated with the production process. On a more general note, we also wish to thank all reviewers of the book. Without their insights and expertise, *An Introduction to Criminal Law* would be a much weaker contribution to the academic discipline.

Daniel K. Maxwell
University of New Haven

Frances P. Bernat
Arizona State University

Gary L. Neumeyer
Arizona Western College

Terrence P. Dwyer
Western Connecticut State University

Maldine B. Bailey
University of North Florida

Michael T. Geary
Albertus Magnus College

Donna Nicholson
Manchester Community College

Again, we sincerely appreciate all who played a role in the development of *An Introduction to Criminal Law*.

# Substantive Criminal Law: Principles and Working Vocabulary

**CHAPTER 1**

## Key Terms

Actual cause
*Actus reus*
Administrative law
Attendant circumstances
Beyond a reasonable doubt
Burden of proof
But-for test
Canon law
Capital felony
Case law
Civil law
Code of Hammurabi
Common law
Compensatory damage
Constitutional law
Constructive intent
*Corpus delicti*
Courts of equity
Crime
Criminal law
Culpable
Declaratory relief
Democracy
Deviance
Ecclesiastical courts
Federalism
Felony
General intent
Gross misdemeanor
Injunctive relief
Intervening cause
Jurisdiction
Kings courts
Law courts
Least restrictive mechanism
Legal cause
Lesser included offense
*Mala in se*
*Mala prohibita*
*Mens rea*
Misdemeanor
Misprision of felony
Natural law
Negligence
*Nulla poena sine lege*
Ordinance
Ordinary misdemeanor
Petty misdemeanor
Positive law
Precedent
Preponderance of the evidence
Procedural law
Property crime
Proximate cause
Punitive damage
Recklessness
Republic
Social contract theory
Specific intent
*Stare decisis*
Statutory law
Strict liability
Substantial factor test
Substantive law
Tort
Tortfeasor
Transferred intent
Uniform Crime Reports
Violation
Violent crime
Wobblers

## Introduction

From the genesis of time, human beings have sought to establish guidelines to govern human behavior. In ancient civilizations, rules were derived from morals, customs, and norms existing within society. Thus, in most societies, modern laws evolved from a loose set of guidelines into a formal system of written laws designed to maintain social order. Because each society—ancient or modern—possesses different moral values, customs and societal norms, laws and legal systems vary.

This chapter explores and describes the foundations of American criminal law. While progressing through its content, readers are informed of the extent to which serious crime occurs in America. Readers will also develop an appreciation for the Republic form of government used in this nation and how social contract theory guides the construction of criminal law. The chapter then explains the differences between civil and criminal law, with a focus on procedural and substantive law. Next, the evolutionary path of criminal law is chronicled by delving into its ancient, religious, and common law heritage while concurrently demonstrating more modern sources (e.g., statutory, case) for regulating societal conduct. Finally, crime is broadly defined, classified (felonies, misdemeanors, violations), distinguished from deviant conduct, explained using an elements approach, and discussed along degrees of social harm.

## The Republic for Which It Stands

The United States is known around the globe for its commitment to democratic values and as such has become regarded, even among its own citizens, as a Democracy. Most people use the term "Democracy" as a generic means to describe America's popular tolerance for free elections and the voice of the people. The reality, however, is that the Unites States was founded as (and continues to support the goals of) a Republic form of government. A simple recitation of the pledge of allegiance highlights this simple truth: . . . *and the Republic for which it stands*.

Republic and democratic forms of government could not be more dissimilar. **Democracy** is a form of government whereby elected leaders make decisions for the populous with no legal safeguards (such as a constitution) to protect the nation (and rights of the people) against the manner in which that power is exercised—an unlimited power of sorts. **Republic**, on the other hand, defines a form of government comprised of elected leaders operating under the umbrella of a Constitution that safeguards the best interest of the nation and its people by limiting power. In this way, it is believed that the right decision, as opposed to the desires of the elite ("snob rule") or majority ("mob rule"), will be achieved regardless of public sentiment or personal favoritism. Without a Republic form of government, our founding fathers were aware that the superfluous whims of the day would take precedent over what is best for the long-term health of the nation. James Madison, in the Federalist Papers, best summarized the dichotomy between these governmental forms:

> Democracy, as a form of government, is utterly repugnant to—is the very antithesis of—the traditional American system: that of a Republic, and its underlying philosophy, as expressed in essence in the Declaration of Independence with primary emphasis upon the people's forming their government so as to permit them to possess only "just powers" (limited powers) in order to make and keep secure the God-given, unalienable rights of each and every Individual and therefore of all groups of Individuals.

## Social Construction of Law

One of the more fundamental tenets of American criminal law is that societal expectations be expressed in writing—through statutory code and/or judicial opinion. This rule is so sacred, in fact, that the American legal system follows the maxim ***nulla poena sine lege***, Latin for "no penalty without a law." Essentially, this legal principle ensures that one accused of wrongdoing cannot be punished unless the behavior is clearly prohibited in written penal law. Given this well-established legal custom, it remains remarkable that social contract theory essentially functions as the fulcrum (hinge) for the American legal system. Although not a written document with contractual obligation, the social contract reflected a sacred trust between American colonists and government authority and continues to be the foundation for contemporary legal transactions. Without fulfillment of this oral agreement, it is safe to conclude that the American judicial process would become suspect in the eyes of its residents.

**Social contract theory** stipulates that American citizens, in certain well-defined circumstances, will voluntarily waive rights, privileges, and liberties guaranteed in the United States Constitution in exchange for government protection. For example, Americans give the government the authority to establish a judicial process that will detect (police), adjudicate (courts), and punish (corrections) persons who commit violations against the peace and dignity of our nation (or state). In exchange, the government agrees to support (through taxation and regulation of commerce) and protect (against foreign and domestic threats) us and vows to do so with tremendous caution. Known as the **least restrictive mechanism**, the agreement includes a binding promise that any government action against citizens, in addition to being necessary, will be implemented with every effort toward minimizing intrusion. For example, government has the right to restrict the freedom of societal members (through incarceration and other means) when violating laws but must do so with an eye toward the minimal incarceration essential to reasonably ensure that an individual, and society as a whole, is sufficiently deterred from committing future crimes. Do you believe that the government has made a good faith effort to abide by this social contract?

## Origins of Law

Historically, law originated from three primary venues: ancient, natural, and common. Though these origins are discussed as distinct, there is overlap among them. For example, much natural law existed within ancient times. Likewise, much common law consisted of natural law. This section will examine those legal origins.

## Ancient

The **Code of Hammurabi** is routinely cited as the first set of written laws developed to govern a society. This code was developed by King Hammurabi of Babylon between 1792 and 1750 BC. In modern times, we think of austere sets of legal reporters and codes when we imagine the location of our laws. In contrast, the Code of Hammurabi was carved onto a black stone monument. The Code of Hammurabi included approximately 300 provisions and addressed both criminal and civil matters. These provisions were believed to have come from the gods. Matters addressed in the code included criminal offenses, punishments, and domestic relations matters such as marriage and divorce.

Other scholars note the existence of an earlier set of written laws discovered in Ur, an ancient city-state in Sumeria. These laws predate the Code of Hammurabi and appear to be approximately 5,000 years old. The existence of both the Code of Hammurabi and the Sumerian code reflects ancient efforts to develop principles through which governance and control of human behavior could occur. Other examples of ancient laws and legal systems can be found by examination of those which existed in ancient Hebrew, Greek, and Roman civilizations. Each system possessed its own unique attributes. Additionally, the development of laws and legal philosophy in these ancient societies significantly influenced the development of modern European and American legal systems.

## Natural

**Positive law** is man-made law enacted into statutes for the protection of people as a whole. Historically, though, positive law was singularly concerned with human activities not addressed within religious circles. It has been argued, however, that one underlying rationale for distinguishing man-made law from religious law was to draw a clear and distinct line between its laws derived from logical, rational human decisions and the more ambiguous and irrational moral distinctions premised on **natural law** (or God's law). Natural law, as defined within Black's Law Dictionary, is based on "... necessary and obligatory rules of human conduct which have been established by the author of human nature as essential to the divine purposes in the universe . . ." (Garner, 2009). It is important, then, to examine the influence of natural law in the construction of positive law.

Dating back to first-century Rome, natural law embodies the beliefs and values based on accepted moral principles derived from a higher power, nature, and/or reason. Religion is the premier natural law source in most world cultures, and without question, American lawmakers have (and still do to some extent) rely heavily on the religious principles of Judaism and Christianity. For example, religious prohibitions embedded in the Old Testament (especially the Ten Commandments) appear (or have appeared) in substantive criminal law. Crimes regarding adultery, murder, theft, and perjury (bearing false witness) are just a small sampling of modern laws grounded in natural law. The historical intertwining of positive and natural law, then, should be readily apparent; their degree of association does seem to be on the decline, however, as certain natural law prohibitions (such as adultery and homosexuality) have for all practical purposes been decriminalized across the nation.

## Common

The origin of modern American law was largely derived from English common law. With the establishment of the American colonies, settlers brought with them existing law as developed in England. English common law developed in contrast to Roman civil law, which was the predominant influence throughout ancient Europe; however, after the fall of the Roman Empire, local communities were left to develop their own systems of justice.

In England, the legal system developed through the influence of monarchs and ecclesiastical authorities, the Catholic Church, and later, the Church of England. Before the Norman Conquest, local communities resolved most legal disputes through reliance on local customs and mores, with penalties for transgressions consisting of harsh physical violence. After the Norman Conquest, however, efforts to centralize power in the monarch provided a more uniform legal system throughout England. The Norman influence is reflected in the efforts of William the Conqueror to vest greater control over the development of law, operation of the legal system, and general business of government in the monarch. The formalization and centralization of the English legal system continued through the reigns of Henry II and Edward I. These efforts marked the transition from a civil law system to the common law system in which judges traveled the countryside (or "rode the circuit") to handle legal matters, a practice formally endorsed and enacted in the Statute of Westminster in 1285.

**Common law** is often referred to as "judge made law." In other words, common law consists of the rulings of judges following the application and interpretation of existing laws, customs, or adherence to prior cases. These judicial decisions were maintained and relied on as precedent for future cases and followed a principle known as ***stare decisis*** ("let the decision stand"). In the English system, then, judges possessed significant authority to identify and define common law crimes and fashion remedies.

The transformed English system also possessed several types of courts distinguished by the nature of their jurisdiction. **Jurisdiction** refers to the authority of the court to hear and decide a case. Courts fell into two categories (law and equity), which essentially differentiate **kings courts** (law courts) from **ecclesiastical courts** (equity courts). Ecclesiastical courts, referred to as Chancery Courts, existed to enforce **canon law** (or the laws of the Catholic Church). The primary distinction between law courts and **courts of equity** was the nature of the remedy that could be ordered. **Law courts** were restricted to an award of monetary damages, whereas courts of equity were vested with much more discretion and flexibility. As such, a court of equity could award extraordinary relief with the goal of achieving a sense of fairness with the award. Although the historical distinction between courts of law and equity eventually disappeared in England and in most of America, a few American jurisdictions have retained the distinction. For example, Mississippi has retained the distinctions, with Chancery Courts possessing jurisdiction over cases involving domestic relations, divorce, child custody, probate matters, and minor's business.

As discussed earlier, American settlers brought with them the influence of English common law; however, although English common law served as a foundation for the development of modern American law, inhabitants of the new world quickly modified this foundation to fit the needs of an emerging nation based on more democratic values. Although many common law legal definitions were retained, many were not. We therefore need to examine the sources of criminal law forming the primary basis from which modern law is derived.

## Primary Sources of Criminal Law

Notwithstanding the three primary origins of law just discussed, criminal law can specifically be traced to five sources: common, statutory, case, constitutional, and administrative. With regard to substantive criminal law (the focus of this text), constitutional and administrative play a lesser role but are nonetheless important (and thus included within the forthcoming discussion).

### Common

As previously discussed, a brief historical examination is sufficient to conclude that American colonists relied heavily on their English culture to form the basis for American criminal law. Without doubt, the laws common to the circuits of England were used to shape the substance of American criminal law. Following our nation's independence campaign against the British, all 13 colonies initially anointed common law as the appropriate foundation for American jurisprudence. Although colonial Americans did not agree with a substantial portion of English practices (hence the American Revolution), they did recognize the logic of many common law prohibitions (such as murder, rape, kidnapping, and burglary). Once the United States was formed, however, states acquired the sovereign power to abolish common law (at its discretion) under a system of **federalism** (nationalized strong central government) negotiated within the U.S. Constitution. Accordingly, most states today have exercised that option, choosing instead to adopt a civil system permitting legislators (on behalf of the people) to declare through statute (**statutory law**) what laws should and will be constructed. It remains true, though, that even in the absence of a common law directive, common law continues to influence the construction of law, as legislative and judicial officials often depend on its heritage of judicial decisions for legal interpretation.

### Statutory

Statutory law currently serves as the prominent source for the establishment of criminal law. The preference for statutory law has its basis in the fundamental nature of democratic values. Statutes are created and enacted by legislative representatives of the people after deliberation and debate, rather than by judges. Essentially, the process of statutory law allows elected legislators to regulate behavior of its constituents based on their beliefs, assuming those beliefs remain within the parameters of constitutional guidelines, regarding what is best for the people within its jurisdiction. Thus, statutory law is thought to represent the will of the citizenry as opposed to what may be the isolated opinion of one individual. What emerged was a unique modern American legal system that was comprised of a vast and complex system of laws originating from a variety of sources. Many of these criminal regulations are new legal constructions designed to protect society from emerging problems (i.e., computer crime), but many have merely been adopted from the historical traditions of old England.

The American legal system today has abandoned the notions of omnipotent monarchs, replacing it with a government structure reliant on the power of its citizens. Respect for the sovereignty of states, limited government, and personal liberties is the hallmark of this new legal system. Because, however, the U.S. Constitution places few restrictions on what can be a crime and does not regulate in any meaningful way labels and definitions attached to crimes, the statutory codes of the 50 sovereign states differ significantly. As such, an attempt to discuss the codes of all states would prove mind boggling at best and monopolize years of time. It is, however, important to be familiar with the three major restrictions regarding law creation. One, legislators must establish a compelling public need for adding to the body of criminal law. Two, the law passed must possess no constitutional infringements on the rights of the people (these constitutional protections are thoroughly discussed in Chapter 2). Three, the legislature must provide the people with fair and adequate notice regarding the passage and implementation of said new laws. The notification of new law is, in practice, fairly simple to accomplish, usually employing techniques associated with billboards and road signs and announcements in the newspaper and on radio and television, as well as other various techniques.

Because statutory codes of independent states vary widely, it is important to develop a familiarity with their respective structures. For this reason, a comparison of the Mississippi (conservative) and New York (liberal) grand larceny statutes is presented to illustrate the importance of common law as a baseline for understanding modern law. Exhibit 1–1 provides a comparison of two statutes from states differing with respect to (1) degrees of grand larceny (one in Mississippi and four in New York), (2) value placed on the property (less in Mississippi), and (3) penalties associated with their violations (greater punishment in Mississippi for the most basic larcenous offense).

---

### Exhibit 1–1 Larceny Statutes

#### Mississippi

**§ 97-17-41    Grand Larceny**
(1) Every person who shall be convicted of taking and carrying away, feloniously, the personal property of another, of the value of Five Hundred Dollars ($500.00) or more, shall be guilty of grand larceny, and shall be imprisoned in the Penitentiary for a term not exceeding ten (10) years; or shall be fined not more than Ten Thousand Dollars ($10,000.00), or both. The total value of property taken and carried away by the person from a single victim shall be aggregated in determining the gravity of the offense. . . .

#### New York

**§ 155.30    Grand Larceny—fourth degree**
A person is guilty of grand larceny in the fourth degree when he steals property where (1): The value of the property exceeds one thousand dollars; or …. Grand larceny in the fourth degree is a class E felony; sentence shall not exceed four years.

**§ 155.42    Grand larceny—first degree**
A person is guilty of grand larceny in the first degree when he steals property and when the value of the property exceeds one million dollars. Grand larceny in the first degree is a class B felony; sentence shall not exceed twenty-five years.

*Source*: MS § 97-17-41; NY § 155.30 & § 155.42

## Case

Federal and state constitutions, through a process of checks and balances, grant the judiciary authority to review and interpret decisions of legislative bodies, subsequently providing judicial officials with an equal opportunity (if not more) to inject belief systems into criminal law. At its core, then, the judiciary possesses the authority to greatly alter the development, growth, and direction of American criminal law through what is referred to as **case law**. Arguably the greatest tool at the disposal of the judiciary is the common law procedure known as *stare decisis*, meaning "let the decision stand." Essentially a system of **precedent**, *stare decisis* requires inferior (lower) courts to abide by the decisions of higher courts and further demands that even higher courts examine all court decisions when addressing complex legal issues. Adherence to the value of precedent promotes stable and predictable court outcomes. Without such dependability, people would regard the legal system as unfair. Meaning? You guessed it—a loss of respect for the law and an increased likelihood of criminality.

Case law represents judicial opinions that impact the constitutionality of criminal laws, lower court rulings, and decisions of executive bodies. When appellate courts issue opinions, four options are at their disposal. First, the judicial decision is affirmed, meaning that the lower court's ruling is supported. Second, the decision is reversed, meaning that the lower court's ruling is overturned. Third, the decision is reversed but remanded back to the lower court with instructions on how to proceed; the case can then come back for a second review (if necessary). Fourth, the decision is reversed and rendered (meaning judgment is immediately proclaimed and entered into the record). One of the most publicly recognized pieces of case law, the U.S. Supreme Court in *Roe v. Wade* (1973), held that a woman's right to an abortion fell within the right to privacy and as such gave women absolute autonomy over pregnancies during the first trimester. Exhibit 1–2 illustrates how case law appears in legal venues.

## Constitutional

**Constitutional law**, though to a lesser degree, also pertains to substantive criminal law. The constitution of the United States and those of the independent states regulate what is required and prohibited in the process of legal enactments. There is little debate that the bulk of constitutional law addresses procedural law, but constitutional principles also protect society from potential abuse stemming from the construction and application of substantive criminal law. Although discussed more thoroughly in Chapter 2, common examples of constitutional protections include the void-for-vagueness doctrine, *ex post facto* prohibition, due process, and equal protection.

## Administrative

Even though criminal law is the most visible deterrent against societal rules violations, there are actually more administrative policies and regulations (thousands in fact), collectively referred to as **administrative law**, that restrict our behavior than contained within criminal codes. Violations of regulatory policies, such as those constructed by the Internal Revenue Service and Environmental Protection Agency, are ordinarily adjudicated in civil courts through fines, economic sanctions, and privilege restrictions. More recently, however, federal and state legislatures have begun to empower administrative agencies increasingly with the backing of criminal sanctions. As such, regulatory policy infractions, once only civil in nature, now carry more legal weight, as prospective violators must now increasingly take into consideration possible referral of said violations to judicial authorities for criminal sanction consideration.

With these understandings, Figure 1–1 outlines the major sources of criminal law.

## Types of Legal Wrongs

There are two recognized forms of legal wrongs: public and private. The content of this book is substantive criminal law and its public focus, and therefore, private wrongs receive minimal attention. Please do not interpret this brevity of coverage as an indictment regarding its value though, for it is an invaluable mechanism for the resolution of dispute between societal members. As such, it greatly reduces the necessity for criminal law intervention in that, among other things, many crimes that likely would have been committed out of retribution or retaliation are not committed due to the availability of civil remedies.

## Private

A private wrong is within the jurisdiction of **civil law** (not criminal law) and is referred to as a **tort** when there is a cause of action, with the person accused of causing the harm (whether intentional or negligent) being the **tortfeasor**. The process entails a complainant making a formal accusation of harm with a court possessing civil jurisdiction and seeks to attain one of three remedies (or combination thereof) for inflicted wrong: monetary award, injunc-

### Exhibit 1–2 *Roe v. Wade*

**SUPREME COURT OF THE UNITED STATES**

410 U.S. 113
*Roe v. Wade*

APPEAL FROM THE UNITED STATES DISTRICT COURT FOR THE
NORTHERN DISTRICT OF TEXAS
No. 70-18 Argued: December 13, 1971—Decided: January 22, 1973
BLACKMUN, J., delivered the opinion of the Court, in which
BURGER, C.J., DOUGLAS, BRENNAN, STEWART, MARSHALL, and POWELL, JJ., joined.
WHITE, J. and REHNQUIST, J. filed dissenting opinions.

**Issue:**

A pregnant single woman (Roe) brought a class action challenging the constitutionality of the Texas criminal abortion laws, which proscribe procuring or attempting an abortion except on medical advice for the purpose of saving the mother's life. . . . A three-judge District Court . . . declared the abortion statutes void as vague and overbroadly infringing those plaintiffs' Ninth and Fourteenth Amendment rights.

**Decision:**

State criminal abortion laws, like those involved here, that except from criminality only a life-saving procedure on the mother's behalf without regard to the stage of her pregnancy and other interests involved violate the Due Process Clause of the Fourteenth Amendment, which protects against state action the right to privacy, including a woman's qualified right to terminate her pregnancy. Though the State cannot override that right, it has legitimate interests in protecting both the pregnant woman's health and the potentiality of human life, each of which interests grows and reaches a "compelling" point at various stages of the woman's approach to term.

*Source*: United States Supreme Court

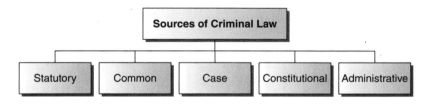

**Figure 1–1** Sources of criminal law.

tive relief, or declaratory relief. Monetary damage is the most common remedy for private harm. There are two forms of monetary damage: compensatory and punitive. **Compensatory damage** seeks reimbursement of actual expenses associated with wrongful conduct. For example, an employee wrongfully fired may sue and receive actual losses stemming from their dismissal, such as back wages, withheld benefits, and emotional distress. **Punitive damage**, on the other hand, aims to deter and punish individual wrongdoers from committing the same act(s) in the future. In essence, the goal of punitive damages is to teach wrongdoers a lesson that will not be forgotten, while concurrently deterring others from contemplating future similar acts. For example, a sexual harassment victim may sue and receive compensatory damages, but punitive damages (sometimes in the millions of dollars) may also be assessed by the court to send a deterrent message. Although securing money from tortfeasors is often the goal, wronged individuals often turn to civil courts for assistance with operational problems, namely in the form of injunctive or declaratory relief. **Injunctive relief** occurs when a court issues an injunction (or order) for someone (or group of persons) to do or stop doing something that is (or may) bring about harm. For example, a building scheduled for demolition may be protected, at least for a period of time, through the securing of a court injunction. **Declaratory relief** describes a judge's determination (called a "declaratory judgment") of parties' rights under a contract or a statute often requested in a lawsuit.

## Public

A public wrong is addressed within the body of **criminal law**: substantive and procedural. **Procedural law** encompasses numerous procedures required of those empowered to carry out the duties of the criminal justice system. The purpose of procedural law is to protect the due process rights of citizens (and illegal aliens) and therefore essentially defines the *do's* and *don'ts* of criminal justice professionals. Fourth amendment search and seizure guidelines and sixth amendment trial rights are but two of a plethora of procedural restrictions. Turning attention to our more fundamental interest, **substantive law** is comprised of the behavioral dictates placed on the people who live within our great nation. The essence of substantive law ensures that members of society are afforded fair notice of what is expected and therefore can be defined as a prescription regarding the *do's* and *don'ts* of societal members. The elements constituting murder, rape, assault, and robbery are just a few examples of what constitutes substantive law. Figure 1–2 summarizes the divergent paths of these two forms of legal wrongs.

## Crime Defined

Generally speaking, a **crime** is a public wrong that causes social harm. On its face, such an ambiguous definition may appear to adequately define criminal behavior. After all, no one reading this text would ever behave in a manner that could be construed as adverse to the public welfare—right? If you believe crime is sufficiently defined in such a generic manner, consider for one moment the person who was adjudicated a criminal for doing little more than being what many regard as the most moral person ever to walk on Earth. You guessed it—Jesus of Nazareth! It should be obvious, then, that who determines what is criminal and how it is determined are of the utmost importance.

Through the years, many crime definitions have been formed within legal circles. For purposes of simplicity, however, we embrace one specific yet encompassing definition to assist our understanding of this legal concept. Crime broadly defined requires three distinct components: (1) the commission of an act prohibited by law or the omission of an act required by law, (2) without defense (excuse or justification), and (3) codified as a felony or misdemeanor.

## Commissions and Omissions

The first component of this crime definition illustrates that criminal punishment is singularly reserved for behavioral conduct (not thoughts alone). The law is clear, however, that behavior consists of both what is **done** (commission) and not done (omission). In other words, even though most criminal regulation proscribes what an individual must refrain from doing (forging, robbing, etc.), it

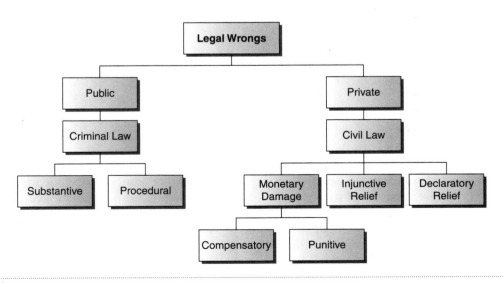

**Figure 1–2** Legal wrongs.

also often demands action of a person (filing taxes, emergency assistance, etc.). Commissions come in a variety of forms (possession, procuring, attempt, etc.) as defined within differing jurisdictions, but omissions are much more narrowly defined. One historical example, although used today only on the federal level, provided that it was a criminal misdemeanor to conceal the commission of a felony committed by another person, an offense known as **misprision of felony**.

## Without Legal Defense

Component two of this crime definition further clarifies that not all persons who consummate such behavioral conduct are criminally accountable. The law aims to punish only those who commit prohibited conducts or ignore (omit) required conduct with no reasonable defense (be it justification or excuse); therefore, an individual is not necessarily guilty of a crime when deviating from legally established behavioral guidelines.

## Codified

The third component of this crime definition mandates that legal proscriptions be codified, meaning that the law must provide written advance notice of its behavioral expectations (referred to as an annotated code) and specifically outline available punishments. Figure 1–3 outlines these essential components to the definition of what a crime must constitute.

## Crime Classifications

Crime is classified in reference to: (1) the degree of punishment and (2) moral turpitude. With respect to authorized punishment, crime is broadly classified as felonies, misdemeanors, and violations. Moral turpitude, on the other hand, is divided into *mala in se* and *mala prohibita* designations.

## Felonies, Misdemeanors, and Violations

A felony at common law was a serious crime for which a person was required to forfeit property to the king as restitution for harm against the crown.

**Figure 1–3** Crime defined.

Common law felonies were subject to a punishment of death and included murder, manslaughter, rape, sodomy, assault, robbery, burglary, larceny, and arson. A common law crime not punishable by death was referred to as a misdemeanor (less serious crime). Although not required, most states today have abandoned the common law guidelines defining felonies and misdemeanors in favor of a quantified approach. Essentially, most states now define a **felony** as a crime for which the authorized punishment is 1 year or more in a federal or state prison, or a fine. Felonious crimes eligible for the punishment of death or life imprisonment without parole are also referred to as a **capital felony**.

A **misdemeanor** is a crime for which punishment is authorized up to, but not including, 1 year in a local (municipal or county) jail. Much like felonies, misdemeanor crimes have been dissected into multiple seriousness scales. Using this classification system, a crime for which punishment ranges from 6 to 12 months in jail is a **gross misdemeanor**. Continuing this logic, an **ordinary misdemeanor** becomes a crime for which punishment ranges from 3 to 6 months in jail. Lastly, a **petty misdemeanor** represents crimes punishable from 10 to 30 days in jail. Each sovereign state is free to penalize criminal offenses according to the needs and values of its jurisdiction, however, and hence uniformly referring to a particular crime as a felony or misdemeanor is not without risk. Further complicating the classification landscape, some states have even designated certain crimes as **wobblers**, meaning the accused can be charged with either a misdemeanor or felony, depending on the circumstances. Modern legal codes often also include a third classification known as a **violation**. These state-sanctioned crimes are punished with fines only and are not administratively recorded as criminal. Finally, a local **ordinance** is a regulation of problematic behavior at the county and municipal level; littering is one example of an ordinance infraction. Ordinances are not sanctioned at the federal or state level; thus, they are not considered a crime.

### *Mala in Se* and *Mala Prohibita*

Crimes are also distinguished along lines of moral turpitude, which refers to immoral or depraved acts, namely behavior that deviates grossly from the accepted standards of a community. As such, crimes of moral turpitude in one community may be considered otherwise in adjacent communities. Moral turpitude crimes are referred to as **mala in se**, meaning "wrong in itself," or inherently evil or bad. All common law crimes were *mala in se*. Similarly, crimes thought to involve no moral turpitude and considered wrong merely because they are legally prohibited are referred to as **mala prohibita**. Likely the most common *mala prohibita* offense is speeding; it is prohibited but not condemned as immoral in society. Figure 1–4 charts the path of these criminal classifications.

## Crime and Deviance Distinguished

Colleges offer courses on crime and deviance spanning full academic terms (and still fail to provide full coverage); therefore, do not consider this section as anything more than a preliminary introduction to these concepts. With that said, it is important to understand that crime and deviance, although interchangeably used in societal circles, do possess separate and distinct qualities within formal criminal justice settings.

Crime (as previously defined) consists of conduct that society agrees to regulate for its own compelling purposes. **Deviance**, on the other hand, is a sociological concept used to describe behavior that (1) breaches (deviates from) societal norms and values or (2) is a statistical anomaly. Vegetarianism, for example, is a statistical aberration from societal norms because it is practiced by a very small percentage of Americans (approximately 3%). Although classified as deviance, it should be commended—not punished—for its health benefits and commitment to values. Essentially, then, conduct may be "deviate" yet not classified as "criminal" when no compelling need to regulate its consequences exists. It remains true, however, that "crime" is often not regarded as "deviance." Pause for Thought 1–1 illustrates the practical difference between a crime and a deviant act.

## Essential Elements of Crime and Liability

The most fundamental of legal requirements pertaining to governmental regulation of criminal

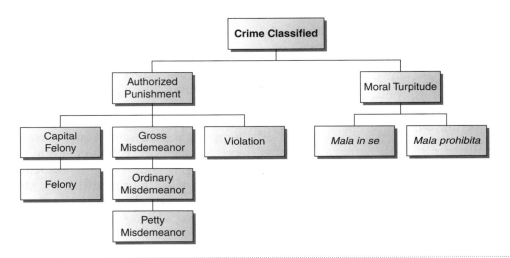

**Figure 1–4** Crime classified.

---

### Pause for Thought 1–1

Consider the following: Kelly is issued a citation for speeding while on her way to work. A colleague witnesses the incident and spreads the word throughout the office. When Kelly arrives, what do you believe the office response will be?

### Scenario Solution

Speeding is a common practice among motorists. Even though most motorists regard themselves as safe drivers, it is undeniable that a large majority have exceeded the speed limit at one time; therefore, speeding is not a statistical anomaly. Speeding also does not qualify as a breach of societal values (or norms) because the practice is considered normal. It is nonetheless regulated as criminal because of the compelling need to protect motorists from each other and themselves.

conduct is that a sanctioned offense (crime or ordinance) must possess an element known as ***actus reus***, translated as "guilty act." Unlike most areas within criminal law, there simply are no exceptions to this legal principle. It is not sufficient, however, to demonstrate merely the likelihood a person committed a prohibited or required conduct; this would be routine in most cases. To hold one **culpable** (or blameworthy) for a legal wrong, the government must at least meet or exceed specified requirements collectively referred to as the **burden of proof**. In criminal cases, this burden is much greater than the **preponderance of the evidence** standard used in civil cases, whereby one need only establish a greater likelihood that harm occurred. With respect to the *actus reus* requirement of a criminal offense, the government must prove to a moral certainty—a standard referred to as **beyond a reasonable doubt**—the presence of *corpus delicti* and proximate cause.

## Phase I of *Actus Reus:* Corpus Delicti

***Corpus delicti*** is translated as "body of the crime" and is best understood when interpreted within the framework of the body of a letter. Essentially, it conveys to all persons engaged in the criminal process that substantial evidence, when examined in its totality, must demonstrate (1) good reason to believe that a crime was committed and (2) good reason to believe that the accused committed the crime. It should be clear that the second component is dependent on the first component, as it becomes impossible to demonstrate that a person likely committed a crime for which there is no good reason to believe was committed in the first place. After the prosecution has successfully established the *corpus delicti* of an offense, it then must address the issue of causation.

In 1959, a California appellate court became the first American court to rule that the *corpus delicti* of murder could be wholly satisfied with circumstantial evidence. In affirming the defendant's murder conviction, the court outlined what they perceived to be the most convincing of the circumstantial evidence:

- The victim was in good physical and mental health before her disappearance and had numerous friends with whom she communicated on a regular basis.
- The victim would not have left home without her eyeglasses and dentures.
- If the victim intended to leave home she would have taken money, baggage, and a wardrobe.
- It would have been impossible for the victim to conceal herself for several years and find a way to live without drawing upon her bank accounts.
- The defendant had a motive for killing his wife to give himself a chance to steal her money through the forgery of her name on many documents.
- The defendant had previously persuaded his wife to convert her securities into cash to make it easier for him to obtain her property through forgeries.
- Every act and every statement of the defendant after the disappearance of his wife were consistent only with knowledge that his wife was dead.

## Phase II of *Actus Reus:* Proximate Cause

The **proximate cause** requirement of a criminal offense demands that the government prove that illegal conduct in question actually caused the harm. For example, suppose one person slaps another in the face (assault) but without apparent harm. Later that night, however, the struck person dies from an apparent heart attack. It is obvious to most reasonable people that the death was not caused by the slap. For the sake of argument, however, it is possible, that a prosecutor could argue that the death was the culmination of a process started with the slap. Given the overzealousness of some prosecutors coupled with jurors unskilled in the rules of law, this person could be convicted of a criminal homicide without having caused the harm at all. It is for reasons such as this that the law aims to protect the criminally accused by requiring the prosecution to prove such a causal connection.

With this understanding, **actual cause** refers to the connection between the harm in question and the actual conduct of the criminally accused. There are two actual cause examination techniques: the but-for test and the substantial factor test. The **substantial factor test** is the preferred prosecutorial tool because it is an easier standard. Essentially, the test requires only that the government establish, without any direct proof, that the person's actions contributed significantly, or were a substantial factor, in the resulting harm. Because of the generalities associated with this test, it is normally permitted by judges in cases in which it would be nearly impossible to establish causation with more certainty. For example, let us presume for one moment that

---

**Actual Cause**

Question 1: Would the harm have been avoided but-for the conduct of the accused?
Finding: Yes
Conclusion: The accused is the actual cause or cause-in-fact.

10 people simultaneously assault another person, resulting in serious bodily harm. Unless the person causing the serious injuries steps forward and accepts responsibility, it would be nearly impossible to determine which of the 10 people should be most accountable; therefore, the prosecution would only have to establish that an accused person was a substantial factor in the sustained injuries. The stricter and more judicially sanctioned approach, the **but-for test**, essentially begs the question: But for the conduct of the accused, would the harm have occurred? If harm to another would not have occurred but for the defendant's conduct, the defendant is said to be the actual cause of the harm.

It must be remembered that actual and proximate cause are not the same. The legal complexities associated with proximate cause often present unique challenges. Proximate cause is premised on **legal cause**, not just actual cause. It recognizes the unfairness of imposing criminal penalties on those who are the actual cause of harm to another, yet should not be criminally accountable for the harm. Where it can be shown that the defendant intended the harm or should have been able to anticipate reasonably dangers associated with certain conduct, a legal cause determination is fairly straightforward. On the other hand, in cases in which the harm is beyond the foreseeable scope of the defendant or in which some independent **intervening cause** severs (or breaks) the connection between the defendant's conduct and its harmful consequence, the defendant's conduct may not be the actual or direct cause of the harm. Keep in mind, however, that the law requires assailants to take victims as they find them, meaning that a lack of awareness concerning victims' health conditions cannot be used to avoid criminal responsibility. Considering that a criminal conviction is prohibited without a proximate cause showing, this legal requirement is of monumental importance. The hypothetical example below illustrates a recipe of sorts for how a proximate cause determination is formulated. Moreover, Pause for Thought 1–2 illustrates the proper legal interpretation regarding proximate cause determinations.

### Role of *Mens Rea*

Most statutes require that prosecutors prove both the *actus reus* (guilty act) and ***mens rea*** (guilty mind) of a criminal offense to hold a person accountable (or culpable) for harmful conduct, a crime generically referred to as a true crime. Although rare, the law does, however, carve out occasional exemptions to this rule. Based on the principle of **strict liability**, the prosecution does not bear the burden of proof. In such a case, the law presumes that the accused is guilty

---

### Legal Cause

Question 1: Was the possibility of harm foreseeable?
Finding: Yes
Question 2: Was there an independent cause intervening between the act and harm?
Finding: No
Conclusion: Accused is the legal cause, and hence the proximate cause.

---

### Pause for Thought 1–2

Consider the following: Driver A becomes enraged at Driver B's aggressive and dangerous maneuvers. Upon arriving at a store and in response to Driver B's callous and cavalier attitude, Driver A punches Driver B in the stomach with no intent to cause serious harm. As a result of a kidney condition unknown to Driver A, Driver B subsequently dies in the hospital from kidney-related complications. Can Driver A be charged with criminal homicide for Driver B's death?

### Scenario Solution

Yes. Driver B would undoubtedly still be alive but for the defendant's conduct. Some might argue that the kidney condition could not reasonably be foreseen and should therefore eliminate the defendant's conduct as the proximate cause of death. Although that perspective makes for interesting debate, the legal requirement that we take victims as we find them makes the condition implicitly foreseeable. Concerning the final element, an intervening cause must be independent. A health condition is not independent, but rather is dependent on the harm. As such, unless Driver A had some lawful justification or excuse to strike Driver B, Driver A is criminally culpable for the death.

without having to prove any mental fault. Drug possession and statutory rape are two common examples of crimes designated as possessing strict liability. The general rule, however, supports the value of treating offenses as true crimes, and as such, forthcoming discussion focuses on those mental fault elements.

## Degrees of *Mens Rea*

Legal codes recognize four forms of mental fault: specific intent, general intent, recklessness, and negligence. Some crimes require states to prove that a criminally accused person possessed **specific intent** to commit the harm in question. To prove this element, states must establish that the accused acted with a willful and intentional mental purpose. With specific intent crimes, a generic belief that the person is at fault, but with no evidence that specific harm was the objective, is not sufficient for a criminal conviction for that particular offense. For example, a first-degree murder conviction ordinarily requires proof of premeditation and deliberation; without evidence to that effect, however, a lesser included form of criminal homicide would be the only permissible prosecutorial avenue. Moving on, most crimes require, at most, proof of **general intent**, meaning some degree of malevolent or wrongful design but with no particularized objective.

Behaviors that possess no mental intent can nonetheless be regulated as criminal to coerce individuals to practice reasonable standards of care; therefore, people whose actions are reckless or negligent are said to have possessed the **constructive intent** to cause harm and thus can be criminally culpable for their harm(s). **Recklessness** is the failure to adhere to a standard of care that a reasonable person knows to exercise, basically behaving in a fashion in which the accused was cognizant of foreseeable danger. **Negligence**, on the other hand, shares a common denominator with recklessness, as it also demonstrates a failure to adhere to a reasonable standard of care, but differs in that the accused was unaware of the anticipatory dangers. One must also keep in mind the doctrine of **transferred intent**, which seals legal loopholes with respect to unsuccessful criminal attempts. Essentially, the principle stipulates that when a person intends to cause harm to any person but instead erroneously inflicts harm on an unintended target, the law can transfer that general intent (but never specific intent) to the party actually harmed. Pause for Thought 1–3 illustrates how to apply the doctrine of transferred intent.

## Attendant Circumstances

In most cases, a person's actions (*actus reus*) and accompanying mental fault (*mens rea*) serve as the essence of what substantive criminal law seeks to eliminate from our midst. With that said, however, it is imperative to keep in mind that many actions (even when mental fault exists) are nonetheless not criminal under laws requiring proof that certain circumstances surrounded the criminal conduct. Referred to as **attendant circumstances**, these legal proscriptions can often mean the difference between freedom and incarceration (or the period of incarceration). For example, the crime of incest, often based on the molestation of a child within the family, often receives greater punishment than the actual crime of child molestation because of its trespass against the sanctity of the family unit—a breach of trust. Statutory rape is another example of the importance of attendant circumstances, in that a female's age can define the difference between criminal sexual intercourse and healthy, adult sexual relationships. Figure 1–5 provides a flow chart to assist with this legal reasoning.

---

### Pause for Thought 1–3

Consider the following: Joe becomes angry with Nicholas. In a moment of rage, Joe throws a knife in the general direction of Nicholas. The knife hits and seriously injures an innocent bystander. Is Joe criminally liable for the unanticipated harm?

### Scenario Solution

Yes, under the doctrine of transferred intent, Joe can legally be viewed as having the general intent to harm the bystander, and as such, the state would be entitled to charge him with a crime, even though Joe held no willful or purposeful intent toward the bystander. Furthermore, it should be obvious that Joe committed his act (at a minimum) with recklessness because he chose not to exercise a standard of care expected of reasonably prudent persons.

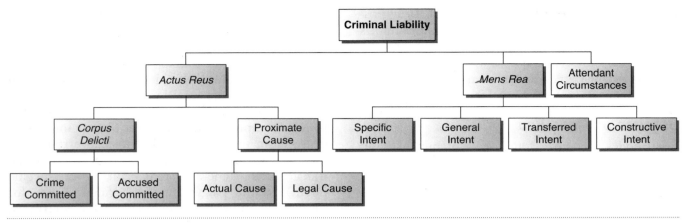

**Figure 1-5** Criminal liability.

## Crime in America

Few people would dispute that human behavior in the United States is highly regulated—some even argue overregulated. The volume of legislation designed to curtail harmful consequences is so great, in fact, that an attempt to organize and discuss all crimes would produce thousands of head-spinning legal pages. For this reason, this book adheres to a blueprint designed to expose students to the most encountered and problematic crimes within criminal justice professions. The logical starting point for the identification of such crimes is with the Federal Bureau of Investigation's (FBI) annual publication *Crime in the United States*. A statistical portrait of crime in America assembled from approximately 17,000 participating law enforcement agencies (representing approximately 95% of the total population), the **Uniform Crime Reports** (UCR) embedded within the annual compilation divide the eight crimes most plaguing the welfare of this nation into two fundamental crime categories: violent crime and property crime. Although the severity of a crime is a major factor in the designation of those crimes included in the publication, frequency, geographic impact, and economic consequences of an offense also serve as major considerations to the selection process.

Murder, forcible rape, aggravated assault, and robbery comprise the **violent crime** grouping. In 2007, law enforcement agencies reported more than 1.4 million such criminal commissions. Aggravated assault was the most frequent violent criminal act (60.8%), with robbery (31.6%), forcible rape (6.4%), and murder (1.2%) representing a decreasing presence. Turning attention to **property crime**, burglary (22.1%), larceny (66.7%), motor vehicle theft (11.1%), and arson (0.65%) were committed more than 9.8 million times in 2007, resulting in estimated economic losses approaching 18 billion dollars (FBI, 2008).

If there is a silver lining, it would be that violent (−0.7%) and property (−1.4%) crime both decreased from the previous year (2006). In addition to the eight offenses comprising violent and property crime, however, their lesser included crimes (or cousins, so to speak) also are discussed throughout the text. A **lesser included offense** is a crime possessing the fundamental elements required of a greater, more serious crime but missing a key component. For example, murder is the most serious form of criminal homicide; manslaughter, however, is a lesser included offense of murder because you cannot commit murder without meeting or exceeding the *actus reus* (guilty act) and *mens rea* (guilty mind) requirements of manslaughter.

## Summary

This chapter sought to outline the principles and working vocabulary (legalese) essential for developing a fundamental understanding of substantive criminal law in the Republic known as the United States. From the formation of social contract theory to the application of law in contemporary society, this chapter aimed to provide students with a comprehensive understanding of the historical evolution and practical application of the rules of substantive criminal law. Students should now have little trouble citing the sources from which law is derived (common, statutory, case, constitutional, administrative) and how crime traditionally is defined and classified (felonies, misdemeanors). Against this backdrop, students now should be armed with the legal tools with which to decipher whether an accused is liable for conduct outlined in substantive codes: *mens rea*, *actus reus*, and attendant circumstances.

# Practice Test

1. The _____ is routinely cited as the first set of written laws to govern society. [written: Code / Hammurabi]
   a. Code of Hammurabi
   b. Ten Commandments
   c. Dead Sea scroll
   d. Babylonian Sacrament
   e. Assyrian Statutory Code

2. _____ law refers to the historical laws of the Catholic Church. [written: Canon]
   a. Ash
   b. Positive
   c. Canon
   d. Common
   e. Papal

3. The FBI annually compiles the _____ to provide data regarding the extent of violent and property crime in America. [written: UCR]
   a. National Crime Survey
   b. American Crime Statistics
   c. Federal Crime Report
   d. Criminal Activity Survey
   e. Uniform Crime Reports

4. _____ defines a government of elected leaders operating under the umbrella of a Constitution which safeguards the interest of the nation through limiting power. [written: Republic]
   a. Constitutionalism
   b. Republic
   c. Sovereignty
   d. Socialism
   e. Democracy

5. _____ stipulates that citizens will voluntarily waive rights, privileges, and liberties guaranteed in the U.S. Constitution in exchange for government protection. [written: Social Contract theory]
   a. Due Process
   b. *Stare decisis*
   c. Equal Protection
   d. Social contract theory
   e. Natural law

6. _____ law originates with legislative bodies and serves as the prominent source for the establishment of substantive criminal law. [written: Statutory]
   a. Common
   b. Statutory
   c. Administrative
   d. Constitutional
   e. Case

7. _____ law is defined as judicial decisions manifesting the customs and traditions practiced throughout the circuits of England.
   a. Criminal
   b. Positive
   c. Ecclesiastical
   d. Common
   e. Canon

8. _____ means "let the decision stand." [written: Stare decisis]
   a. *Mala prohibita*
   b. *Actus reus*
   c. *Mala in se*
   d. *Mens rea*
   e. *Stare decisis*

9. A(n) _____ is defined as a crime punishable from 3 to 6 months in jail. [written: Ordinary misde]
   a. ordinary misdemeanor
   b. strict liability crime
   c. gross misdemeanor
   d. petty misdemeanor
   e. true crime

10. A(n) _____ represents the legislative efforts of local government (county and/or municipal) to regulate problem behaviors within its jurisdictional boundaries. [written: Ordinance]
    a. misdemeanor
    b. crime
    c. felony
    d. administrative policy
    e. ordinance

11. Moral turpitude crimes are referred to as _____, meaning wrong in itself. [written: mala in se]
    a. *mala prohibita*
    b. *actus reus*
    c. *mala in se*
    d. *corpus delicti*
    e. *mens rea*

12. _____ is translated as "guilty act." [written: Actus Reus]
    a. *Actus reus*
    b. *Mala prohibita*
    c. *Corpus delicti*
    d. *Mala in se*
    e. *Mens rea*

13. The _____ standard is used in civil cases, whereby one need only establish a greater likelihood that harm occurred. [written: Preponderance of the evidence]
    a. beyond a reasonable doubt
    b. civil scale
    c. civil injury
    d. preponderance of the evidence
    e. incurred harm rule

Summary 15

**14.** _____ is translated as "body of the crime."
   a. *Mens rea*
   b. *Mala prohibita*
   c. *Corpus delicti*
   d. *La cosa de criminal*
   e. *Actus reus*

**15.** In order to hold an accused person liable for harm, the _____ for a criminal offense must be sufficiently proven.
   a. burden of proof
   b. actual cause standard
   c. but-for test
   d. substantial factor test
   e. proximate cause standard

**16.** _____ refers to the connection between harm in question and the actual conduct of the criminally accused.
   a. Uniform Determination
   b. Actual cause
   c. Contractual relation
   d. Legal cause
   e. Proximate cause

**17.** Drug possession and statutory rape are two examples of crimes often exempt from the *mens rea* requirement, meaning that they possess _____.
   a. injunctive relief
   b. declaratory relief
   c. strict liability
   d. general intent
   e. specific intent

**18.** _____ is defined as having some degree of malevolent or wrongful design but with no particularized objective.
   a. General intent
   b. Aimless intent
   c. Specific intent
   d. Malicious design
   e. Criminal design

**19.** The doctrine of _____ stipulates that when a person intends to cause harm but instead erroneously inflicts harm on an unintended target, the law can presume general intent was present with respect to the party actually harmed.
   a. actual cause
   b. transferred intent
   c. federalism
   d. legal cause
   e. proximation

**20.** Referred to as _____, these legal elements must accompany *actus reus* and *mens rea* for most crimes to be punished.
   a. *corpus delicti*
   b. incarceration factors
   c. substantial components
   d. attendant circumstances
   e. extenuating circumstances

## References

Federal Bureau of Investigation. (2008). *Crime in the United States, 2007: Uniform Crime Reports*. Retrieved July 14, 2009, from http://www.fbi.gov/ucr/cius2007/index.html

Garner, B. A. (Ed.). (2009). *Black's law dictionary* (9th ed.). Eagan, MN: West Group.

# Crime and Punishment: Constitutional Limitations and Protections

**CHAPTER 2**

## Key Terms

Aggravating circumstance
Bail
Bifurcated proceeding
Checks and balances
Civil forfeiture
Compulsory process
Criminal forfeiture
Death-qualified jury
Determinate sentence
Deterrence
Doctrine of overbreadth
Dual sovereignty
Due process
Eminent domain
Equal protection clause
Excessive bail
Exclusionary rule
Executive branch
Federal Sentencing Guidelines
Fine
Forfeiture
General deterrence
Grand jury
Habitual offender statute
Incapacitation
Incorporation
Indeterminate sentence
Information
Intensive supervision probation
Judicial activism
Judicial branch
Legislative branch
Mitigating circumstance
No bill
*Parens patriae*
Predicate crime
Privilege against self-incrimination
Probable cause
Probation
Procedural due process
Proportionality of punishment
Rehabilitation
Restitution
Restorative justice
Retribution
Selective incorporation
Sentencing disparity
Sentencing Reform Act of 1984
Separation of powers
Specific deterrence
Speedy trial
Substantive due process
True bill
Void for vagueness
*Voir dire*
Wergild

## Introduction

This chapter explores the function of the United States Constitution as a principled measure to avoid uncontrolled government intrusion into the lives of citizens. While progressing through its content, readers will be exposed to the role the Constitution plays in (1) protecting the integrity of the law making function and (2) mediating the relationship between sovereign and citizen. Specific attention is given to those constitutional amendments which impact the criminal justice process. In addition to constitutional notions of fair play that impact the legal process, this chapter will also explore prevailing theories of punishment, constitutional limitations regarding the nature and extent of punishment, and the many alternative forms of punishment.

## United States Constitution and Criminal Law

In light of the oppressive system of government which existed in England, those who settled the United States designed a government that would be "by the people and of the people." The U.S. Constitution and The Bill of Rights provide the framework for the balance between government power and personal liberty. Rights are guaranteed and may not be taken away or limited without certain protections. The U.S. Constitution proscribes a three-pronged system of government with limited powers. Government power is therefore delineated among three distinct branches. The doctrine of **separation of powers** reflects a concerted effort by the drafters to avoid the concentration of government power in one individual (i.e., monarch, dictator, or one branch of government). Remember that these individuals fled a monarchy to avoid such situations. The U.S. Constitution not only separates powers among the three branches of government but also distinguishes between power granted to the federal or national government and that which is reserved to the states. Thus, the doctrine of separation of powers exists on two levels within the Constitution.

The effort to strengthen the national government through identifying matters that fall within its exclusive power is referred to as federalism; on such matters, federal law always prevails. Principles of federalism are entrenched in the American legal system and serve to maintain a strong central government while respecting state sovereignty. Subjects within the purview of the federal government are the powers to declare war, regulate interstate commerce, and operate the national government. In contrast, certain matters are specifically reserved for the states. Police power, or the protection of the health, safety, and welfare of citizens, is an example of a power specifically reserved for the states. Thus, states have the primary authority to enact laws affecting the health, safety, and welfare of their citizens. The power to tax citizens, however, seems to be one possessed by all levels of government.

The presence (or absence) of power within each branch of government serves to regulate its power. This system of **checks and balances** is designed to prevent tyranny by any singular branch. Thus, each branch is kept "in check" by specific powers vested to the other two branches. The legislative branch is vested with the power to make law, whereas the executive branch enforces and judicial branch interprets those laws. This triumvirate form of government also is used by the states; however, state governments are proscribed by state constitutions.

In the federal government, the **legislative branch** (United States Congress) possesses the constitutional authority to enact laws regarding the areas discussed previously. The **executive branch**, however, consists of many entities, with the President serving as its chief executive. Finally, the **judicial branch** possesses the authority to interpret both the U.S. Constitution and laws enacted by Congress. In principle, the judicial branch should not be in the business of making law; however, in many cases, critics contend that federal judges have overstepped and attempt to legislate by way of court decisions. This is referred to as **judicial activism**. An activist judge is one who relies on personal ideology to guide judicial decisions as opposed to the facts of the case and rule of law. Critics argue that activist judges fail to rule in an objective and impartial manner in furtherance of their own agenda.

## Constitutional Principles and Limitations

The U.S. Constitution provides several provisions designed to limit the nature and extent of government intrusion into the lives of American citizens. In the following discussion, we focus on those provisions that may apply to the application of criminal law or impact criminal procedure and punishment. Before examining the constitutional amendments related to criminal law and procedure, it is important to understand certain overarching legal doctrines restricting the manner in which laws may be drafted or applied. First, laws must be sufficiently specific so that an average or reasonable person is able to determine what conduct is or is not prohibited. If a law is not sufficiently specific or clear, a court may conclude that the law is "**void for vagueness**." In other words, the law is so unclear that the average person is not able to determine what conduct is legal or illegal. Vague laws are not constitutional and may violate the due process clauses of the Fifth or Fourteenth Amendments.

Although the creation and application of criminal law have historically been matters reserved to the states, the jurisdiction of the federal government with regard to crime continues to expand. Thus, criminal offenses may exist on both the state and federal levels. Given the significant and increasing areas of overlap among federal and state crimes, we address constitutional principles that may apply to each. Constitutional provisions regarding the application of criminal law are generally found in four amendments to the U.S. Constitution: Fourth, Fifth, Sixth, and Eighth. These provisions, all contained in the Bill of Rights, are made applicable to the states through the Fourteenth Amendment.

The Bill of Rights was originally intended only to apply to the federal government. The U.S. Supreme Court affirmed this interpretation in *Barron v. Baltimore* (1833). The underlying purpose of this limited applicability was to assure states that the federal government would not encroach on state issues, hence the enduring states' rights debate. Most state constitutions possessed comparable provisions to protect individual rights; however, after the Civil War, it became apparent that constitutional limitations on government power must be extended to the states to protect the newly freed slaves from states seeking to infringe on individual liberties.

## Fourteenth Amendment

> All persons born or naturalized in the United States, and subject to the jurisdiction thereof, are citizens of the United States and of the State wherein they reside. No State shall make or enforce any law which shall abridge the privileges or immunities of citizens of the United States; nor shall any State deprive any person of life, liberty, or property, without due process of law; nor deny to any person within its jurisdiction the equal protection of the laws.

In 1868, the Fourteenth Amendment was enacted and included three central provisions: the privileges and immunities clause, the due process clause, and the equal protection clause. Over the next several decades, the Amendments in the Bill of Rights were, through the process of incorporation, applied to the states. **Incorporation** refers to the process wherein the protections contained in the Bill of Rights are applied to the states via the Fourteenth Amendment. Incorporation occurred incrementally and primarily through the use of the due process clause of the Fourteenth Amendment.

The position of the U.S. Supreme Court regarding incorporation has fluctuated over the years. Although some justices favor total incorporation of all rights contained in the Bill of Rights, others opt for selective incorporation. **Selective incorporation** refers to the process of applying only those rights fundamental in nature. Determination of what rights are fundamental has been a long and arduous process for the Court. Today, however, only two provisions in the Bill of Rights have not been incorporated and applied to the states: (1) the Fifth Amendment right to grand jury indictment and (2) the Eighth Amendment prohibition against excessive bail.

**Due process** refers to the requirement that government follow certain procedures before infringing on the life, liberty, or property of a private citizen. A precise definition of due process is difficult to provide, as its parameters have proven elusive. Interpretation of the due process clause is more thoroughly discussed in the section devoted to the Fifth Amendment. The **equal protection clause** prohibits states from making arbitrary and unreasonable distinctions among people in terms of their rights and freedoms. Although the equal protection clause does not prohibit all distinctions, states must be able to demonstrate sufficient justification for classifications. For example, states may not enact laws or regulations that only allow Native Americans to drive. This would be a distinction based on race or ethnicity.

Because the distinction is based on an immutable characteristic (one the person cannot control) that has nothing to do with one's ability to drive, this law would be unconstitutional. All race-based classifications are treated as suspect classifications by the U.S. Supreme Court and are illegal.

Cases involving gender, age, and illegitimacy have not necessarily been deemed unconstitutional but are subject to heightened scrutiny by the Court. The state must establish the existence of an important state interest and that the law that draws such distinctions is substantially related to that interest. If so, the law or policy may be upheld. For example, age requirements for obtaining a marriage license may be upheld if the state can demonstrate an important state interest that underlies this requirement and that the law is substantially related to that state interest. This law would be upheld. Underlying laws of this nature reflect the state's interests in protecting minors from the consequences of immature decision making and from the consequences of early marriages, and to promote the stability of marriage. These have all been deemed important state interests. Moreover, the law requiring that individuals be of a certain age before a marriage license is granted is substantially related to the *parens patriae* function of the state. ***Parens patriae*** (as defined within Black's Law Dictionary), literally interpreted as the "king is the father," refers to the ability of the state to serve as the ultimate parent or guardian of persons with legal disabilities such as children (Garner, 2009). As you can see, resolving equal protection challenges can be difficult and has evolved into a complex area of the law.

## Fourth Amendment

> The right of the people to be secure in their persons, houses, papers, and effects, against unreasonable searches and seizures, shall not be violated, and no Warrants shall issue, but upon probable cause, supported by Oath or affirmation, and particularly describing the place to be searched, and the persons or things to be seized.

The Fourth Amendment is extremely important in terms of individual rights and liberties. This Amendment protects citizens by limiting the authority of government actors to intrude on the privacy of private citizens when searching for evidence. Given the unbridled ability of the police to interfere in the lives of private citizens in England, the drafters of the U.S. Constitution wanted to ensure that American citizens were protected from unreasonable searches and seizures of their homes and other places where there is a reasonable expectation of privacy (*Mapp v. Ohio*, 1961). The task of determining whether a search or seizure is unreasonable has been an arduous one for state and federal courts. Some scholars suggest that all searches conducted with less than probable cause are unreasonable; however, the U.S. Supreme Court has allowed a lesser standard known as reasonable suspicion to be used as a basis for searches in limited circumstances. For example, searches in public schools (*New Jersey v. T.L.O.*, 1985) and stop and frisk (*Terry v. Ohio*, 1968) are governed by the reasonable suspicion standard.

The drafters also included a provision that requires probable cause to exist before an arrest or search warrant may be issued. **Probable cause** is a legal standard that requires a judicial determination that a strong probability exists that a crime has been committed, the individual named in the warrant application committed the crime, or in the case of search warrants, that evidence of a crime will be located in the area(s) described in the application. Probable cause requires a threshold level of proof and is not as exacting as the level of proof required for other standards such as beyond a reasonable doubt, clear and convincing evidence, or preponderance of the evidence.

When applying for a warrant, the Fourth Amendment requires law enforcement to describe the areas or persons to be searched or arrested with "particularity." The particularity requirement was included to prevent the use of the general warrant. The use of the general warrant was common in England and, once in hand, essentially allowed the police to search any place for anything. Unrestricted access to the homes, persons and effects of private citizens created significant opportunities for abuse. Thus, the drafters of the U.S. Constitution included the particularity requirement to ensure that limits exist when searches and/or arrests occur. Finally, all applications for warrants must be made under oath or affirmation.

Without a mechanism for enforcement and accountability, Fourth Amendment protections are meaningless. Thus, the U.S. Supreme Court affirmed the use of the **exclusionary rule**, which prohibits the use of evidence obtained in violation of the Fourth Amendment in criminal trials (*Weeks v. United States*, 1914; *Mapp v. Ohio*, 1961). Thus, if a search is conducted without probable cause, the evidence may not be used at trial. The risk of losing the use of evidence in a criminal trial is intended to deter law enforcement from knowingly violating the Fourth Amendment.

## Fifth Amendment

No person shall be held to answer for a capital, or otherwise infamous Crime, unless on a presentment or indictment of a Grand Jury, except in cases arising in the land or naval forces, or in the Militia, when in actual service in time of War or public danger; nor shall any person be subject for the same offence to be twice put in jeopardy of life or limb; nor shall be compelled in any criminal case to be a witness against himself, nor be deprived of life, liberty, or property, without due process of law; nor shall private property be taken for public use, without just compensation.

When most Americans think of those protections contained in the Fifth Amendment, they immediately think of the privilege against self-incrimination; however, this protection is one of several which are contained in this important amendment.

### Grand Jury

The Fifth Amendment also guarantees citizens the right to be indicted by a **grand jury**—a body of citizens drawn from the rolls of registered voters who evaluate whether there is sufficient proof to go forward with criminal charges and have a trial. Thus, after criminal charges are filed against an individual, the prosecutor will present the state's case to the grand jury. The grand jury hears only the prosecutor and does not hear from the defense. A defendant may testify if he or she would like, but this would be extremely rare given the potential for a defendant to incriminate themselves. The purpose of the grand jury is to protect citizens from arbitrary prosecutions. Thus, use of the grand jury procedure is another check on the power of prosecutors to bring citizens to trial for criminal offenses.

Grand juries operate in secret, and their deliberations are closed to the public. After hearing the evidence presented by the prosecution, the grand jury may return a true bill of indictment or a no bill. A **true bill** indicates that there is sufficient evidence to go to trial. A **no bill** means the opposite—that there is insufficient evidence to go to trial. Grand juries may also serve as an investigatory body. In this capacity, a grand jury may subpoena witnesses and compel testimony or the production of documents. In *Hurtado v. California* (1884), the U.S. Supreme Court held that the right to be indicted by a grand jury is not binding on the states. Thus, in all federal cases, the defendant is entitled to have his or her case presented to the grand jury for review, but this is not required in state prosecutions. Most states do, however, use grand juries even though not constitutionally required, whereas others use the information as an alternative to indictment by a grand jury. The **information** is a formal charging document filed by the prosecutor with the court.

### Double Jeopardy

The U.S. Constitution, as well as most state constitutions, contains a prohibition against double jeopardy.

During medieval times, there were no limits on the number of prosecutions or punishments that a criminal defendant might endure. As such, the drafters of the U.S. Constitution were careful to include provisions that eliminated any risk of these practices occurring in their new legal system. **Double jeopardy** occurs when a citizen is twice put in jeopardy of prosecution or punishment for the same offense. Thus, the prohibition against double jeopardy applies on two levels: multiple prosecutions and multiple punishments. The underlying purpose of the double jeopardy clause is to prevent the government from repeatedly placing citizens in jeopardy of conviction or loss of liberty. The prohibition against double jeopardy serves to shield citizens from the extreme physical and psychological distress that may result from multiple prosecutions or punishments. In *Green v. United States* (1957), the U.S. Supreme Court held that

> [t]he underlying idea, one that is deeply ingrained in at least the Anglo-American system of jurisprudence, is that the State, with all its resources and power, should not be allowed to make repeated attempts to convict an individual for an alleged offense, thereby subjecting him to the embarrassment, expense and ordeal and compelling him to live in a continuing state of anxiety and insecurity, as well as enhancing the possibility that even though innocent he may be found guilty.

Many of us have heard statements such as "the government only gets one bite at the apple." This is actually a reference to double jeopardy. The government may prosecute an individual only once for a particular crime. If the jury finds the defendant "not guilty," he or she may not be retried for the same offense; however, if there is a mistrial or if the defendant wins an appeal, he or she may be tried again. Multiple punishments are also prohibited by double jeopardy. If a defendant is tried for murder and found guilty, he or she may only receive one sentence for the crime of murder; however, if the defendant is convicted of two murders, he or she may receive a sentence on each crime. Thus, when a defendant is charged with multiple counts or multiple crimes arising out of the same circumstance or transaction, he or she may receive separate punishment on each count or crime.

There are a few exceptions to the general prohibition against double jeopardy. For example, **dual sovereignty** refers to multiple prosecutions by different governments (i.e., different states, or federal and state). In such situations, the authority to prosecute, convict, and punish is derived from different sovereigns. Pause for Thought 2–1 illustrates how to interpret this legal issue.

In *United States v. Lanza* (1922), the U.S. Supreme Court upheld the prosecution and conviction in federal court of a defendant convicted in state court for the same offense. Later, in *Heath v. Alabama* (1985), the Court also held that the dual sovereignty doctrine allowed successive prosecutions by different states for the same conduct.

### Self-Incrimination

The **privilege against self-incrimination** provides that no man or woman can be forced to be a witness against himself or herself. Force may include physical or psychological coercion. Inclusion of this provision in the U.S. Constitution was necessary to protect Americans from the widespread use of physical and mental torture, which was commonly used in England to obtain confessions. In light of this history, the drafters of the Constitution sought to forbid expressly the use of such tactics. Thus, if questioned, a suspect may refuse to speak to law enforcement about a crime. Also, the privilege allows defendants to refuse to testify at trial, and

### Pause for Thought 2–1

Consider the following: Kix the Kidnapper abducts a convenience store clerk from a small town in Louisiana and then transports his victim to Florida before releasing him. Kix is later apprehended and charged with the crime of kidnapping by the FBI. This a federal charge resulting from a federal crime. The state of Louisiana, however, also charges Kix with kidnapping pursuant to a state statute. In a pretrial motion, Kix's lawyer argues that both federal and state (Louisiana) charges for the same offense violate the double jeopardy clause. Is the attorney's argument constitutionally valid?

### Scenario Solution

No, the charges are by different governments (i.e., state of Louisiana and federal government), and thus, the dual sovereignty exception applies. There is no double jeopardy violation.

the prosecution cannot comment on this refusal (*Griffin v. California*, 1965).

The privilege against self-incrimination applies only to testimonial evidence, that is, verbal admissions. Thus, a person may be compelled to provide a writing sample, blood sample, fingerprints, or other forms of nontestimonial evidence without violating the Fifth Amendment. Another criterion is the requirement that the admissions must be incriminating. Thus, the defendant may invoke the privilege only when admissions may provide incriminating evidence. In order for a statement to be incriminating, it must in some way provide information that subjects a declarant to the possibility of loss of liberty. If the disclosure would only result in a civil action such as a claim for monetary damages, embarrassment, or humiliation, the privilege may not be invoked.

Privilege against self-incrimination gained national attention when it became the central issue in *Miranda v. Arizona* (1966). In *Miranda*, the U.S. Supreme Court addressed the need for verbal warnings regarding self-incrimination (and other rights). The Court acknowledged the psychological coercion, which often occurs when suspects are in custody of law enforcement and interrogated. If these two circumstances exist, law enforcement officers must read the *Miranda* warnings to suspects before interrogation. The warnings state the following:

1. You have the right to remain silent.
2. Anything you say can and will be used against you in court.
3. You have the right to speak to an attorney before questioning and to have your attorney present during questioning if you wish.
4. If you cannot afford a lawyer, one will be appointed free of charge before questioning.
5. You can decide at any time not to answer any questions or make any statements.

## Due Process

The due process guarantee is first found in the Fifth Amendment. Because the U.S. Supreme Court in *Barron v. Baltimore* (1833) held that the Bill of Rights only restricts the actions of federal government, a second due process clause was included in the Fourteenth Amendment. The clauses are virtually identical and interpreted by courts in a similar manner. As noted earlier in this chapter, due process refers to the requirement that government follow certain procedures before infringing on the life, liberty, or property of a private citizen. A precise definition of due process is difficult to provide as the exact parameters of due process have proven elusive. In *Solesbee v. Balkcom* (1950), the U.S. Supreme Court held

> It is now settled doctrine of this Court that the Due Process Clause embodies a system of rights based on moral principles so deeply imbedded in the traditions and feelings of our people as to be deemed fundamental to a civilized society as conceived by our whole history. Due Process is that which comports with the deepest notions of what is fair and right and just.

The U.S. Supreme Court has spent decades interpreting the due process clause, producing two distinct dimensions: substantive due process and procedural due process. **Substantive due process** refers to certain freedoms and protections that are inherent in the concept of liberty. In other words, there are certain areas of life that are so central to a free society that government interference in those areas should be restricted. For example, the freedom of choice regarding an abortion (*Roe v. Wade*, 1973), the freedom of choice regarding conception (*Griswold v. Connecticut*, 1965), and the freedom to parent one's child as seen fit, as well as others, fall within the guarantee of substantive due process. **Procedural due process** refers to the requirement that a fair process must be in place before a person is deprived of life, liberty, or property. For example, in order to be deprived of one's liberty, a person must be provided notice of charges, the opportunity to be heard, and a fair trial. Such guarantees ensure that individuals accused of crimes will not be persecuted in an arbitrary and capricious manner.

The U.S. Supreme Court has held that individuals should not be required to "speculate" as to the meaning of laws (*Lanzetta v. New Jersey*, 1939). Vague laws (both state and federal) are not constitutional and violate the due process clause. A vague state law violates the due process clause of the Fourteenth Amendment, whereas vague federal laws violate the due process clause of the Fifth Amendment. Vagrancy, curfew, and loitering statutes have been particularly problematic under the void for vagueness doctrine. The Court has addressed problems with these types of statutes in several important cases, including *Papachristou et al. v. City of Jacksonville* (1972), *Chicago v. Morales* (1999), and *Coates v. City of Cincinnati* (1971). In many cases, a law is argued as both vague and overbroad. Laws that are vague are not specific enough, whereas overly broad laws tend to be so general that they could apply to and criminalize both legal and illegal behavior. The **doctrine of overbreadth** typically is raised in cases involving First Amendment protections such as freedom of assembly and speech.

## Eminent Domain

The final protection provided by the Fifth Amendment is that of **eminent domain**, a requirement that citizens must be given just compensation when the government takes private property for personal use. Although this provision has little to do with criminal law or procedure, it provides a remedy for citizens when their property is needed for public use. This ensures that the government may not seize private property at will without compensating the owner. Eminent domain has evolved into a complex area of the law with many avenues for challenging the taking of private property as well as the reasonableness of the compensation.

## Sixth Amendment

> In all criminal prosecutions, the accused shall enjoy the right to a speedy and public trial, by an impartial jury of the State and district wherein the crime shall have been committed, which district shall have been previously ascertained by law, and to be informed of the nature and cause of the accusation; to be confronted with the witnesses against him; to have compulsory process for obtaining witnesses in his favor, and to have the Assistance of Counsel for his defense.

Like the Fifth Amendment, the Sixth Amendment contains many different protections that apply to criminal procedure. These include the right to a speedy and public trial, the right to an impartial jury drawn from the venue where the crime occurred, the right to receive notice of the charges brought by the government, the right to confront witnesses at trial, the right to compel witnesses to appear at trial, and the right to assistance of counsel.

### Right to Speedy and Public Trial

In felony matters, criminal defendants are entitled to a speedy and public trial. A **speedy trial** is one that occurs without unnecessary delay. All familiar with the criminal justice system understand that delays are inevitable. Thus, all delay is not prohibited by the Sixth Amendment, only that which is unreasonable or unnecessary. This protection serves to provide criminal defendants with the opportunity to have their cases heard and disposed of in a reasonable amount of time. It is imperative, however, that criminal defendants assert their right to a speedy trial.

What is reasonable is defined on two levels. First, state statutes establish timelines for criminal trials. Thus, each state will set forth a time period during which a criminal trial must be held. For example, a state statute may require that a trial be held within 270 days from indictment. As such, trials that are not held within these time frames may be in violation of state statute. Second, reasonableness is determined by reference to the Sixth Amendment as interpreted by the U.S. Supreme Court in *Barker v. Wingo* (1972) and other cases. In *Barker*, the Court established a four-prong balancing test to evaluate claims that the government had denied the defendant a speedy trial. These factors are as follows: (1) length of the delay, (2) reason for the delay, (3) the defendant's assertion or non-assertion of the right to speedy trial, and (4) prejudice to the defendant resulting from the delay. A criminal defendant cannot allow the clock to run and then claim that his or her rights were violated. There is a clear obligation to demand a speedy trial. If a violation of the speedy trial provision is claimed, the appellate court will apply the *Barker* balancing test to the facts of the case. During this analysis, the conduct of the prosecution and the defense is weighed. If the court concludes that a violation has occurred, the court may dismiss the indictment or reverse the case so that the trial court may dismiss the indictment. As such, violations of the Sixth Amendment may result in a dismissal of the case against the defendant. Such a result is warranted as the Sixth Amendment's right to a speedy trial is considered to be a fundamental constitutional right.

Jury trials also must be public. The requirement that trials be public assures the criminal defendant that the proceedings will be open for public scrutiny. Sunlight, according to many, is the best antiseptic. As such, this right serves to protect criminal defendants from persecution by the government in secret proceedings. The U.S. Supreme Court stated that "[t]he knowledge that every criminal trial is subject to contemporaneous review in the form of public opinion is an effective restraint on possible abuse of judicial power" (*In re Oliver*, 1948).

### Right to Impartial Jury

With the exception of petty offenses, those facing criminal charges are entitled to have their case heard by a jury. This requirement was not applied to the states until 1968 when the Court, in *Duncan v. Louisiana* (1968), incorporated the right via the due process clause of the Fourteenth Amendment; however, most states provided the right to a jury in their own state constitutions or statutes prior to *Duncan*. The right to a jury trial attempts to ensure that criminal defendants are shielded from overzealous prosecutors and/or judges. Thus, the central issue in criminal trials, resolution of factual matters, is a matter left solely to the discretion of the jury.

The Sixth Amendment also guarantees criminal defendants the right to an impartial jury. Again, this protection is an attempt to ensure cases are resolved by objective jurors. Juries should be chosen from a cross-section of the community in which the crime occurred. In order to ensure objectivity, *voir dire* is used to assess the competence of jurors, bias, previous knowledge of the case or the actors involved. **Voir dire** is a process in which the prosecutor and defense attorney are allowed to question prospective jurors. Assessment of juror responses allows the attorneys to determine which jurors they would accept on the jury. Attorneys may exclude or challenge jurors. A challenge for cause is one based on the juror's responses to questions posed during *voir dire*. Challenges for cause may be made where the juror has preexisting knowledge of the case, is related to or knows the defendant, judge, or attorneys, has a conflict of interest in the case, or knows other facts that may undermine the juror's ability to be impartial in the case.

Peremptory challenges may also be used by attorneys to exclude prospective jurors. Unlike challenges for cause, the peremptory challenge, in theory, may be used for any reason. For example, perhaps the defense attorney does not like the hairstyle of a particular juror. The attorney may use one of the allotted peremptory challenges to exclude that juror. The juror will be dismissed. The continued use of peremptory challenges has been the subject of much debate. Given the potential for abuse, legal scholars and commentators have suggested that peremptory challenges no longer be allowed. In *Batson v. Kentucky* (1986), the U.S. Supreme Court held that the due process clause of the Fourteenth Amendment prohibited the use of peremptory challenges to exclude jurors solely on the basis of their race. In a later case, the Court, in *J.E.B. v. Alabama* (1994), extended the logic of the *Batson* holding to the use of peremptory challenges to exclude jurors solely on the basis of gender. Thus, peremptory challenges may not be used as a tool to perpetuate gender or racial discrimination. Although peremptory challenges may not be used in a discriminatory manner, criminal defendants are not entitled to have a jury of any particular racial or gendered makeup, only one that is impartial.

## Notice of Charges

The Sixth Amendment requires that criminal defendants receive notice of the charges against them. Notice of the nature and cause of the accusation is required to ensure that the defendant is able to formulate a meaningful defense to the allegations. Notice, for purposes of the Sixth Amendment, is typically in the form of an indictment or information that contains written notice of the specific allegations. In order to satisfy the Sixth Amendment, the indictment must be served on the defendant.

## Right to Confront Witnesses at Trial

The Sixth Amendment right to confrontation includes the right to confront adverse witnesses and cross-examine those witnesses during trial. Confrontation of adverse witnesses has been a central issue in many cases before the U.S. Supreme Court. The Court has held on many occasions that the confrontation clause guarantees a criminal defendant the right to a face to face meeting with his or her accuser (i.e., right to cross-examine witness at trial).

In *Crawford v. Washington* (2004), the U.S. Supreme Court reviewed the historical bases for the confrontation clause. In essence, the Framers sought to prohibit the use of *ex parte* affidavits and depositions, as opposed to in-court testimony, at trial. The Court specifically addressed the case of Sir Walter Raleigh. During his trial for treason, the prosecution presented an affidavit from his accuser. The accuser did not appear at trial, but rather, the affidavit was read to the jury. Raleigh had no opportunity for cross-examination. This process merely allowed the reading of hearsay evidence to a jury in a capital case. Such procedure deprived the defendant of an opportunity to confront his or her accuser and cross-examine regarding his or her recollection, credibility, and motives. Thus, the accuser went untested by the "... crucible of cross-examination." As a result, the Framers included the right to confrontation in the Constitution to preclude such opportunities in criminal trials. As such, prior testimonial evidence may not be produced at trial unless the prosecution can establish that the witness is no longer available and the defendant had a previous opportunity for cross-examination.

## Right to Compel Witnesses

During criminal trials, the defendant has the right to **compulsory process** requiring the presence of witnesses who may offer favorable testimony. Defendants may need witnesses to testify regarding character, alibi, or to contest facts of the case. Although many witnesses voluntarily attend court proceedings when needed, others may not. In those cases, a defendant may subpoena the witness to court. The subpoena power of the courts enables a defendant to compel the presence of an individual in court. If the individual fails to attend court, law enforcement may secure their presence, or alternatively, they may be held in contempt of court.

In *Washington v. Texas* (1967), the U.S. Supreme Court applied this Sixth Amendment right to states through the Fourteenth Amendment. There, the Court held that

[t]he right of an accused to have compulsory process for obtaining witnesses in his favor stands

on no lesser footing than the other Sixth Amendment rights that we have previously held applicable to the States.

The Court held that the right is a fundamental element of due process, as the right to present testimony of witnesses is the linchpin of the ability to present a sound defense.

## Right to Counsel

The Sixth Amendment guarantees criminal defendants the right to counsel. Counsel must be afforded at all critical stages of the criminal process (*Kirby v. Illinois*, 1972). Thus, after formal charges have been filed and the machinery and resources of the government are focused on the defendant, the Sixth Amendment right to counsel is triggered. After indictment, counsel must be present at interrogations, lineups, and any other legal proceeding. The requirement of counsel is to ensure that proceedings against the defendant are fair and that the rights of the defendant are protected. Given that few criminal defendants possess the requisite legal knowledge to represent themselves, counsel is afforded to them. In an adversary system, the defendant requires counsel to ensure that the playing field is level. For indigent defendants, counsel must be appointed by the court and compensated with government funds.

In *Gideon v. Wainwright* (1963), the U.S. Supreme Court addressed the issue of appointed counsel for indigent defendants. There, the Court held that lawyers in criminal courts are ". . . necessities, not luxuries." Thus, regardless of financial status, all defendants should have access to assistance of counsel. This requirement applies when defendants are facing charges that could result in imprisonment for 6 months or more. The mere appointment or presence of counsel does not fulfill the obligations imposed by the Sixth Amendment. Rather, counsel must provide "effective" assistance (*Strickland v. Washington*, 1984). Thus, attorneys in criminal cases, appointed or retained, must provide competent legal assistance for their clients. When failure to provide effective assistance results in prejudice to the defendant, it may be deemed a violation of the Sixth Amendment.

# Eighth Amendment

> Excessive bail shall not be required, nor excessive fines imposed, nor Cruel and unusual punishments inflicted.

The Eighth Amendment provides three rights for those charged with criminal offenses. These protections prevent excessive bail or fines as well as punishments that are cruel and unusual. Inclusion of these rights in the Bill of Rights reflects the drafters' intent to protect individuals charged with crimes from the severe punitive measures that were often exerted on criminal defendants before and after conviction of a crime. Thus, the Eighth Amendment provides the primary source of constitutional limitation on criminal punishment.

In *Atkins v. Virginia* (2002), the U.S. Supreme Court explained that the Eighth Amendment guarantees individuals the right not to be subjected to excessive sanctions. The right flows from the basic "precept of justice that punishment for crime should be graduated and proportioned to [the] offense" (*Weems v. United States*, 1910). By protecting even those convicted of heinous crimes, the Eighth Amendment reaffirms the duty of government to respect the dignity of all persons.

## Excessive Bail

**Bail** refers to a court-determined amount of money or property to be deposited before a defendant may be released pending trial. The sole purpose of bail is to ensure the presence of the defendant at criminal proceedings. Bail is not intended to punish the defendant for an alleged wrongdoing. At common law, bail was guaranteed, but was often set at an amount so high that the defendant could not obtain release. Although the Eighth Amendment does not guarantee that bail will be set in any particular case, it does require that when bail is set, the amount should not be excessive; however, what constitutes "excessive" bail? In *Stack v. Boyle* (1951), the U.S. Supreme Court provided some guidance in holding that **excessive bail** is an amount that exceeds what is necessary to assure reasonably the presence of the defendant at trial. In cases in which the defendant is charged with a capital crime (i.e., eligible for the death penalty), many states do not even allow bail. As such, excessive is not an issue; however, in noncapital cases, a criminal defendant is entitled to bail unless the prosecutor can demonstrate the defendant is a threat to public safety, to witnesses, or to self or that the defendant is a flight risk.

States differ somewhat regarding the process to set bail. In some jurisdictions, a bail amount may be set by the court during arraignment. Once a bail is set, the defendant can then request a hearing to reduce its amount. In other jurisdictions, a full hearing is required to set a bail amount. At bail hearings, each side may present evidence regarding the following factors:

- Nature of the offense
- Criminal history of the defendant
- Risk of flight or failure to appear in court
- Financial ability of the defendant
- Any threat posed by the defendant (to self or others)

Following consideration of these factors, the court may then determine an initial bail amount or reduce the amount originally set by the court.

## Excessive Fines

The Eighth Amendment also places constitutional limitations on the amount of fines that may be imposed by the federal government. Again, this provision simply requires that fines imposed must not be "excessive" in nature; however, the U.S. Supreme Court has not held that this particular provision applies to state governments through the process of incorporation and the Fourteenth Amendment.

## Cruel and Unusual Punishment

Punishments after conviction of a crime were extremely severe in England. As such, the drafters of the Constitution sought to ensure that constitutional protections were in place to prevent use of torture, maiming, and other forms of cruelty as punishment. This provision applies to punishment that is disproportionate to the crime. In *Weems v. U.S.* (1910), the U.S. Supreme Court held that the Eighth Amendment requires ". . . that punishment for a crime should be graduated and proportioned to the offense." Examination of the **proportionality of punishment** is most apparent in cases in which the death penalty is imposed. In capital cases, the court must conduct a proportionality review to ensure that capital punishment is not disproportionate to the crime. In reviewing other murder cases, the court attempts to determine whether a punishment other than death has been ordered for similarly situated defendants. Such a process is designed to ensure that the death penalty is not used in an arbitrary manner (*Walker v. Georgia*, 2008).

An exact and precise meaning of "cruel and unusual" does not exist, but the U.S. Supreme Court has addressed the issue in many cases. The prohibition against "cruel and unusual punishments," like other expansive language in the Constitution, must be interpreted according to its text, by considering history, tradition, and precedent, and with due regard for its purpose and function in the constitutional design (*Trop v. Dulles*, 1958). In evaluating whether particular punishments are so disproportionate to be cruel and unusual in nature, the High Court uses a standard that draws on the "prevailing standards of decency that mark the progress of a maturing society" (*Trop v. Dulles*, 1958). In order to assess society's standards of decency, the Court reviews legislative enactments, state practices, and jury behavior.

Recently, in *Roper v. Simmons* (2005), the U.S. Supreme Court concluded that use of the death penalty was cruel and unusual as applied to juveniles who were under the age of 18 years at the time they committed the offense. A few years earlier, the Court held that the use of the death penalty was cruel and unusual as applied to offenders who were mentally retarded (*Atkins v. Virginia*, 2002). In each of these opinions, the Court reviewed legislative enactments, state practices, and jury behavior. These cases provide an interesting review of the process that the High Court uses to determine whether certain punishments violate the Eighth Amendment; in other words, whether our society appears willing to tolerate such measures.

Having reviewed the constitutional limitations on the restriction of personal liberty and punishment, we turn to a more general discussion of punishment to examine the goals and purposes of criminal punishment as well as the many forms of punishment available after conviction of a criminal offense.

## Criminal Punishment

In general, the underlying purpose of punishment is ultimately to achieve social order and control. Punishment or the threat thereof exists to prevent individuals from violating the law and thereby creating social harm. There are many goals or theories that guide the use of punishment for criminal offenses. Here, we focus on the several goals that will help us understand the use of punishment. These are retribution, deterrence, incapacitation, and rehabilitation. We also briefly address the emerging use of restorative justice.

### Retribution

**Retribution** is often referred to as the "just desserts" model of punishment. The central theme is that offenders deserve punishment for their wrongful acts. Moreover, society has a responsibility to inflict punishment on those who violate its norms. Biblical notions of punishment and the sentiment of The Code of Hammurabi reflect the notion of retribution, or *lex talionis*, meaning "an eye for an eye, a tooth for a tooth," as found in the Book of Exodus 21:23–27. These principles, both biblical and nonbiblical versions, suggest that wrongdoers must be punished in a manner that fits the crime. Thus, the crime and the resulting punishment should be proportionate to each other. Such ideas have been debated for centuries.

### Deterrence

**Deterrence** is a principle that evolved from the philosophy of the Classical School that refers to the effect that threat of punishment has on criminal offenders and society at large. Deterrence suggests that the existence of punishment will prevent individuals from engaging in illegal acts. Punishment must be swift, certain, and consistent to have a meaningful effect. Deterrence, however, requires one to embrace the notions of free will and rational choice. To subscribe to deterrence, one must acknowledge that humans possess free will and make rational choices about their behavior, criminal or otherwise. Such choices are a result of a process wherein the offender weighs the pros and cons or advantages and disadvantages of engaging in the behavior under consideration. If the cons or disadvantages (such as punishment) outweigh the advantages, the actor will not engage in the behavior; however, if the cons or disadvantages do not outweigh the advantages, the actor will engage in the behavior. Thus, in order to be meaningful in terms of preventing social harm, punishment must be proportionate to the crime. In other words, the punishment must be severe enough to thwart criminal behavior.

Deterrence can be divided into two categories: general deterrence and specific deterrence. **General deterrence** refers to the effect of the punishment for criminal offenders on the greater community. Although it is too late to impact the choice made by the criminal offender, society at large sees the punishment being meted out on the convicted and will be deterred or prevented from engaging in similar behavior. In contrast, **specific deterrence** refers to the impact that the existence of certain punishments has on the offender. Thus, if the punishment is severe enough, offenders will be deterred or prevented from commission of the crime.

## Rehabilitation or Reformation

**Rehabilitation** suggests that society should use punishment or sanctions for the purpose of correction or reform. Thus, the ultimate purpose of punishment should be to assist the offender with reformation so that after the punishment has concluded the offender will be transformed into a meaningful member of society. Education, mental health and drug abuse treatment, and vocational training are examples of rehabilitative interventions. Rehabilitation was the prevailing justification for criminal corrections in the 1960s and 1970s; however, with the political shifts that occurred in the late 1970s and early 1980s, rehabilitative ideals quickly lost their appeal for legislators and policy makers who were faced with increasing crime rates and staggering numbers of drug-related offenders. In general, the tolerance for rehabilitative programs declined and public sentiment demanded accountability from its correctional system.

## Incapacitation

**Incapacitation** refers to the removal of the offender from society after conviction of a criminal offense. Thus, the focus is to avoid the potential for future harm by the offender. There are many ways to incapacitate criminal offenders, but the prevailing method is incarceration. Once incarcerated, the offender is removed from society and unable to engage in acts that damage society for a proscribed number of years.

## Restorative Justice

Restorative justice is an emerging model through which the relationship between victim and offender is reconciled. **Restorative justice** refers to the process whereby victims and offenders, with the assistance of a trained mediator, identify and address the consequences of crime. Restorative justice focuses on healing. Rather than viewing crime as an offense against the state, restorative justice embraces the consequences of crime on the victim and the offender and attempts to work through those. Restorative justice is not always appropriate, but those who elect to participate usually experience positive outcomes.

## Types of Criminal Punishment

In the modern American criminal justice system, various sanctions exist for those convicted of criminal offenses. These range from monetary fines to capital punishment. Congress or state legislatures determine the appropriate penalty for the offense. Typically, specific penalties are described in the statute wherein the offense is defined. This provides notice to the public regarding possible penalties for violations. In many statutes, legislatures provide for fines and/or terms of imprisonment. Constitutionally speaking, the primary limitations on criminal punishment are as follows: due process before the imposition of punishment (5th/14th), no excessive fines (8th), and no cruel and unusual punishment (8th).

## Fines

A **fine** is an order by the court for the offender to pay a fixed sum of money as penalty for the criminal offense. Modern fines are descendants of the wergild. In common law England, **wergild** required an offender to pay compensation to the state and the victim. The amount of a fine varies with the severity of the crime. Fines for misdemeanors may be as low as $25.00, whereas certain felonies may include fines of hundreds of thousands of dollars. The amount of fines for criminal offenses is an issue largely left to the discretion of the legislative body.

## Forfeiture

**Forfeiture** refers to the taking or seizure of real or personal property that was used to commit or to facilitate a criminal offense. Forfeitures, like fines, were also allowed at common law. After conviction of a felony at common law, the king could seize real or personal property as punishment. Modern-day forfeitures may be civil or criminal in nature. **Civil forfeiture** refers to the loss of property due to a civil proceeding. **Criminal forfeiture** is the loss of property that results as a penalty during a criminal proceeding. The use of forfeitures has increased significantly over the last several decades and is now commonly used by both federal and state authorities. Forfeitures are typically used in cases involving the drug trade, white collar crime, conspiracy, and pornography. Several cases challenging the use of forfeitures have been heard by the U.S. Supreme Court.

## Incarceration

Incarceration (or imprisonment) is also a common form of punishment. The United States currently has one of the highest incarceration rates among industrialized nations. A variety of reasons may account for the high incarceration rate. For example, politicians frequently run for election on platforms with a "tough on crime" message. The public seems to prefer tough approaches to crime. Many are more comfortable with offenders being locked behind bars as opposed to remaining in the community. Thus, the use of incarceration contributes to a feeling of public safety and societal well-being. Finally, despite the high costs associated with building and operating prisons, there is a lack of resources to develop alternative programs.

If the crime is a felony, the offender will serve time in a state penitentiary, whereas a misdemeanor offender will serve time in a county jail. Correctional facilities may be either public or private. An increase in privatization has considerably changed the American system of corrections over the last 30 years. Incarceration or imprisonment comes in many forms. There are different models that exist to determine the number of years that offenders receive. The following discussion focuses on several key concepts that will assist in understanding how terms of imprisonment occur.

### Indeterminate Sentences

An **indeterminate sentence** occurs when legislatures proscribe minimum and maximum incarceration periods but where trial judges, correctional authorities, or parole boards ultimately determine when an offender will be released. For example, the legislature may set the sentence for burglary at 10 to 20 years in the penitentiary. After conviction and a sentencing hearing, the trial court will sentence the offender to a term of imprisonment not less than 10 and not more than 20 years. The actual sentence served, however, will be determined by the correctional system using a number of factors to include inmate behavior while incarcerated, nature of the offense, criminal history, and participation in activities or programs indicative of rehabilitation. In the end, the correctional system simply wants to ensure that an inmate is no longer a threat to public safety and has been rehabilitated.

### Determinate Sentences

**Determinate sentences** regained their popularity in the 1970s. The public and policy makers were frustrated with the vast discretion of trial judges to impose sentences. In certain cases, trial judges imposed little or no prison time for heinous crimes. Alternatively, in other cases, trial judges imposed extremely severe prison terms for minor offenses. Thus, to limit the discretion of the trial judge and to reduce sentencing disparity, many states turned to determinate sentencing schemes. A determinate sentence exists when legislatures more specifically set forth the term of imprisonment for particular crimes. Rather than the wide range of imprisonment available in an indeterminate sentencing scheme, the judge is limited to a more specific range. This, in turn, limits the discretion of judges in sentencing.

### Sentencing Guidelines

Sentencing guidelines are an example of determinate sentencing used in many states and in the federal government. The chief purpose of sentencing guidelines is to reduce **sentencing disparity**, which occurs when individuals receive markedly different sentences for similar offenses. With sentencing guidelines, a complex grid containing offenses and recommended sentences is created. Judges are restricted to sentences that are commensurate with the recommended sentence in the grid; however, some judicial discretion is allowed. Judges may consider the pre-sentence report with its summary of the offender's criminal, psychological, employment, educational, family, and social history to either reduce or increase the number of points, thereby influencing the sentence. Consideration of these factors must be specifically addressed by the judge in his or her sentencing order.

In passing the **Sentencing Reform Act of 1984**, Congress created the Federal Sentencing Commission responsible for developing the **Federal Sen-**

tencing Guidelines for federal courts. The guidelines were designed to reduce sentencing disparity in the federal court system, reduce judicial discretion in the imposition of sentences, and provide an objective standard on which to base federal sentences. Initially, federal judges were required to strictly adhere to the guidelines when imposing sentences; however, in *United States v. Booker* (2005), the U.S. Supreme Court held that the provisions of the federal statutes that made the guidelines binding on federal judges violated the Sixth Amendment guarantee of trial by jury. After *Booker*, the guidelines became advisory rather than mandatory for federal judges.

## Mandatory Sentences

Mandatory sentences are another example of determinate sentencing. The move toward the greater use of mandatory sentences by the states stems from public demand for truth in sentencing. Convicted offenders who received substantial sentences at the time of their convictions often received significant credit or good time while in prison and therefore were often released after having served a fraction of their original sentence. This result angered the public and policy makers and resulted in a shift toward truth in sentencing measures such as mandatory minimum sentences.

Mandatory sentences remove all discretion from the trial judge. The legislature sets the mandatory sentence for its imposition on convicted offenders. For example, a statute stating that anyone convicted of murder shall be sentenced to a term of life imprisonment in the penitentiary is an example of a mandatory sentence. Mandatory sentences and sentencing guidelines in state courts have come under intense scrutiny by the U.S. Supreme Court. In *Apprendi v. New Jersey* (2000) and *Blakely v. Washington* (2004), the Court invalidated certain provisions of mandatory sentencing schemes and indicated a preference for more individualized decision making during sentencing.

## Habitual Offender and Three Strikes Laws

Habitual offender (or "three strikes") laws are a reflection of the public's growing intolerance with recidivism and career offenders. These laws specifically target offenders who fail to cease their criminal behavior after an initial conviction. Most of these statutes become effective after conviction of a third felony, hence the name "three strikes laws." Thus, a **habitual offender statute** requires the imposition of a specific sentence after conviction of a third felony. In the majority of habitual offender sentencing schemes, the penalty is life imprisonment if all prerequisites are met.

The number of previous offenses is the first prerequisite that the state must establish to sentence a person as a habitual offender. Although not all states use three as the magic number, many do. States may also specify the type of felony that qualifies as a **predicate crime**, defined as previous offenses for which a defendant has been convicted. For example, a state may limit predicate crimes to violent felonies. Thus, the prosecution would have to demonstrate that the offender had committed three violent felonies before he or she could be sentenced pursuant to the habitual offender statute. Finally, the statute may require that the offenses occur within a particular time frame (such as a conviction for three violent felonies within a 10-year period); however, time limitations are rare among most habitual offender statutes. Thus, conviction of a third felony offense, regardless of the time frame, would result in a mandatory sentence of life imprisonment.

Critics contend that these sentencing schemes are too harsh and result in life sentences for many offenders who could be rehabilitated. The research on habitual offender sentences has produced mixed results regarding their deterrent effect. Although the use of such sentences is costly, these measures appear to sit well with the voting public, and thus policy makers and politicians are unlikely to abandon their use anytime soon. The use of habitual offender sentencing was upheld by the U.S. Supreme Court in *Ewing v. California* (2003) and *Lockyer v. Andrade* (2003). Each of these cases involved the application of California's three strikes law, which is routinely cited as one of the most severe in the United States. Under California's statute, even relatively minor offenses can serve as predicate offenses.

## Capital Punishment

Other than abortion, the death penalty is one of the most controversial issues in the American discourse. Advocates argue the death penalty is appropriate for crimes because of its deterrent and retributive value; however, critics contend that the death penalty does not deter crime, and with the inherent difficulties of fairly applying the death penalty, its use should be abolished. Support for abolition of the death penalty (or at least a moratorium on its use) has increased in recent years as a result of the publicity surrounding several cases in which innocent people have been wrongly convicted and later released from death row through the use of forensic science tools such as DNA analysis. The debate has been an enduring one and is not likely to end soon.

The U.S. Supreme Court has recently addressed the use of the death penalty for offenders who were juveniles at the time the crime was committed (*Roper v.*

*Simmons*, 2005) and for the mentally retarded (*Atkins vs. Virginia*, 2002). In each of these cases, the Court held that use of the death penalty for such offenders violated the Eighth Amendment. Central to the disposition of these cases was that juveniles and mentally retarded offenders do not possess the level of criminal culpability necessary to impose the ultimate punishment. Moreover, the Court reviewed the practices used by states and juries to assess whether use of the death penalty in such cases was in accordance with the "evolving standards of decency of a maturing society" (*Trop v. Dulles*, 1958). After review, the Court concluded that it did not.

Currently, 38 states have the death penalty. In those states, the death penalty is reserved for cases involving murder. In states where a jury determines or recommends the sentence, death penalty cases may only be heard by a **death-qualified jury**, defined as one in which the jurors have undergone *voir dire* regarding their ability to impose the death penalty. Jurors who are unable to impose the death penalty under any circumstances are excluded from service. These jurors commonly cite religious or personal beliefs for their opposition to the death penalty.

In death penalty cases, proceedings are bifurcated. A **bifurcated proceeding** occurs in two distinct phases. The first phase of the proceedings is the guilt phase, during which the defendant is tried for the murder as alleged in the information or indictment. After the guilt phase concludes, the sentencing phase may begin. During the sentencing phase, the judge or jury focuses solely on whether the defendant should be sentenced to death. In order to sentence a defendant to death, the state must establish the existence of one or more aggravating circumstances that outweigh the presence of mitigating circumstances. Evidence is presented by both the prosecution and the defense regarding the presence or absence of these factors. An **aggravating circumstance** is a fact that distinguishes one murder as more serious than others. Examples of aggravating factors include multiple victims, murder for hire, felony murder, special status victims (such as police officers and judges), and offenders with extensive criminal histories. A **mitigating circumstance**, on the other hand, is a fact that serves to lessen a defendant's culpability. Examples of mitigating factors include a lack of criminal history, mental illness or deficiency, troubled childhood, child abuse or neglect, or a substance abuse history.

## Alternatives to Incarceration

Although incarceration is quite popular within political circles and appeals broadly to the retributive spirit among many citizens, it simply is not essential with respect to many criminal offenders. Think back to the social contract discussed in Chapter 1. Within that construct, it has been well settled that only the least restrictive mechanism should ever be used by government to encroach on the liberties and freedoms of its people. With this in mind, it becomes clear that incarceration of some criminal offenders goes well beyond that which is essential to protect society. The following discussion examines the variety of non-incarceration options available to the criminal justice system for people who stray from the boundaries of legal parameters.

### Probation

The most popular form of punishment in the American system is probation. This is especially true for juvenile and nonviolent adult offenders. **Probation** is a nonsecure method of providing supervision for criminal offenders while maintaining them in the community. When compared with incarceration, there are several advantages to probation. First, offenders may remain in their communities. Issues associated with prolonged institutionalization that face incarcerated offenders are avoided. Probation is also significantly less costly to administer. The cost is further offset by the payment of supervision fees by probationers. Another advantage of probation is its ability to provide services to offenders while in the community. Often, rehabilitative programs offered within prison systems have lengthy waiting lists. As such, inmates may spend years waiting for acceptance, whereas in the community, a greater variety of programs and more alternatives are available. Remaining in the community also avoids the transition or reintegration issues that may occur when incarcerated offenders are released back into communities. There is no entitlement to probation; rather, offenders must meet certain qualifications. Most states restrict probation to nonviolent offenders and to those with no significant criminal history. Probation, then, is viewed as an opportunity, not a right, to rehabilitate within the community; however, for those offenders who do pose a greater risk to the community because of the nature of their offense or a significant criminal history, the use of **intensive supervision probation** (ISP) is increasing because it allows these offenders to remain in the community under much more stringent supervision and additional conditions.

Probationers are supervised by probation officers to ensure adherence to the conditions of their release. Under the terms of probation contracts, freedom is conditional. In order to remain abroad in the community, an offender must adhere to certain terms and conditions. In order to create an individualized probation contract, conditions are designed to meet the needs of each offender and are therefore constructed on a case by case basis: examples of probation conditions include no criminal behavior, sobriety, drug treatment, drug testing, employment, avoiding criminal associates, curfew, attendance at all meetings with probation officers, and payment of all fees, fines, and restitution. Probation conditions may also include house arrest and electronic monitoring, each of which is an additional alternative to incarceration; however, special conditions may be tailored to meet unique characteristics of the offender. For example, pedophiles may be forbidden to have contact with children under the age of 18 years. Failure to adhere to probation conditions will result in a probation violation, which then will be reported to the sentencing court by the probation officer assigned to the case. Violations may result in a variety of sanctions, including warnings, additional conditions, temporary incarceration in a local jail, or imprisonment in a penitentiary for the original term of years contained in the sentencing order.

Probation may be ordered by the court in two ways. Probation can be the original sentence imposed by the court. Alternatively, an offender may receive a sentence of incarceration that is immediately suspended and placed on probation. In the latter scenario, if the offender violates his or her probation, incarceration can then be ordered under the original sentencing decision. In the first scenario, however, an offender who violates a probation condition may be held in contempt of court and sanctioned on the contempt charge.

To revoke an offender's probation, a revocation hearing must be held. The revocation hearing must meet minimal expectations of due process. In *Morrissey v. Brewer* (1972), the U.S. Supreme Court addressed offender due process rights during parole revocation proceedings. These rights also apply, however, to probationers. In *Morrissey*, the Court held that minimal due process requires notice and an opportunity for the offender to be heard, contest allegations, and confront adverse witnesses. One year later in *Gagnon v. Scarpelli* (1973), the Court held that probationers and parolees have a limited right to counsel in revocation proceedings.

### Restitution

Restitution also is commonly used as punishment. Typically, restitution is ordered in conjunction with other forms of punishment. Rarely, and usually only in very minor cases, it is imposed as a stand-alone punishment. The purpose of restitution is to provide accountability and to compensate victims. An order of **restitution** requires an offender to provide services or money as compensation for their wrongdoing. Monetary restitution may be ordered to the victim. Generally, all payments are made through the court so that an official record of payments as well as their timeliness is documented. Offenders may also be required to perform a certain number of hours of community service in order to satisfy a restitution order. Community service can be useful when offenders are unable to pay monetary amounts.

## Summary

An understanding of the foundation of the American form of government and individual liberties is necessary when exploring modern criminal law. Criminal laws are designed to protect the safety and security of citizens from individuals who violate our notions of public order and the social contract; however, the U.S. Constitution and Bill of Rights each extends necessary protections to those charged with criminal acts. These limitations are intended to serve as a barrier between the individual and the government and provide a fair and just process through which criminal behavior may be adjudicated. The goal is a balance between the power possessed by the government and its ability to impose such power on individual citizens. As such, constitutional limitations on governmental power were explored. The nature and types of punishment for criminal offenses also were explored to provide an understanding of the measures that may be employed to penalize individuals for law violations. Prevailing theories regarding punishment too were discussed. The theoretical rationale for punishment sheds light on whether certain measures should be employed given American notions of liberty.

## Practice Test

1. The doctrine of _____ powers reflects a concerted effort by the drafters of the U.S. Constitution to avoid the concentration of power within one branch of government.
   a. dispersion of
   b. division of
   c. separation of
   d. equal
   e. mutual

2. The effort to strengthen the national government by identifying certain matters that fall within its exclusive power is referred to as _____.
   a. sovereign power
   b. nationalism
   c. exclusive authority
   d. incorporation
   e. federalism

3. A system of _____ within our system of government is designed to prevent tyranny by any one branch.
   a. division of powers
   b. checks and balances
   c. dispersion of powers
   d. retribution
   e. mutual governance

4. The _____ branch of government is vested with the power to enforce laws.
   a. judicial
   b. enforcement
   c. executive
   d. legislative
   e. authoritarian

5. The _____ provides numerous provisions that limit the nature and extent of government intrusion into the lives of American citizens.
   a. U.S. Constitution
   b. Federalist Papers
   c. Declaration of Independence
   d. Intrusion Prevention Bill
   e. Government Limitation Statute

6. _____ refers to the requirement that government must follow certain procedures before infringing on the life, liberty, or property of a private citizen.
   a. Federalism
   b. Restorative Justice
   c. Equal Protection
   d. Selective Incorporation
   e. Due process

7. The _____ prohibits states from making arbitrary and unreasonable distinctions among people in terms of their rights and freedoms.
   a. Ninth Amendment
   b. due process clause
   c. dual sovereignty doctrine
   d. equal protection clause
   e. predicate power bill

8. The _____ Amendment includes the privilege against self-incrimination.
   a. 5th
   b. 6th
   c. 8th
   d. 4th
   e. 14th

9. The _____ Admendment provides numerous trial rights, including the rights to a speedy and public trial, impartial jury, and to confront witnesses at trial.
   a. 5th
   b. 6th
   c. 8th
   d. 4th
   e. 14th

10. The _____ Amendment provides protection from excessive bail or fines, as well as punishments that are cruel and unusual.
    a. 5th
    b. 6th
    c. 8th
    d. 4th
    e. 14th

11. _____ is an emerging model through which the relationship between victim and offender is reconciled.
    a. Deterrence
    b. Restorative justice
    c. Retribution
    d. Incapacitation
    e. Rehabilitation

12. _____ suggests that the existence of punishment will prevent individuals from engaging in illegal acts.
    a. Incapacitation
    b. Rehabilitation
    c. Retribution
    d. Restorative justice
    e. Deterrence

13. _____ refers to the taking or seizure of real or personal property used during the commission or facilitation of a criminal offense.
    a. Incapacitation
    b. Forfeiture
    c. Apprehension
    d. Confiscation
    e. Wergild

14. _____ occur(s) when legislatures proscribe minimum and maximum years of imprisonment, but others in the system determine when an offender will be released.
    a. Sentencing disparities
    b. Discretionary sentences
    c. Authoritarian incarceration
    d. Indeterminate sentences
    e. Legislative incarceration

15. _____ occurs when individuals receive markedly different sentences for similar offenses.
    a. Sentencing disparity
    b. Predicate sentencing
    c. Prejudicial guidelines
    d. Discriminatory sentencing
    e. Imbalanced punishment

16. A grand jury will return a(n) _____ when the evidence is insufficient for sending a person to trial.
    a. motion *in limine*
    b. no bill
    c. information
    d. true bill
    e. affidavit

17. _____ crimes refer to past offenses for which a defendant has been convicted.
    a. Predicate
    b. Strict liability
    c. Habitual
    d. Determinate
    e. True

18. A(n) _____ refers to facts that serve to lessen a defendant's culpability.
    a. aggravating circumstance
    b. information
    c. mitigating circumstance
    d. wergild
    e. true bill

19. _____ is the most popular form of American punishment.
    a. House arrest
    b. Incarceration
    c. Supervision
    d. Probation
    e. Parole

20. In common law England, _____ required offenders to pay compensation to the state and victim.
    a. ex post facto laws
    b. biven actions
    c. bills of attainder
    d. wergilds
    e. civil forfeiture

## References

Garner, B. A. (Ed.). (2009). *Black's law dictionary* (9th ed.). Eagan, MN: West Group.

# CHAPTER 3

# Theft Offenses and Fraudulent Practices

## Key Terms

Abandoned property
Alteration
*Caveat emptor*
Claim of right
Constructive asportation
Constructive taking
Continuing trespass
Conversion
Counterfeiting
Creation
Custody
Direct asportation
Direct taking
Document
Embezzlement
False pretenses
Forgery
Grand larceny
Instrument
Intangible property
Larceny
Larceny by trick
Legal efficacy
Misappropriation
Of another
Ownership
Personal property
Petit larceny
Possession
Real property
Receiving stolen property
Service
Shoplifting
Stealth
Superior right of possession
Surety
Tangible property
Treble damages
Unauthorized use
Uttering

## Introduction

This chapter guides the reader through the historical evolution of theft law, including theft-related fraudulent practices. The journey begins with an examination of the legal simplicity with which theft was viewed in the earliest of times and concludes with the complex codes of modern substantive law. The crimes of larceny, embezzlement, and false pretenses are defined and differentiated as functions of custody, possession, and ownership. Then the crime of receiving stolen property and other modern theft offenses will be explored with an eye toward discussing contemporary state efforts to consolidate what has become a voluminous body of criminal offenses ultimately regulating the same thing—theft of property. Next, the chapter defines and explains the crimes (and elements) of forgery and uttering. Finally, federal legislation for each of these crimes is introduced at the conclusion of each topical section.

## Theft

Even though the Uniform Crime Reports (FBI, 2008) isolate motor vehicle theft into its own category, it fundamentally is nothing more than a specialized form of larceny-theft. As such, motor vehicle theft will be integrated with larceny-theft crime data to present a more comprehensive portrait of the extent of theft in America today. With that understanding, approximately 7.7 million incidents of theft (including 1.1 million motor vehicle thefts) were committed in the United States in 2007. Theft crimes accounted for nearly 78% of all property crime and produced staggering annual losses in the neighborhood of $13 billion. Specifically, motor vehicle theft averaged $6,755 per incident ($7.4 billion), whereas other forms of larceny-theft yielded substantially lower gains (approximately $900 per incident). When compared with crime data from the previous year (2006), the prevalence of theft as a whole decreased, with motor vehicle theft declining 8.1% and larceny-theft declining by 0.6%. Long-term trends reveal even more optimism as both theft categories demonstrated substantial double-digit declines over the previous decade (1998).

## Theft in General

Theft in the earliest years of common law differed greatly from contemporary theft legislation. In those times, larceny was the only legal mechanism that protected property from being stolen. Larceny at that time was limited in its application, too, as it protected property only when taken through force (actual or threatened) or **stealth** (sneaking away with property without permission). Essentially, common law larceny sought mostly to protect livestock (e.g., cattle and horses), which was necessary to their livelihood. In a nutshell, the principle of **caveat emptor**, "let the buyer beware," reigned supreme and served to exempt all voluntary surrender of property from criminal designation as theft. Over time, societal growth and economic development necessitated enhanced protection against trickery, deceit, and fraud. Robbery also evolved as a means to protect property stolen through forceful means (actual or threatened). Eventually, legislative bodies passed additional statutes to compete against new and emerging theft forms, primarily in the areas of embezzlement, false pretenses (fraud), and lesser included larcenous acts such as receiving stolen property, unauthorized use, and theft of services.

## Differentiating Custody, Possession, and Ownership

Thirty-seven states have expanded larcenous statutes through differentiating circumstances surrounding theft. Thus, the framework for understanding the evolution of theft law begins with mastering the legal complexities associated with the terms custody, possession, and ownership. **Custody** occurs when persons do not possess title or ownership of property but do have a limited right to use property (but with little real discretion regarding how the property is exercised). In custodial cases, property is said to be within one's care and control (or dominion) and hence aligns with larceny because the person committing the theft has done so while trespassing against one's right of possession. Pause for Thought 3–1 illustrates the principle of custody.

With this understanding, one who commits theft against the property of another while in possession (but not ownership) has committed the crime of embezzlement. **Possession** refers to having broad discretion regarding the use of property within control; again though, possessors do not hold title or ownership to said property. Grocery store managers and bank tellers are examples of persons clearly entrusted with possession of the money and property of others. Meanwhile, a person who deprives another of ownership to property through fraudulent means has committed the crime of false pretenses. **Ownership** is defined as possessing title to property. Figure 3–1 diagrams the relationship between custody, possession, and ownership with respect to the crimes of larceny, embezzlement, and false pretenses.

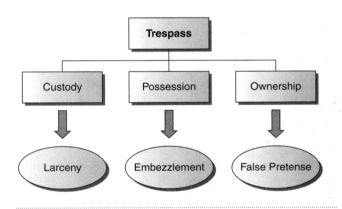

**Figure 3–1** Trespass.

---

### Pause for Thought 3–1

Consider the following: While shopping at a community store, Jack removes merchandise from a shelf and leaves the store without paying for the item. Did Jack (at the time of the taking) have custody, possession, or ownership of the property?

### Scenario Solution

Jack was in custody of the property because the store never agreed to more than the transportation of the item around the store in consideration of its purchase. Larceny is the appropriate charge, as it represents a trespass against the store's possessory right.

## Larceny at Common Law

Common law **larceny** was defined as the taking and carrying away of the personal property of another with the intent to deprive permanently. It is important to understand that larceny is a trespass against another's custodial rights to personal property. As such, trespass against another's right to possess or own was not within larceny's jurisdiction. Common law larceny consisted of five *corpus delicti* elements: two *actus reus*, one *mens rea*, and two attendant circumstances. Figure 3–2 outlines the elements required for the crime of larceny at common law.

### Taking and Carrying Away

The first *actus reus* element for the crime of larceny—that of "taking"—is defined along the same lines as robbery. Larceny's taking element can be consummated in one of two ways: direct or constructive. **Direct taking** occurs when one acquires physical custody of money or personal property for a brief period of time, whereas **constructive taking** (often referred to as an indirect taking) occurs when one causes an innocent third party to take custody of money or personal property which the accused never physically touched. Pause for Thought 3–2 illustrates the difference between a direct and constructive taking.

Often misinterpreted by legal novices, the second *actus reus* element requires the desired items be carried away. The law does not require money or personal property be carried away in the traditional sense, but rather only that it be moved from its original position. The movement need not be extensive; even the slightest movement is legally sufficient. Without movement, however, a charge of larceny will not withstand legal challenge. Also known as asportation, the carrying away element occurs through one of two avenues: direct or constructive. **Direct asportation** refers to the physical moving (through touching) of money or personal property, whereas **constructive asportation** (also known as indirect asportation) refers to situations in which one causes an innocent third party to move money or personal property without ever having to personally carry away an item.

Examination of our previous scenario should assist with understanding the application of asportation. It has heretofore been established that Bill constructively took the bicycle when he sold it to an unsuspecting third party without touching the stolen property; however, Bill also committed a constructive asportation the moment he caused the bicycle to be moved from its original position, even though he never physically carried away the property. Once again, without such an indirect path to consummating

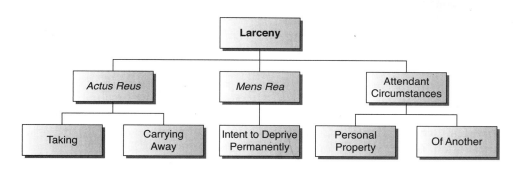

**Figure 3–2** Larceny.

---

### Pause for Thought 3–2

Consider the following: Standing at a bicycle rack on a college campus, Bill stops a passerby and offers to sell another's bicycle. Without touching the bicycle, Bill collects the money, grants permission for the bicycle to be removed from the rack, and then watches the student ride away on the stolen bicycle. Does Bill's action constitute a direct taking or constructive taking?

#### Scenario Solution

Bill's action constitutes a constructive (or indirect) taking because he never directly (or physically) touched the personal property. Without such a rule, all larcenous thefts absent some physical contact would be legally protected against larceny charge.

an asportation, all larcenous thefts absent some physical carrying away of money or property would escape criminal punishment.

## Personal Property of Another

**Real property** refers to anything attached to the ground (such as real estate, trees, and even farming crops). Against this backdrop, we can now better understand the first attendant circumstance required for the crime of larceny. Basically, larceny at common law required that items "taken" and "carried away" must be **personal property**, defined as property that is not affixed to real estate. Two avenues are used for understanding the context of personal property—both of which are housed within the heading of what is referred to as **tangible property**. First, property must possess a concrete quality (capable of being physically touched). Second, property must be capable of being moved (or carried away). Property possessing value but no actual concrete qualities is referred to as **intangible property**. For such cases, legislative bodies often create additional crimes (such as theft of services) to deter and punish theft.

With regard to tangible property, the value of stolen property is used in most states to distinguish the seriousness of larcenies. In short, larceny against personal property whose value meets or exceeds a predetermined value (usually at least $100) is called **grand larceny**, whereas property with lesser value constitutes **petit** (or petty) **larceny**. Moreover, personal property must be of some value, meaning that it must have economic value on the market. Items with only sentimental value do not qualify for protection under traditional larceny statutes. Many states do, however, disregard the monetary value argument when involving certain kinds of property. For example, states often consider theft of credit cards, livestock, and motor vehicles (among other things) as felonious grand larceny, regardless of the property's actual value or the monetary intent of the person who took the property.

The second attendant circumstance for larceny at common law was that personal property must belong to another. Defined much the same as burglary and robbery (see Chapter 4), **of another** refers to rightful possession, not ownership. Rightful possession refers to one who holds the **superior right of possession**, defined as the first-order authority (or right) to possess personal property in cases where multiple parties have a lawful right to possess. In such cases, it is possible for one rightful possessor (usually an owner) to commit larceny against his or her own property when taken and carried away from the one holding the superior right of possession. Conversely, retrieving one's own property to which no other holds a superior right of possession is not larceny because the owner has the singular right to possess. Thus, the rightful possessor can retrieve property under the legal guideline **claim of right**. This right to reclaim property, however, does not excuse other crimes that may be committed in the recovery process (such as trespass or assault). Pause for Thought 3–3 illustrates the legal interpretation for a claim of right and superior right of possession.

## Intent to Permanently Deprive

Common law larceny required that an accused intend to deprive another of personal property permanently. Thus, one who satisfied all larcenous elements but had only intent to temporarily deprive was not guilty of larceny. Other crimes have been created to deal with these forms of takings (discussed later in the chapter). It is important to understand, however, that permanent intent and taking

### Pause for Thought 3–3

Consider the following: A car dealership leases an automobile to Ricardo. Before expiration of the lease contract, the dealership (for no valid reason) retracts the agreement and hires a repossession company to reclaim the car against Ricardo's wishes. Can the dealership be charged with larceny even though it is the rightful owner?

### Scenario Solution

Yes, the "of another" circumstance is satisfied because Ricardo has the superior right of possession. The bank had no legitimate "claim of right" and therefore would be subject to larceny's provisions given that they did take and carry away the car with the intent to deprive permanently.

need not be contemporaneous (within the same scope of time). Basically, a person's continued unauthorized use of another's personal property could be larcenous even when the intent to permanently deprive was formed at some point beyond the taking and carrying away. This concept is referred to as **continuing trespass** and contradicts common law burglary standards (see Chapter 4) requiring that entry and intent co-exist.

Another important issue to keep in mind is that stealth (sneaking away with property) is not always required to constitute larceny. In some cases, a person can be convicted of larceny even when granted permission to take and carry away personal property of another. Referred to as **larceny by trick**, the law refuses to recognize such consent as valid when obtained through fraud (trickery or deceit). Pause for Thought 3–4 illustrates what is known as a continuing trespass.

## Lost, Mislaid, and Abandoned Property

Have you heard the expression "finders' keepers, losers' weepers"? How about "possession is nine-tenths of the law"? More importantly, do you agree with their basic premises? If so, you must belong to the small group of people fortunate enough not to have lost something of value. Most of us have lost property, however, and as such recognize the unfairness of such an outcome; therefore, regardless of one's position on this issue, the law recognizes a person's continuous right to own and possess property that has been lost or mislaid.

People routinely lose property, and as a consequence, it will usually be others who find the property. In many such cases, the value of the property may appear too small to even justify an effort at locating its rightful owner. On the other hand, it may seem equally unreasonable to leave such a piece of invaluable property in its "lost" state due to some fear that a theft accusation might be alleged. The central question then is what effort must be expended by the locator of lost or mislaid property to avoid being regarded (socially and legally) as a thief? The answer to this question is simple and consistent with most other legal conclusions.

One who finds lost or mislaid property must make a "reasonable" effort to locate its owner. What constitutes a reasonable effort, however, is dictated by the circumstances surrounding the finding of lost or mislaid property. Thus, if a reasonable person would conclude in a like situation that locating the owner of the property was improbable, then no effort is expected under the law. Remember, however, that the rule is based on what a "reasonable" person would conclude, not what a careless, reckless, or greedy person actually concludes. Moreover, the degree of effort placed on the finding the property's owner is to a great degree premised on the value of the lost or mislaid property. Pause for Thought 3–5 provides a hypothetical scenario pertaining to lost or mislaid property.

People sometimes abandon property too. **Abandoned property** refers to items left behind (often by a tenant) intentionally and permanently when it appears that the former owner (or tenant) does not intend to claim ownership; essentially, then, the owner has no intent to come back, pick it up, or use it. In such cases, others who come across the property are not legally expected to make efforts at finding the owner because the owner has already expressed the voluntary desire to rid himself or herself of the property. Those who find property perceived to be abandoned, however, must use extreme caution in concluding its abandoned nature; for if wrong, the property reverts to lost or mislaid status

### Pause for Thought 3–4

Consider the following: With every intention to sell Elijah's riding lawn mower, Isaac convinces Elijah to loan him the mower for the weekend. Elijah then sells (as planned) the mower without permission from Isaac. Has Elijah committed the crime of embezzlement or larceny?

### Scenario Solution

Elijah has committed the crime of larceny. The law invalidates the consent from Isaac because it was based on fraudulent information; therefore, the crime is not embezzlement because the mower was not legally entrusted. Elijah only had custody of the mower, the trespass upon which equates to larceny (also referred to as larceny by trick).

> ### Pause for Thought 3–5
>
> Consider the following: While walking across a parking lot en route to her car, Veronica spots a diamond ring on the ground. She picks up the ring and scans the area for possible owners, at which point she notices one person who appears to be looking for something. Concluding that no one witnessed her picking up the ring, Veronica places the object in her pocket and drives away. A bystander noticed both parties' actions, however, and gives the owner the license plate number for Veronica's car. The owner reports these facts to the police, who in turn locate Veronica and retrieve the ring without incident. Can Veronica be charged with the crime of larceny?
>
> ### Scenario Solution
>
> Yes, there is no question Veronica took and carried away the personal property of another. The most important issue is whether she intended to permanently deprive the ring's rightful possessor. Regarding this mental state, it seems apparent that Veronica possessed such intent. She made no effort to locate the owner. Moreover, a bystander testified that Veronica became evasive when she noticed another looking for the ring. Based on the letter of the law, Veronica's actions do illustrate the crime of larceny.

and would then be protected under larceny guidelines. Pause for Thought 3–6 examines how theft law applies to the taking of abandoned property.

## Other Custodial Theft Statutes

The crime of **shoplifting** refers to theft of merchandise from store merchants. Essentially, shoplifting is nothing more (or less) than the crime of larceny, differing only with respect to where the theft must occur. There are two primary reasons for segregating the crime of shoplifting from its larcenous host. First, it allows crime statistics to accurately portray the extent of the damage associated with merchandise theft. Second, states often enhance the penalties for shoplifting with each successive conviction. In so doing, that which ordinarily would have been nothing more than just another misdemeanor larceny offense could become felonious when dealing with a habitual shoplifting offender.

There are other circumstances where theft is committed but without one or more of the essential larcenous elements. As such, a theft perpetrator would go unpunished (and hence undeterred) without some legislation sealing those legal loopholes. Three common examples of such legislative actions among states include receiving stolen property, unauthorized (or unlawful) use, and theft of services.

### Receiving Stolen Property

Mere possession of stolen goods does not necessarily constitute a crime. It is equally true, however, that

> ### Pause for Thought 3–6
>
> Consider the following: While driving to work each morning for two weeks, Corey views an old car parked along the side of the highway. After one week, he begins to speculate about how great it would be to refurbish the car for his son. After the second week, Corey decides the car has been abandoned and hauls it to his home. After the third week, the owner returns for the car to realize it has been taken. The owner assumes that the police impounded the car and inquires accordingly. After confirmation that someone other than the police took the car, a theft complaint is filed. The police ultimately locate the vehicle in Corey's possession. Has Corey committed the crime of larceny?
>
> ### Scenario Solution
>
> Corey has not committed the crime of larceny because he did not intend to deprive the owner of the property permanently. Although an argument could be made that a reasonable person would not have confiscated the car without consulting the police, most would concur that Corey was well within reasonableness in believing the car was abandoned.

continuous possession of stolen goods under circumstances where a reasonable person should be aware of its wrongful acquisition does equate to criminal action. Actions of this nature presented a problem historically because the taking required of larcenous statutes was absent. As such, the law had no mechanism (other than misprision of felony) through which it could punish those who purposely chose to deprive others of personal property permanently. In time, jurisdictions created a separate theft offense to seal this legal loophole (of sorts). Now a felony in most states, **receiving stolen property** is generically defined as acquiring property of another with knowledge of its stolen origin and with no intent of returning the property to its rightful owner. Pause for Thought 3–7 illustrates the proper legal interpretation for this crime.

## Unauthorized Use

The law had further problems deterring and punishing custodial thefts committed with no intent to deprive permanently. Once again, a separate offense evolved to remedy this legal oversight. Most jurisdictions created a crime called unauthorized (or unlawful) use. Essentially, the crime is nothing more than larceny minus the permanent intent to deprive. As such, **unauthorized use** is ordinarily defined as the taking and carrying away of the personal property of another without permission (or consent). Pause for Thought 3–8 illustrates the proper legal interpretation regarding this issue.

## Theft of Services

Common law larceny prohibited taking and carrying away the personal property of another. When the object of theft was intangible property, the crime could not be larcenous. In contemporary society, the most common example of such thefts occurs with respect to **service**, which constitutes paid work performed by others. One could argue that deterring such theft is even more important in an economically driven and service-oriented society. Without protection, businesses and entrepreneurs would be at grave risk of economic catastrophe. For example,

### Pause for Thought 3–7

Consider the following: Francis purchases a car stereo worth $2,000 from a roadside van for the unbelievable price of $300. He suspects the stereo is stolen but falls prey to the price temptation. To his horror, he is pulled over by the police about a mile down the road for suspicion of receiving stolen property (police were conducting surveillance). Francis argues he had no knowledge of the property's origin but is arrested and charged accordingly. Has Francis committed the crime of receiving stolen property?

### Scenario Solution

Yes, Francis can be charged with receiving stolen property because a reasonable person would have been aware of the property's stolen origin. In this case, a car stereo purchased from an unlicensed roadside stranger at 15% of its retail value would be enough to constitute such knowledge.

### Pause for Thought 3–8

Consider the following: Knowing his neighbor (Tom) would be on vacation for 2 weeks, Jimmy "hot wires" Tom's motorcycle for some leisure riding. Tom then returns home early to find his motorcycle missing. While the police are at Tom's home investigating the disappearance, Jimmy returns with the motorcycle. Tom wishes to file criminal charges against Jimmy for the theft. Has Jimmy committed a crime?

### Scenario Solution

Yes, Jimmy committed the crime of unauthorized use. It is obvious that Jimmy had no intent to permanently deprive Tom of his property, and as such, the crime of larceny would not withstand legal challenge. It is important, however, that people not be allowed to use property at will without permission.

most any television viewer likely has seen commercials that warn of the consequences for cable theft. To ensure these kinds of thefts can be successfully prosecuted, some states have altered the language of their existing statutes. Other states, however, created separate and distinct statutes called theft of services. As with larceny, intent to deprive another of remuneration (payment) must be present at the time a service is rendered (or taken advantage of). Other examples of service theft include (but not limited to) unlawfully watching a movie in a theater (not paid) and refusing to pay for lawn service.

## Embezzlement

Larceny at common law required one to trespass against the possessory rights of another and thus failed to provide criminal sanctions for thefts committed by those in lawful possession of property at the time of misappropriation or conversion. As societies continued to industrialize, choices regarding whether to entrust property to third parties became nearly unavoidable with the advent of the local bank, mercantile, and other commercial establishments. In response, the crime of embezzlement evolved as a necessary legal construction to deter misuse of entrusted property. Simply put, **embezzlement** at common law was defined as the unlawful conversion or misappropriation of another's money or property by one to whom such items were entrusted. Even though theft is fundamentally the object of embezzlement, one simply cannot ignore the importance of the breach of trust component. It is not difficult to recognize that the breach of trust is being punished as much as (if not more than) the theft itself. Figure 3–3 outlines the elements required for the crime of embezzlement at common law.

### Conversion or Misappropriation

The *actus reus* for the crime of embezzlement does not require that an accused, as with larceny, took and carried away personal property. Rather, embezzlement substitutes one of two other acts for those elements: conversion or misappropriation. **Conversion** refers to the transforming of property into something other than its original status. Although it may seem insignificant in the broader scheme of things, even the unauthorized exchange of two 5-dollar bills for one 10-dollar bill is an example of a conversion.

**Misappropriation**, on the other hand, refers to the unauthorized use of unconverted property. For example, let us say I loan you my riding lawnmower for the expressed purpose of cutting your grass. Before its return, you decide to cut the lawns of other family members. Although not likely to receive much judicial attention, this action nonetheless constitutes a misappropriation under our legal guidelines.

### Entrusted with Property of Another

A major hallmark of an industrial nation is its reliance on economic transactions. Thus, it is essential that a **surety**—defined as a person entrusted with personal property of others—be sufficiently deterred from misappropriating or converting items conveyed to them. Remember that it was for this reason (as discussed earlier) that the crime of embezzlement emerged as necessary, as the crime actually aims to punish the breach of trust more so than the theft itself. One need examine only two fundamental differences between the crimes of embezzlement and larceny to conclude this breach-of-trust emphasis. First, consider that the crime of embezzlement is always a felony, whereas larceny can be a misdemeanor when the value of the property is below a certain threshold. Second, the authorized punishment for the crime of embezzlement is substantially greater than (usually twice as much) that established for grand larceny.

Keep in mind that embezzlement statutes require the trespass be against another's right to ownership but where the offender has lawful possession. Once again, however, the personal property must have actual market value. It is not possible to commit embezzlement when one does not lawfully possess the

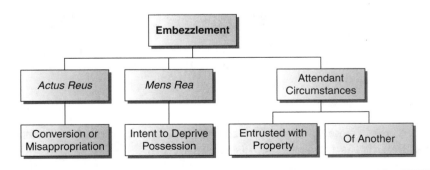

**Figure 3–3** Embezzlement.

property of another. In short, then, the crime of embezzlement cannot be committed against one's own property. Finally, some states do not allow the charge of embezzlement pertaining to circumstances involving personal relationships, instead reserving the crime for breach of trust within business transactions.

### Intent to Deprive Possession

Even though larceny required permanent intent to deprive, the intent required for embezzlement need not rise to such an elevated degree. Embezzlement merely purports that an offender intend to deprive the rightful possessor of possession. Moreover, the misappropriation or conversion within embezzlement need not even be permanent. Thus, embezzlement can occur even when the accused had intent to return the personal property. In some cases, too, embezzlement remains a viable charge even when one returns money or property of equal value (but not the same money or property) before being discovered. Pause for Thought 3–9 illustrates how embezzlement is processed within legal circles.

## False Pretenses

Heretofore, theft crimes in which personal property has been misappropriated while in custody (larceny) and possession (embezzlement) have been examined. Neither of those crimes, however, defines a situation where one actually deprives another of title (or ownership) to personal property without a "taking" and without being in "possession." In response, the law sought to deter such fraudulent practices with the creation of yet another statute. In short, **false pretenses** at common law represented the acquiring of title to the personal property of another through false representations made with intent to defraud. False pretenses was (and often remains) a felony that mandated **treble damages**, meaning that an offender had to pay restitution in an amount three times the worth of the items fraudulently acquired.

The crime of false pretenses at common law required five essential elements. The attendant circumstances were identical to those contained within larceny and embezzlement statutes and thus necessitate no further explanation: personal property of another. The *mens rea* component—intent to defraud—also has been sufficiently explained. Keep in mind, however, that the intent to defraud had to occur simultaneous with the making of any false representation. If the intent formed only after realizing an error was made, no false pretenses occurred.

Two *actus reus* elements are required for the crime of false pretenses. First, an accused actually had to acquire title (or ownership) to the personal property. A mere unsuccessful attempt was insufficient to consummate false pretenses. Second (and a bit more complex), false pretenses required some false representation. The nature of those representations had strict requirements, however. One such parameter mandated that the false representations had to be directed toward past or present circumstances. It was historically accepted (though less today) that false representations about future transactions were not within the scope of false pretenses. Although some disagreed with such an approach, the primary reason for excluding future transactions rested with the difficulties associated with determining that a person intended to honor the promise were it not for unexpected intervening circumstances. False representations about past or present material facts were satisfied when victims transferred title to personal property in reliance on a false impression that was created or reinforced by the accused or when an accused failed to correct an existing false impression. Originally, false pretenses

---

### Pause for Thought 3–9

Consider the following: A local pizzeria store manager supplements his weekly income by opting not to ring up three sales per day and then pocketing the cash. A store employee does not regard this conduct as acceptable and informs the store owner of the practice. The owner installs a surveillance camera to document such off-the-record transactions. The police are notified of the thefts. What crime (if any) has the store manager committed?

### Scenario Solution

The store manager has committed the crime of embezzlement because he misappropriated cash acquired from the lawful conversion of pizza. The crime is much more serious than larceny because the store manager was entrusted with the property. Thus, it is impossible to trespass on another's possessory right when one is in lawful possession of said property.

followed the reasonableness standard regarding a victim's reliance on misrepresentations, meaning that naïve victims were not protected from their inability to use logic; states have since abandoned that philosophy. Figure 3–4 outlines the elements required for the crime of false pretenses. Pause for Thought 3–10 additionally illustrates how to legally resolve its elements.

## Modern Consolidation of Theft Statutes

Without question, larceny as defined at common law proved insufficient to combat the cunning methods that soon emerged from those seeking to become theft entrepreneurs within an industrializing nation. With its narrow foci regarding deprivation—taking and carrying away—the common law crime of larceny became an antiquated weapon in the battle against serious forms of theft involving breach of trust (embezzlement) and fraud (false pretenses). As a consequence, the number of statutes designed to deter theft activities began to mount, ultimately causing more confusion than benefit. With that understanding and to remedy other problems discussed within this text, the American Law Institute developed the Model Penal Code in an effort to reduce legal chaos. Many states use the code; others do not.

## Federal Theft Law

Federal theft legislation is divided into three categories. First, Title 18, Chapter 31 of the U.S. Criminal Code combines embezzlement with other larcenous offenses to form Embezzlement and Theft. The section essentially defines a multitude of conducts and penalties for nonfraudulent theft committed under circumstances of federal concern. A large number of federal laws (§§641-669) have been created to combat theft and embezzlement. Unlike state legislation with customary one-size-fits-all codes, federal theft laws are extraordinarily delineated. For example, there are codes singularly reserved for defining, among other things, receiving loans from court officers (§647), theft of livestock (§667), and theft of major artwork (§668). The most generic of the fed-

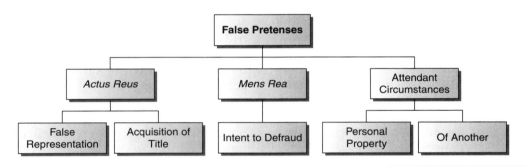

**Figure 3–4** False pretenses.

### Pause for Thought 3–10

Consider the following: Wendi wishes to sell an antique table. She believes that the table has substantial value but is uncertain of its actual worth. She visits a local antique store and requests an appraisal for the table. The store owner concludes the table is worth $3,000 (fair market value) but tells Wendi it is worth $600. Based on the store owner's expertise, Wendi sells the table to the antique store for $600. Is this a legal transaction, or has the store owner committed the crime of false pretenses?

### Scenario Solution

The store owner has committed the felonious crime known as false pretenses. Wendi certainly has the right to voluntarily sell the table for $600. In this case, however, the voluntariness of the sale was negated when the store owner created a false impression concerning the table's value. As such, the store owner deprived Wendi of title to the table through false representation.

eral theft laws pertains to thefts of money, property, or records of the United States (§641) and states that

> whoever embezzles, steals, purloins, or knowingly converts to his use or the use of another, or without authority, sells, conveys or disposes of any record, voucher, money, or thing of value of the United States or of any department or agency thereof, or any property made or being made under contract for the United States or any department or agency thereof; or . . . receives, conceals, or retains the same with intent to convert it to his use or gain, knowing it to have been embezzled, stolen, purloined or converted—Shall be fined . . . or imprisoned not more than ten years, or both; but if the value of such property does not exceed the sum of $1,000, he shall be fined . . . or imprisoned not more than one year, or both.

Second, Title 18, Chapter 47 of the U.S. Criminal Code creates a body of legislation entitled Fraud and False Statements (§§1001-1037). Once again, federal fraud legislation is well delineated, including (but not limited to) possession of false papers to defraud the United States (§1002), insurance fraud (§1007, 1014, 1033, 1035), identity theft (§1028), and credit/debit card fraud (§1029). Other fraudulent offenses regulated within federal law also include (but not limited to) wire fraud (47 USC §1343), telemarketing fraud (26 USC §113A), and securities fraud (15 USC §2D).

Third, Title 18, Chapter 113 of the U.S. Criminal Code creates legislation entitled Stolen Property (§§2311-2322). Ranging from transportation of stolen vehicles (§2311) to chop shops (§2322), crimes of stolen property are numerous and designated as felonies. For example, the sale or receipt of stolen vehicles (§2312a) stipulates that

> whoever receives, possesses, conceals, stores, barters, sells, or disposes of any motor vehicle, vessel, or aircraft, which has crossed a State or United States boundary after being stolen, knowing the same to have been stolen, shall be fined under this title or imprisoned not more than 10 years, or both.

## Forgery and Uttering in General

Forgery and uttering at common law protected society from fraudulent documents. The fundamental difference between the two offenses hinged on whether one made (forgery) or merely passed (uttering) a forged document. Both behaviors were subject to legal penalties, but the classification separation made forgery felonious and uttering a misdemeanor.

## Forgery Defined

**Forgery** at common law was defined as the unlawful making or alteration of a document possessing legal significance (or legal efficacy) and with the intent to defraud. Essentially, the aim of a forger was to produce a false writing to bring about some advantage (usually economic). Forgery was a felony but now is often a misdemeanor when the value of the document is below a certain threshold. Common law forgery possessed four *corpus delicti* elements: one *actus reus*, one *mens rea*, and two attendant circumstances. Figure 3–5 outlines the elements for the crime of forgery at common law.

### Creating and Altering

The *actus reus* element for common law forgery was creation or alteration. Legally, **creation** (or making) refers to the manufacture of a document or instrument, whereas **alteration** refers to the any addition, deletion, or manipulation of an existing document or instrument. Keep in mind, however, that merely creating or altering a document does not constitute a forgery, as the *mens rea* and attendant circumstances must also be present.

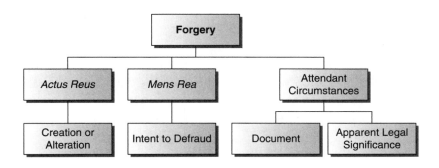

**Figure 3–5** Forgery.

## Documents with Legal Efficacy

The circumstances that must attend a forgery charge at common law were twofold. One, the object of the forgery must be a **document**, which can be anything with writing on it. And two, the document must possess **legal efficacy** (apparent legal significance), meaning that it must create a legal obligation that could be enforced in civil courts (such as a birth certificate, death certificate, or educational diploma). Some documents are not initially recognized as possessing legal significance but can be identified as such through the nature of its use. For this reason, an **instrument**, defined as a written legal document (such as a contract, deed, will, or currency), is the preferred target of forgery investigation. Today, some states have eliminated the legal efficacy requirement and allow all documents within forgery statutes. Pause for Thought 3–11 illustrates how to determine if a document possesses legal efficacy.

## Intent to Defraud

The *mens rea* for the crime of forgery requires intent to defraud. It is important to understand that the intent need not be directed at a specific victim; general intent to defraud is sufficient. It also matters not whether the intent is put into motion, as the passing of such fraudulent writings are addressed under a separate criminal offense known as uttering. Furthermore, the intended advantage sought through forgery is not restricted to monetary (or pecuniary) gain. Forgery law also cares not whether the intended fraud was fruitful, meaning that unsuccessful attempts to defraud others are nonetheless criminal violations because it serves to undermine the authenticity (and thus confidence) of documents in general. Simply put, after a legally significant document (or instrument) is created (or altered) with a mental forecast to defraud others of money or property, the crime has been consummated regardless of its successfulness. Pause for Thought 3–12 illustrates the proper legal interpretation regarding intent to defraud.

Currency is usually the mental image one conjures when hearing the term counterfeiting, but such statutes extend well beyond money. **Counterfeiting** actually refers to the making (or altering) of any government obligation (such as food stamps and bonds). Much like with burglary, counterfeiting cannot be accomplished without special equipment and tools; therefore, statutes also have been created which criminalize the possession of such counterfeiting paraphernalia (such as plates, paper, and stones).

## Uttering

**Uttering** at common law was defined as unlawfully passing a forged document with intent to defraud. Uttering often is consolidated with modern forgery statutes, but as a separate offense possesses three *corpus delicti* elements: one *actus reus*, one *mens rea*, and two attendant circumstances. Figure 3–6 outlines the elements required for the crime of uttering.

The *actus reus* element for uttering is "passing" a forged document. It is not necessary within uttering statutes that one create or alter a document. The mere passing of a document known to be forged is sufficient. Uttering also requires an offender possess intent to defraud when passing forged items. It is quite possible for persons to unknowingly possess

### Pause for Thought 3–11

Consider the following: Sally is a recent college graduate applying for a job with a prominent criminal justice agency. The organization requires submission of a recommendation letter from a major professor or department chair, but both decline her request. Fearing that she will not get the job, Sally writes a recommendation letter from the department chair and signs the appropriate name. Does this document possess legal significance? If so, does this action constitute the crime of forgery?

### Scenario Solution

Yes, any document or instrument with a pecuniary (financial) interest, even when victims are unaware of its value, possesses legal efficacy. In this scenario, the fraudulent recommendation can cause pecuniary harm to the criminal justice organization while also subjecting the department head to possible reputation damage for the potential hiring of a poor employee; therefore, forgery would be an appropriate charge.

> **Pause for Thought 3–12**
>
> Consider the following: The police stop a motorist for speeding. During the course of the stop, the officer smells alcohol emanating from the car and receives permission from the motorist to search the automobile. During the lawful search, the officer discovers a series of checks belonging to others. Several checks had been completed "to the order of" differing community merchants. The motorist explains he found the blank checks and did write in the information but had no intent to pass them. Further investigation determined that the motorist owed those exact sums of money to the merchants. Are these facts sufficient to prove the elements for the crime of forgery?
>
> **Scenario Solution**
>
> Yes, the first step in this evaluation requires the checks be altered. In this case, the motorist admits the alteration (handwriting analysis likely could verify this action as well). Second, bank checks are unquestionably regarded as possessing legal efficacy; therefore, the only remaining issue relates to whether the motorist had intent to defraud. Given the circumstances, it can be reasonably inferred that such intent exists, and as such, the crime of forgery was committed.

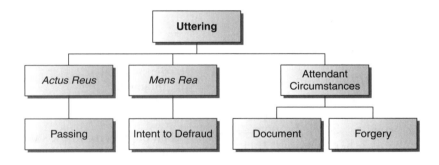

**Figure 3–6** Uttering.

forged items and subsequently pass them. Uttering was not intended to penalize such innocent transactions. Pause for Thought 3–13 provides a hypothetical scenario for the crime of uttering.

## Modern Forgery and Uttering Examined

Forgery at federal law combines common law forgery and uttering. Title 18, Chapter 25 of the U.S. Criminal Code, entitled Counterfeiting and Forgery, defines a multitude of conducts and penalties for forgery offenses (§§470-514). A large portion of federal forgery legislation aims to curtail counterfeiting schemes, whereas others seek also to combat the forging of money orders, bank notes, postage stamps, and other government instruments. The most generic of federal forgery law pertains to the making and uttering of government obligations or securities (§472). The making of fraudulent government documents and instruments (§471) states this:

> Whoever, with intent to defraud, falsely makes, forges, counterfeits, or alters any obligation or other security of the United States, shall be fined under this title or imprisoned not more than 20 years, or both.

Meanwhile, the uttering of counterfeit obligations or securities (§472) stipulates this:

> Whoever, with intent to defraud, passes, utters, publishes, or sells, or attempts to pass, utter, publish, or sell, or with like intent brings into the United States or keeps in possession or conceals any falsely made, forged, counterfeited, or altered obligation or other

> **Pause for Thought 3-13**
>
> Consider the following: Alex finds a series of counterfeit $100 bills and subsequently purchases computer equipment. If he knew the bills were forged, has he committed the crime of forgery, uttering, or both?
>
> ### Scenario Solution
>
> Alex has not committed a forgery because he did not create or alter the $100 bills. With that settled, however, he did commit an uttering offense because (1) Alex passed the bills when purchasing the computer equipment, (2) the currency possessed legal efficacy, and (3) Alex had the intent to defraud the store merchant.

security of the United States, shall be fined under this title or imprisoned not more than 20 years, or both.

## Summary

People have always cherished (and still do) their property (and money). This chapter sought to chronicle how the many theft crimes came into existence, while also tracing their contemporary evolution and consolidation. Specifically, this chapter aimed to provide students with a fundamental understanding of the elements associated with the crimes larceny, embezzlement, false pretenses, receiving stolen property, and other custodial theft crimes (unauthorized use, theft of service, and white-collar activities). The fraudulent theft-related crimes of forgery and uttering also were examined. The chapter also outlined how these crimes are defined within federal law, as well as their statistical presence in the United States today.

## Practice Test

1. _____ occurs when a person has lawful physical control of property but no discretion regarding how the property is handled or exercised.
   a. Custody
   b. Indirect ownership
   c. Direct ownership
   d. Transparent custody
   e. Invalid claim

2. _____ occurs when one causes an innocent bystander to take possession of money or personal property of which the accused never controlled.
   a. Direct asportation
   b. Constructive asportation
   c. Direct taking
   d. Constructive taking
   e. Secondhand theft

3. _____ occurs when one causes an innocent bystander to move money or personal property.
   a. Direct asportation
   b. Constructive asportation
   c. Direct taking
   d. Constructive taking
   e. Secondhand theft

4. _____ property possesses value but no actual concrete qualities.
   a. Undetermined
   b. Sentimental
   c. Intangible
   d. Fluctuating
   e. Real

5. Theft of property whose value meets or exceeds a predetermined value is called _____ larceny.
   a. petit
   b. simple
   c. grand
   d. felony
   e. first degree

6. _____ is defined as first-order authority to possess property which multiple parties have a lawful right to possess.
   a. Superior right of possession
   b. Rightful claim
   c. Primary heir
   d. Custody
   e. Ownership

7. _____ refers to the continued unauthorized use of another's property and can constitute larceny even when intent to deprive permanently is formed after taking.
   a. Perpetual theft
   b. Second degree larceny
   c. Deprivation after the fact
   d. Continuing trespass
   e. Delayed intent

8. _____ is a form of theft in which the law does not recognize consent as valid when obtained through fraud.
   a. Continuing trespass
   b. Consent fraud
   c. Invalid consent
   d. Fraudulent coaxing
   e. Larceny by trick

9. _____ refers to items left behind intentionally and permanently when it appears that the former owner does not intend to come back, pick it up, or use it.
   a. Forfeiture of property
   b. Abandoned property
   c. Ownership abandonment
   d. Property desertion
   e. Property forfeiture

10. _____ is ordinarily defined as the taking and carrying away of the personal property of another without permission but with no intent to deprive permanently.
    a. Unauthorized use
    b. Larceny
    c. Burglary
    d. Robbery
    e. Uttering

11. _____ constitutes paid work performed by others.
    a. *In limine*
    b. Usury
    c. Equity
    d. Service
    e. Contract

12. Common law _____ was the unlawful conversion or misappropriation of another's money or property by one to whom such items were entrusted.
    a. larceny
    b. embezzlement
    c. false pretenses
    d. forgery
    e. counterfeiting

13. _____ refers to the unauthorized use of unconverted property.
    a. False pretenses
    b. Illegal use
    c. Misappropriation
    d. Receiving stolen property
    e. Theft of services

14. _____ refers to transforming property from its original status.
    a. Transformation
    b. Metamorphosis
    c. Inducement
    d. Conversion
    e. Manipulation

15. A(n) _____ is a person entrusted with money and property of others.
    a. *ad litem*
    b. advisor
    c. supervisor
    d. guardian
    e. surety

16. _____ is a punishment that requires an offender to pay restitution in an amount three times the worth of items fraudulently acquired.
    a. Triple restitution
    b. Treble damages
    c. Victim compensation
    d. Fraudulent reimbursement
    e. Collateral compensation

17. Common law _____ was the acquiring of title to money or property of another through representations made with intent to defraud.
    a. fraud
    b. forgery
    c. false pretenses
    d. deception
    e. identity theft

18. _____ is the unlawful creation or alteration of a document or instrument possessing legal significance and with the intent to defraud.
    a. Tampering
    b. Embezzlement
    c. Fraud
    d. Forgery
    e. Counterfeiting

19. _____ refers to the addition, deletion, or manipulation of a document.
    a. Alteration
    b. Revision
    c. Creation
    d. Tampering
    e. Forgery

20. _____ is defined as the unlawful passing of a forged document or instrument with the intent to defraud.
    a. False documentation
    b. Uttering
    c. Forgery
    d. Fraudulent intent
    e. Counterfeiting

## References

Federal Bureau of Investigation. (2008). *Crime in the United States, 2007: Uniform Crime Reports*. Retrieved July 14, 2009, from http://www.fbi.gov/ucr/cius2007/index.html

# CHAPTER 4

# Crimes Against Habitation, Robbery, and Assault

## Key Terms

Actual breaking
Actual entry
Actual possession
Aggravated assault
Anti-Car Theft Act of 1992
Armed robbery
Arson
Assault
Attempted battery
Battery
Burglary
Burning
Carjacking
Constructive breaking
Constructive entry
Constructive possession
Crimes against habitation
Disablement
Dismemberment
Domestic violence
Dwelling
Extortion
False imprisonment
Harassment
Jostling
Kidnapping
Malicious intent
Mayhem
Menacing
Mutual affray
Nighttime
Parental Kidnapping Prevention Act of 1980
Possession of burglar tools
Right of locomotion
Robbery
Serious bodily injury
Simple assault
Stalking
Strong-armed robbery
Structural degradation
Threatened battery
Uniform Child Custody Jurisdiction Act of 1968

## Introduction

This chapter guides the reader through the historical evolution of laws designed to protect the sanctity of the home as well as violence in general. Known as **crimes against habitation**, common law arson and burglary each sought to deter and punish those who brazenly violated a person's place of last resort—the home (or dwelling). Our journey in this chapter begins with an examination of the narrow elements defining arson and burglary at common law, as well as explanations regarding their expansion in modern definitions. The chapter also examines lesser included offenses such as criminal trespass and possession of burglar tools. Discussion then turns to violence against persons, with a detailed examination of how aggravated assault and robbery were defined at common law and how modern definitions of these offenses take into consideration expanding concerns. Crimes associated with assault and robbery also are explored, with special attention paid to extortion and kidnapping. Finally, the extent to which these major crimes occur in America is examined. Federal legislation pertaining to each of these crimes is introduced at the conclusion of each topical section.

## Arson

In 2007 (FBI, 2008), law enforcement agencies reported the commission of 57,224 arsons, accounting for less than 1% of all property crime in the United States. Most arson (42.9%) was committed against structures (residential, storage, etc.), but a large portion of arson also targeted mobile property (27.9%) and other miscellaneous property (29.2%). Although the volume of arson is sparse when compared with other property crimes, its costliness to society is of extreme importance. Specifically, the dollar value associated with an average arson was $17,289. Arsons of industrial and manufacturing structures, however, produce substantially higher losses ($114,699 per offense).

## Arson at Common Law

Arson at common law was a serious crime that was punishable with death through burning. Common law **arson** was defined as the malicious burning of the dwelling of another. Essentially, then, the crime of arson specified four *corpus delicti* elements: one *mens rea*, one *actus reus*, and two attendant circumstances. Even though common law arson was not nearly as broad as its modern coverage, it nonetheless was considered a serious violation against the sanctity of one's home. Figure 4–1 outlines the elements that were required for arson at common law.

## Burning

At common law, the *actus reus* for the crime of arson was **burning**—defined as structural degradation caused by fire. **Structural degradation** referred to permanent change in a structure's material composition. Fire involved with acts of arson did not have to destroy totally or even cause substantial damage to the dwelling in question. Rather, any structural degradation sufficed when caused with fire. The most common test used for the determination of structural damage was whether fire caused charring. Mere blistering or blackening (caused by smoke) to a surface was insufficient to constitute arson. Even though explosions clearly pass the test for causing structural degradation, they too were not eligible for arson consideration at common law because the harm was not the product of a burn. Keep in mind, too, that the burning of items within the dwelling could also constitute arson when those items were permanently affixed to the dwelling structure (such as a lighting fixture). Personal property, however, was never eligible for protection within common law arson.

## Dwelling of Another

Arson at common law required that two attendant circumstances accompany an act of burning. First, the object of the burning had to be a dwelling. Historically, **dwelling** was defined as the primary safe haven where one habitually sleeps. Even though the traditional portrait of a dwelling conjures an image of a residential home with four constructed walls, the legal definition for a dwelling encompasses structures of all kinds (such as a boat, recreational vehicle, or tent). As we soon discuss, modern arson statutes have expanded beyond dwellings, but no structure other than a dwelling was included within the scope of common law arson. The value of the dwelling was not relevant to the crime of arson, nor was it necessary that the dwelling be used as such at the time of the actual burning. In short, it mattered only what was reasonably known to the public, meaning that a structure perceived to be occupied was legally a dwelling even when the structure was unoccupied. For example, a structure would still be regarded as a dwelling even if the arsonist was aware that its inhabitants were away on an extended vacation. Likewise, a structure known to be occupied was a dwelling even when its habitation was illegal.

The second attendant circumstance for common law arson was that the dwelling be "of another." It was well established that arson could not be committed against one's own dwelling. As noted in Chapter 3, however, "of another" refers to possession, not ownership, and as such, the burning of one's own dwelling could be arson when another person shared a proprietary (ownership) interest in the dwelling. For example, burning one's home that is lawfully leased to another would be arson because the tenant actually has the superior right of possession during the contractual period. Pause for Thought 4–1 illustrates the proper legal interpretation for resolving the dwelling debate.

## Malicious Intent

The law has always presumed fire to emanate from an act of God (natural occurrence) or accident. As such, arson at common law was classified as a general intent crime, meaning that its use as a criminal

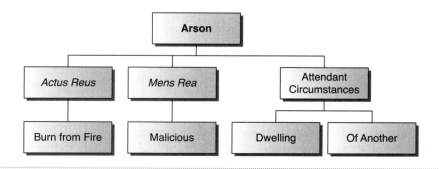

**Figure 4–1** Arson.

### Pause for Thought 4–1

Consider the following: Contractor Tad has been unable to sell his homes because of a failing real estate market. Fearing that he may lose his business, Tad decides to generate revenue by burning one of his houses for insurance money. Tad is aware that a homeless person unlawfully occupies the house, but proceeds with his plan. Under the common law standard, has Tad committed the crime of arson?

### Scenario Solution

Yes, even though the house should not have been occupied, Tad's knowledge that the homeless person was using it as a dwelling would constitute arson because he burned the dwelling of another with malicious (or purposeful) intent. Conversely, the act would not have constituted common law arson if the home had not been used as a dwelling. Keep in mind, too, that many states today include insurance fraud as arson; thus, modern law often would classify this act as arson regardless of property designation.

---

charge was unjustified when the fire was the consequence of any mental state less than intentionality. Conversely, arson's designation as a general intent crime equally made clear that the accused did not have to possess specific intent to set a fire. Rather, the *mens rea* element required only that the setting of fire be committed with malicious intent. With arson, **malicious intent** was consummated the moment a person voluntarily (or willfully) set a fire without legal justification or excuse; therefore, even though the crime of arson existed to deter harm associated with fire setting, the mental state of the offender was most important when considering criminal sanctions. An offender's motive, however, is not relevant. With that understanding, no proof is necessary that the accused held any ill will toward the victim or intended to cause personal harm to the victim as a consequence of the burning. Pause for Thought 4–2 illustrates the proper legal interpretation regarding the crime of arson.

## Modern Arson Examined

Arson statutes today have far outgrown their narrow origins. Compared with arson's common law blueprint, substantial changes have evolved over the years. To begin, arson now protects more than just dwellings. Most modern arson statutes extend protection to all structures (usually buildings and motor vehicles) and sometimes incorporate personal property (such as clothes) and woods. Some states even go so far as to include insurance fraud within arson statutes when policy claims are submitted for property that was intentionally burned. A second expansion among arson laws is to include all fire setting with the definition of a burn, even if the fire never reaches the targeted property. Contemporary arson statutes also choose to include explosions within burn interpretations.

A third change from common law arson relates to the classification of arson offenses across levels of severity. The most common approach segregates arson into offenses committed under extraordinarily dangerous circumstances (usually called first-degree arson, or aggravated arson) from those without such added danger. Grounded in common law, one factor that often aggravates the severity of arson is the burning of a dwelling (because of its heightened threat to human life). Aggravated arson (of sorts) also often occurs when certain people (such as firefighters,

---

### Pause for Thought 4–2

Consider the following: Kevin is upset with his wife's adulterous relationship with Jimmy. Kevin then proceeds to pour gasoline over Jimmy's property and sets it on fire with a match. Fortunately, a neighbor sees smoke emanating from the point of origin and suppresses the smoldering before a fire could form—thus negating any structural damage. Has Kevin committed arson as defined at common law?

### Scenario Solution

A charge of arson would not be appropriate at common law because the *actus reus* element (burning) was missing. Kevin definitely set out to maliciously burn the dwelling of another. The problem with a criminal charge of arson is that a "burn" never occurred; the property suffered no structural degradation.

police officers, and other emergency personnel) are harmed while attempting to combat a fire stemming from an act of arson. Conversely, burning unoccupied buildings and structures is an example of arson that ordinarily comprises the lesser degrees of arson. Finally, many states now include as arson burnings that result from the reckless handling of fire. Quite controversial, arson statutes of this nature allow for an arson charge against persons who had a legal right to start the fire but failed to use a reasonable standard of care in its management.

## Federal Arson Law

Arson committed against property on federal land is within the jurisdiction of the federal government. Arson at federal law is regulated by Title 18, Chapter 25 of the U.S. Criminal Code (§81). The statute stipulates that

> whoever, within the special maritime and territorial jurisdiction of the United States, willfully and maliciously sets fire to or burns any building, structure or vessel, any machinery or building materials or supplies, military or naval stores, munitions of war, or any structural aids or appliances for navigation or shipping, or attempts or conspires to do such an act, shall be imprisoned for not more than 25 years, fined the greater of the fine under this title or the cost of repairing or replacing any property that is damaged or destroyed, or both.

The statute aggravates the crime up to life imprisonment if the building is a dwelling or if the life of a person is placed in jeopardy.

## Burglary

In 2007 (FBI, 2008), law enforcement agencies reported the commission of nearly 2.2 million burglaries in the United States. Accounting for 22.1% of all property crime, burglaries cost society an estimated $4.3 billion in property losses (average dollar value, $1,991). When compared with the previous year (2006), the number of burglaries decreased marginally (–0.2%), but actually decreased significantly (6.6%) over the past decade (1998). Residential burglaries accounted for the majority of burglaries (67.9%) and are primarily committed in the daytime (63.6%). Conversely, however, nonresidential burglaries are committed primarily in the nighttime (56.4%). Finally, 61.1% of burglaries involved forcible entry.

### Burglary at Common Law

**Burglary** at common law was defined as the breaking and entering of the dwelling of another in the nighttime with intent to commit a felony therein. Burglary was a crime against the habitation of another, or the physical security of the home. Historically, little was more important than protecting the sanctity and safety of the home. Six *corpus delicti* elements comprised the definition of common law burglary: two *actus reus*, one *mens rea*, and three attendant circumstances. The absence of even one of the six elements negated the charge of burglary (although it likely still qualified for a lesser property crime). Figure 4–2 outlines the elements required at common law for the crime of burglary.

### Breaking and Entering

Common law burglary possessed two *actus reus* elements: breaking and entering. Breaking could be consummated through two avenues: actual breaking and constructive breaking. **Actual breaking** occurred when physical force was used to effect entry. The degree of force did not have to be substantial, nor did it have to cause damage. The elements for breaking simply required any degree of physical force (no matter how slight) used against some portion of a structure to create an opening to allow entry. One classic example of an actual breaking with the slightest use of force is the opening of an unlocked door through turning the doorknob. **Constructive breaking**, mean-

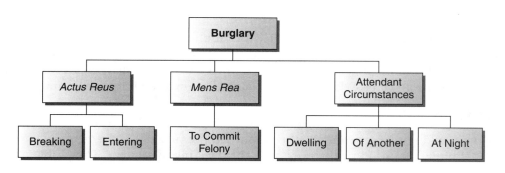

**Figure 4–2** Burglary.

while, occurred when one caused an unsuspecting third party to create an opening for the purpose of allowing the accused to enter unlawfully. In these situations, then, an accused never makes physical contact of any sort with the structure in question. On another note, common law was clear that a breaking had to occur just before the entry. In cases in which the breaking occurred while exiting the dwelling, the law did not recognize the breaking as sufficient to consummate the *actus reus* requirement for the crime of burglary. Pause for Thought 4–3 additionally illustrates the proper legal interpretation regarding this issue.

Much like breaking, the second *actus reus* element—entering—also could be satisfied through two mechanisms: actual entering or constructive entering. **Actual entry** referred to the physical insertion of a body part into the prohibited structure. Also, insertion of any body part (or portion thereof) satisfied the actual entry standard regardless of the extent of the insertion. For example, a perpetrator caught with only one arm inside a window would be sufficient to constitute an actual entry. Conversely, **constructive entry** occurred when one caused entry without physical insertion of a body part but through the use of some mechanical instrument or tool. Pause for Thought 4–4 illustrates the proper legal interpretation regarding this issue.

### Dwelling of Another in the Nighttime

Three attendant circumstances were required at common law to consummate the crime of burglary. The first of these necessities required that the breaking and entering be of a dwelling (defined in the same manner as arson). Pause for Thought 4–5 illustrates the dwelling concept.

The second attendant circumstance required the dwelling be "of another." As noted within the arson discussion, the central issue surrounding "of another" was that of possession, not ownership; therefore, it was possible to commit burglary against property to which one held ownership rights. Pause for Thought 4–6 illustrates the proper interpretation regarding this legal issue.

The third attendant circumstance required that the breaking and entering occur in the nighttime. The "at night" requirement was implemented because it was

---

### Pause for Thought 4–3

Consider the following: Paul suspects his wife (Brenda) is cheating. While driving home one afternoon, Paul sees his wife's car parked in the driveway of his best friend's home. Convinced that his friend would never open the door with knowledge of his true identity, he knocks and pretends to be a lost pizza delivery man. Deceived by the ruse, the best friend opens the door, at which time Paul enters the home and assaults him. Does this constitute an actual breaking or constructive breaking?

### Scenario Solution

Paul's actions do not describe an actual breaking because he used no force to enter the home. Paul did use deception to cause the opening of the door, however, and thus constitutes a constructive breaking. If not for the constructive breaking rule, common law burglaries without actual force would have been protected from a burglary charge.

---

### Pause for Thought 4–4

Consider the following: George uses a crowbar to open a window of a neighbor. Does this constitute an actual or constructive entry?

### Scenario Solution

The crowbar is an extension of George and therefore constitutes a constructive entry. No portion of George's body physically entered the home and as such cannot be construed as an actual entry. If not for the constructive entering rule, burglaries absent the insertion of an actual body part would have been exempt (at common law) from criminal prosecution under burglary's felonious designation.

> **Pause for Thought 4–5**
>
> Consider the following: A housing contractor recently completes the building of a home and then places the home for sale. During this time period, a homeless person has been staying in the home without permission. Even the community has no knowledge of the homeless person's habitation. Considering that the homeless person is using the home as a dwelling, should this home be considered a dwelling for burglary purposes?
>
> ## Scenario Solution
>
> No, the community has no reasonable grounds on which to believe the vacant home was being used as a dwelling; therefore, the home would not be included within the dwelling definition for burglary even though it was actually being used as a dwelling.

> **Pause for Thought 4–6**
>
> Consider the following: John leases an apartment. While at work, the owner of the apartment complex uses a master key to enter (without permission) the leased apartment for the purpose of taking valuables belonging to John. Given that the apartment was entered by its owner, has the owner committed burglary?
>
> ## Scenario Solution
>
> The owner has committed the crime of burglary. There is no question that the landlord owned the apartment, but the law defines "of another" as a possession issue; therefore, John, not the landlord, had the superior right of possession, and as such, the owner had no right (without permission or negotiated purpose) to enter the apartment. Thus, the dwelling is that "of another" and would be within the parameters of burglary definitions.

the time of day universally regarded as the most dangerous for an unlawful breaking and entering, as it was the primary window during which inhabitants were in the dwelling and sleeping. Although an exact time frame for what constituted "nighttime" was subject to some debate, it generally encompassed that time of day when an offender was not recognizable. In time, more precise definitions emerged, with most jurisdictions defining **nighttime** as the period between dusk (usually one hour after sunset) and dawn (usually 1 hour after sunrise). Application of the nighttime element was quite stringent, however, with any doubt regarding the timing of an unlawful breaking and entering resolved in favor of the accused. The meaning is this: no burglary charge would withstand legal challenge when nighttime was not conclusive.

### Intent to Commit a Felony

The *mens rea* for the crime of burglary required "intent to commit a felony therein." Intent to commit a misdemeanor is usually sufficient to constitute burglary at modern law, but was not adequate at common law. Ordinarily, the intended crime was theft (or larceny). Burglaries can be grounded, however, on the intent to commit other crimes as well. Moreover, because no single crime serves as the underlying crime within burglary definitions, both the crime committed therein and the broader burglary offense can be charged. Pause for Thought 4–7 illustrates the proper legal interpretation regarding this issue.

It also is important to understand that the intent to commit a felony had to be formed at or before the moment of unlawful entry. Formation of such intent at some point beyond the unlawful entry was considered a separate and distinct crime, and thus not burglary. Pause for Thought 4–8 provides an opportunity to test emerging understanding of this issue.

### Modern Burglary Examined

Today, the states have significantly expanded the scope of common law burglary. In general, modern burglary is defined as the unlawful entry of the building or structure of another with the intent to commit a

### Pause for Thought 4-7

Consider the following: Seth sneaks into Julie's home for the purpose of committing rape. After completion of the rape, Julie files a complaint that leads to Seth's apprehension. Can Seth be charged with burglary and/or assault in addition to the crime of rape?

### Scenario Solution

Seth cannot be charged with assault because that crime is a lesser included offense of the rape crime (meaning that rape cannot be committed without an underlying assault). Burglary is not a lesser included offense of rape, however, and thus, even though assault cannot be charged, both rape and burglary charges can be pursued.

### Pause for Thought 4-8

Consider the following: Drew borrows assorted items from his neighbor Donnie. Wanting to keep the items, Drew fabricates a story that the valuables were stolen in a recent burglary. Donnie does not believe the tale but has no alternative but to accept his loss. Soon, however, Drew leaves town for a 1-week vacation, during which time Donnie decides to enter his neighbor's home unlawfully to search for the supposed missing items. While inside the home, he comes across a diamond ring that clearly belongs to Drew. Donnie takes the ring as compensation for his own losses. Did Donnie commit the crime of burglary?

### Scenario Solution

Donnie did not commit the crime of burglary because he had no intent to steal the diamond ring when he entered the home. He merely wanted to retrieve property to which he had a valid "claim of right." He did commit criminal trespass when he unlawfully entered the home, however, as well as the crime of larceny when he took and carried away his neighbor's diamond ring with intent to deprive permanently.

---

crime. Examination of this generic definition reveals four distinct changes regarding the fundamental requirements for the crime of burglary. First, the breaking requirement has been eliminated in most states. Unlawful entry is now sufficient to constitute burglary without a corresponding breaking. Second, the "dwelling" requirement has been expanded in most states to include any building or structure (which, of course, continues to include dwellings). This expanded definition has been used by some states to even include automobiles within burglary statutes. The states that retain the dwelling distinction (like Rhode Island) create other statutes to adjudicate burglary-like offenses against nondwelling structures.

Third, the "at night" common law requirement has been eliminated. Burglary now can be committed in the daytime or nighttime. Fourth, even though a handful of states (such as Indiana and North Carolina) continue to use the "felony therein" restriction, most states have expanded their burglary statutes to include the intent to commit any "crime" therein: felony or misdemeanor (but not ordinance infractions).

Notwithstanding the general abandonment of the four common law elements referenced previously, states continue to use such common law distinctions as aggravating factors to differentiate the severity (or degree) of burglary offenses. Essentially, then, burglaries committed through the use of breaking, or against a dwelling (especially when occupied), or at night, or with intent to commit a felony are often the basis by which states enhance penalties for burglary crimes. Furthermore, committing a burglary while armed with a deadly weapon or an explosive also routinely serves to aggravate the crime.

## Burglary and Criminal Trespass Distinguished

The crime of criminal trespass is well known in America. To people of this independent-minded nation, little is more sacred than privacy rights to owned property. Thus, it has long been regarded a civil violation when one intentionally enters another's property without consent. Trespass can also be a crime, too, under circumstances in which the unlawful entry was known to be prohibited. Often referred to as malicious (or willful) trespass, the crime of trespass (usually a misdemeanor) actually is a lesser included element of the felonious crime of burglary. As such, trespass must occur before a burglary can even be considered under law. In short, the only real difference between burglary and trespass is the intent to commit a crime therein. Pause for Thought 4–9 illustrates the difference between burglary and trespass as criminal charges.

## Possession of Burglar Tools

The emphasis placed on deterring burglary can easily be inferred from the laws of many states that criminalize even the possession of tools ordinarily used to commit burglaries. By now, it should be clear just how important the legal process regards proof that an accused had intent (or *mens rea*) to commit the crime charged. With statutes criminalizing the **possession of burglar tools**, however, the intent to commit burglary can be inferred from the mere possession of such instruments without some legitimate reason for their possession. A locksmith, for instance, would be licensed to carry such an assortment of tools to perform assigned job duties and thus would not be guilty of having violated a crime of this sort. For example, Montana's criminal code (45-6-205) states:

> (1) A person commits the offense of possession of burglary tools when he knowingly possesses any key, tool, instrument, device, or explosive suitable for breaking into an occupied structure or vehicle or any depository designed for the safekeeping of property or any part thereof with the purpose to commit an offense therewith.

The severity of the punishment for possession of burglar tools varies, with some states classifying the offense as a misdemeanor and others as a felony. Montana, for example, regards the crime as worthy of imprisonment not to exceed 6 months. Alabama, on the other hand, stipulates the crime is a felony warranting classification as a Class C felony (1 to 10 years of imprisonment) (13A-7-8).

## Robbery

In 2007 (FBI, 2008), law enforcement agencies reported the commission of 445,125 robberies in the United States. Even though robbery is a theft-based offense, it is categorized within the Uniform Crime Reports as a violent offense because of its reliance on violence. Firearms were used in 42.8% of robbery incidents. Moreover, robbery accounted for 31.6% of all violent crime. Robberies cost victims an estimated $588 million in property losses (average dollar loss, $1,321), with bank robbery representing the greatest target ($4,201 per offense). When compared with the previous year (2006), the number of robberies decreased marginally (–0.5%) but actually increased 7.5% over the previous 5-year period (2003).

### Robbery at Common Law

Robbery is somewhat unique in that it is a felonious offense regardless of the circumstances surrounding the nature and severity of the theft (against property) and assault (against persons). Without the lesser included assault offense, the conduct would be nothing more than larcenous. Moreover, the larcenous violation likely would be a misdemeanor considering the average value of property taken during a robbery is far less than what is ordinarily required for grand larceny. On the other hand, the elimination of theft from robbery leaves only the residue of assault. Once again, too, it is possible that many such assaults would only amount to a misdemeanor given that most robberies do not cause (or intend to cause) serious bodily harm.

**Robbery** at common law was defined as the felonious taking of money or personal property from the

---

### Pause for Thought 4–9

Consider the following: Kyle is going on vacation and gives his neighbor (Josh) the key to his home in case of an emergency. Josh then chooses to use the key to enter the home to watch sports on Kyle's satellite system. Another neighbor notices the unexpected home activity and reports a possible burglary to the police. The police respond and arrest Josh. Has Josh committed the crime of burglary?

### Scenario Solution

Josh has not committed burglary because he possessed no intent to commit a crime within Kyle's home. Nonetheless, even though Kyle likely would not support a criminal charge, Josh is guilty of criminal trespass against Kyle's privacy rights in that he did unlawfully enter the home.

person or presence of another and through the use or threat of force or violence. Six corpus delicti elements comprise robbery at common law. Figure 4–3 outlines those elements required for the crime of robbery at common law.

## Felonious Taking Through Use or Threat of Force

The *mens rea* for common law robbery required an accused possess the intent to commit a theft through the use or threat of force. The intent need not exist in advance; formation of intent even seconds before the taking is sufficient. On another note, the first *actus reus* element of taking required an accused to gain physical control (for any amount of time) over the money or personal property of another and attempt to abscond (leave) with it. Possession can be consummated in one of two ways: actual possession or constructive possession. **Actual possession** occurs when one acquires physical custody of money or property, whereas **constructive possession** occurs when one causes money or property to be possessed without physical interaction. Regardless of the nature of the possession, it nonetheless constitutes the taking element required of robbery. Pause for Thought 4–10 illustrates the difference between an actual and constructive taking.

The second *actus reus* element required the use or threat of force or violence, the interpretation of which is the same as that for assault. In short, threats of force or violence must be imminent, directed at a person, made with at least the apparent ability (some states use a present ability standard) to carry out the threat, and place the victim in actual fear of harm. Keep in mind that the force (be it use or threat) need not be extensive. Any degree of force designed to effect a taking—no matter how slight—satisfies the robbery requirement when allowing for the removal of money or property from a person (or in a person's presence) and against his or her will. The degree of force does matter, however, when differentiating the severity of the robbery.

Robbery is traditionally divided into two major categories dependent on the nature of the force. **Armed robbery** occurs when a deadly weapon is used or threatened and ordinarily is punished up to life imprisonment (mandatory minimum in the neighborhood of 20 years normally applies). At common law, the present ability to carry out the threat was essential. Most states today, however, use an

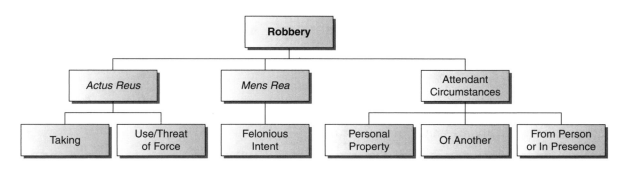

**Figure 4–3** Robbery.

---

### Pause for Thought 4–10

Consider the following: Ted enters a convenience store and orders the clerk (at gunpoint) to remove cash from the register. The clerk complies with Ted's order and transports the money from the register to the countertop. Before attaining actual possession of the money, however, the police intervene. Has Ted completed the crime of robbery?

### Scenario Solution

Yes, even though Ted never physically acquired possession of the money, he nonetheless has committed the crime of robbery because the taking element was satisfied through constructive possession. If not for the constructive possession rule, all robberies absent physical possession would be legally protected from a robbery charge.

apparent ability standard, meaning that simulations of being armed (such as a toy gun) are regarded as actually being armed. **Strong-armed robbery** (which most states refer to as "robbery") refers to forceful takings committed without assistance of a deadly weapon—hence, reliance on personal weapons (hands, feet). Pause for Thought 4–11 illustrates the proper legal interpretation for the use of force within the crime of robbery.

The force requirement stipulates only that a victim be handled in some way. Thus, a pickpocket (removing valuables from a person through stealth) would not be guilty of robbery because the force was limited to removing the valuables and in no way was directed or used against the person of the victim. At common law, the force had to be applied simultaneously with the taking. Modern statutes (for the most part) have abandoned this antiquated rule. Today, then, robbery is normally consummated when force is used contemporaneously (within the scope of the crime) with a taking. As such, even when no force is used against the victim to accomplish the taking, the offense can quickly and unexpectedly evolve into a robbery when a victim becomes aware of a theft attempt. Pause for Thought 4–12 illustrates the proper legal interpretation regarding this point of law.

## Money or Property of Another from Person or Presence

Robbery at common law (and today) required three attendant circumstances. First, the crime of robbery stipulated that the fruit of a felonious taking be money or personal property—the interpretation of which should be well understood at this advanced stage of our criminal examinations. The second attendant circumstance, too, has received in-depth coverage. Similar to burglary and larceny, the "of another" element prohibits charging a person with the crime of robbery when using or threatening force to retrieve items to which a "claim of right" is recognized. Regardless of this robbery exemption, however, the law still requires people to use civil mechanisms to reclaim property. Essentially, then, a person who uses or threatens force to reclaim money or property may not be guilty of robbery, but would be guilty of assault.

---

### Pause for Thought 4–11

Consider the following: Ted yanks a purse from an older woman's arm. Is this action sufficient to constitute the crime of robbery?

### Scenario Solution

The force used by Ted to remove the purse from the older woman's arm is sufficient to constitute the crime of robbery. It should now be clear that the force required for robbery can be of the slightest nature as long it overcomes bodily resistance (natural or overt).

---

### Pause for Thought 4–12

Consider the following: Mitch the Pickpocket, with technical precision, removes a wallet from the pants of an unsuspecting tourist with no force whatsoever to the body of the tourist. A bystander notices the theft by stealth and alerts the tourist. In response, the tourist commences to chase Mitch. After a short pursuit, the tourist tackles Mitch. Not wanting to give back the wallet, Mitch pushes the tourist to the ground and escapes with the coveted wallet. Has Mitch committed the crime of larceny with a subsequent assault, or has he committed a robbery?

### Scenario Solution

Mitch has committed the crime of robbery. It is true that Mitch did not intend to commit a robbery against the tourist at the time of the taking, but the law does not require that force and taking co-exist. Thus, Mitch the Pickpocket's use of force (push) to maintain possession of the wallet contemporaneous to the taking is sufficient to consummate the force requirement. To avoid a robbery charge, Mitch would have needed to discard the wallet to avoid the necessity of resorting to force.

The third attendant circumstance requires that any taking be from the body of a person or in the person's immediate presence. Thus, any person within the vicinity of the felonious taking who reasonably fears use or threat of force constitutes the crime of robbery. It also is important to remember that threats of force do not need to be verbally expressed; such threats can be implied from actions of an accused. Keep in mind, too, that theft without force is larceny, not robbery. Pause for Thought 4–13 explains how to interpret the presence requirement within the crime of robbery.

## Extortion

**Extortion** at common law occurred when public officials while acting under color of law demanded money or property to which there was no lawful entitlement. Most times, the act of extortion constituted a demand for money not owed, not yet due, or more than the amount owed (discussed more in Chapter 8). More recently, state (and federal) governments define extortion more along the lines of making a threat of future harm against persons or property when certain demands are not met (called blackmail). Figure 4–4 outlines the elements required of modern extortion statutes.

### Threat to Cause Harm with Intent to Cause Fear

Most modern extortion statutes possess a *mens rea* requirement that an accused actually intend to place one in fear of harm through the making of a threat. Thus, idle comments about the future that a reasonable person would not construe as a real threat do not constitute extortion even when a particular person (though unreasonably) interprets the same comments as such. For this reason, extortionist threats usually follow the threat guidelines set forth in assault law. Namely, the accused must have at least the apparent ability to carry out the threat, and the victim must actually be placed in fear that the threat will be carried out. Otherwise, thousands (if not millions) of persons daily would be subject to extortion arrest as a consequence of comments made in anger but where no one (including the

### Pause for Thought 4–13

Consider the following: Bryan enters a convenience store late one night and immediately notices that the clerk is alone. He waits until the clerk goes into a back room and then locks the door to that room so that the clerk is out of the picture. Bryan then commences to take money from the cash register while the clerk continuously bangs on the door. Does this action negate a robbery charge as the money was not taken from the person or presence of the clerk?

### Scenario Solution

Most jurisdictions classify Bryan's actions as robbery because force (locking of the door) was used to segregate the clerk from the taking; as such, he took the cash register's contents in the immediate presence of the clerk. Any other conclusion serves only to reward Bryan for the false imprisonment of the clerk. Bryan did commit the crime of larceny, but it serves only as an underlying element to what constitutes robbery.

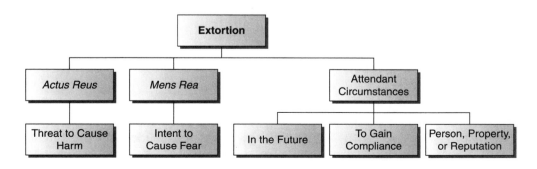

**Figure 4–4** Extortion.

threat recipient) believed the threat was legitimate. Pause for Thought 4–14 illustrates the proper legal interpretation regarding this issue.

## Future Harm to Gain Compliance from Objects of Threat

Threats addressed within extortion statutes can be against persons, properties, or reputations. Generally, threats must suggest future harm. One common exception to this future requirement relates to threats against persons (not property). Imminent threats against persons constitute assault and thus need no extortion protection. Threats against property and reputation, however, are not protected within assault statutes and thus can be (and often are) protected within extortion definitions. Moreover, the intent of the threat must be to gain compliance from the target, although actual compliance from the victim is not necessary. Extortion laws also do not require actual damage result from threats. Furthermore, threats against reputations need not possess truth. In short, the threat to harm one's reputation can pertain to something true or fabricated. Pause for Thought 4–15 illustrates the proper way to legally process these extortion issues.

## Burglary, Robbery, and Extortion at Federal Law

Burglary and robbery are combined within federal legislation. Title 18, Chapter 103 of the U.S. Criminal Code, Robbery and Burglary defines a multitude of actions (§§2111-2119) that attempt to regulate offenses against or on federal facilities (such as banks, post offices, and other territories and installations). Federal efforts to combat robbery of personal property is found within §2112: "Whoever robs or attempts to rob another of any kind or description of personal property belonging to the United States, shall be imprisoned not more than fifteen years." More specifically, §2113 outlines the prohibitions against bank robbery and bank burglary:

> Whoever, by force and violence, or by intimidation, takes, or attempts to take, from the person or presence of another, or obtains or attempts to obtain by extortion any property or money or any other thing of

---

### Pause for Thought 4–14

Consider the following: Andrew is upset about comments that Kyle made regarding his "nerdy" disposition. Out of anger, Andrew (5'5", 135 pounds) yells at Kyle (6'2", 220 pounds) that an apology best be forthcoming by the end of day. If the apology is not received, he promises to "beat up" Kyle and damage his car. Do these threats constitute the crime of extortion?

### Scenario Solution

Extortion would be a legitimate charge regarding the threat against Kyle's car because Andrew possessed the ability to cause the threatened damage. Conversely, however, Andrew does not have the apparent ability to physically harm Kyle in the manner threatened, nor is it likely that Kyle actually feared his threat. On this issue, then, Andrew would not be guilty of extortion.

---

### Pause for Thought 4–15

Consider the following: Professor Jones has never sexually harassed a student. Suzie the Student nonetheless threatens to file a false sexual harassment complaint against the professor unless she receives an A in the course. Professor Jones reports the threat. Has Suzie committed the crime of extortion?

### Scenario Solution

Suzie has committed the crime of extortion. Without question, Suzie made a threat to cause future harm (with the ability to carry out the threat). Moreover, the threat was made with the intent to place the professor in fear of the threatened harm, the purpose of which was to gain compliance from the professor—to issue an A in the course.

value belonging to, or in the care, custody, control, management, or possession of, any bank, credit union, or any savings and loan association; or Whoever enters or attempts to enter any bank, credit union, or any savings and loan association, or any building used in whole or in part as a bank, credit union, or as a savings and loan association, with intent to commit . . . any felony affecting such bank, credit union, or such savings and loan association and in violation of any statute of the United States, or any larceny—Shall be fined under this title or imprisoned not more than twenty years, or both.

Congress also passed legislation aimed at reducing armed auto theft. Entitled the **Anti-Car Theft Act of 1992**, the law made it a federal crime to commit a **carjacking**—taking a motor vehicle from an occupant through use or threat of force. Subsequently, passage of the Violent Crime Control and Law Enforcement Act of 1994 put teeth into the law by permitting the punishment of death for those who cause death in the midst of a carjacking.

Meanwhile, federal extortion is addressed within Title 18, Chapter 41 of the U.S. Criminal Code. Extortion and Threats (§§871-880) combats threats ranging from against the President and successors to the Presidency (§871) to receiving proceeds from extortion (§880). Likely the most well known form of extortion is blackmail (§873):

Whoever, under a threat of informing, or as a consideration for not informing, against any violation of any law of the United States, demands or receives any money or other valuable thing, shall be fined under this title or imprisoned not more than one year, or both.

Finally, anyone interested in working as an officer or employee of the United States should be aware of §872:

Whoever, being an officer, or employee of the United States or any department or agency thereof, or representing himself to be or assuming to act as such, under color or pretense of office or employment commits or attempts an act of extortion, shall be fined under this title or imprisoned not more than three years, or both; but if the amount so extorted or demanded does not exceed $1,000, he shall be fined under this title or imprisoned not more than one year, or both.

## Assault

In 2007 (FBI, 2008), law enforcement agencies reported 855,856 aggravated assaults in the United States. Accounting for 60.8% of all violent crime, aggravated assault is especially dangerous given that 21.4% of its incidents involve the use of a firearm. When compared with the previous year (2006), the number of aggravated assaults decreased marginally (–0.6%), but has actually declined significantly (–12.4%) over the past decade (1998).

## Assault and Battery at Common Law

Assault and battery were separate and distinct crimes at common law. **Assault** was generically defined as an attempted battery or threatened battery. An **attempted battery** referred to an overt act designed to cause harm through physical contact. For example, an attempt to punch another unlawfully but where the effort was unsuccessful constituted an attempted battery. On the other hand, an assault also occurred through a threat to batter. As expected, a **threatened battery** required a threat to batter another person with some degree of mental fault. More importantly, threatened battery outlined four attendant circumstances. First, the threat must be imminent (meaning very soon). Second, the threat must be directed at a person (not property). Third, the person who made the threat must have had the present (or actual) ability to carry out the threat. Keep in mind, however, that most states now require only an apparent ability to carry out a threat, which is guided by a reasonableness standard wherein actual ability is not relevant. Rather, it only need appear that the threatening party could accomplish the threatened harm. Fourth, the victim must actually be placed in fear of harm. Evidence that the victim was not fearful of the threat would negate the crime of assault. Most complaining victims, however, likely will protest that such fear existed if for no other reason than to ensure that the threatening party is punished—revenge. Figure 4–5 outlines the elements required of common law assault. Pause for Thought 4–16 further illustrates the proper legal interpretation regarding the criminal elements for a threatened-battery assault.

**Battery** at common law was defined as the nonlethal culmination of an assault. Essentially, then, a successful attempt to batter (defined as assault) constituted battery when no death resulted (because death would produce the broader crime of homicide). Thus, battery was a lesser included offense of assault, meaning a criminal charge of assault was not valid once the crime of battery was consummated. In a nutshell, battery was the unlawful application of force against a person. To constitute battery, an accused must have intended to use force without justification, or with justification but where the force was unreasonably applied (as in excessive self defense). For example, common law carved out a niche called **mutual affray** to exempt what otherwise would constitute a battery were it not for the

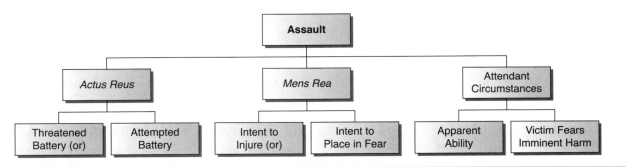

**Figure 4-5** Assault.

---

### Pause for Thought 4-16

Consider the following: Dean insults Raymond in front of friends. Embarrassed at the comment, Raymond threatens to punch Dean's "lights out" if he opens his mouth again. Dean laughs and encourages Raymond to "bring it on." Has Raymond committed an assault?

### Scenario Solution

Raymond likely has committed the crime of assault. Raymond threatened Dean with imminent bodily harm. It also is undeniable that Raymond possessed the apparent ability to carry out the threatened punch. Thus, deciding the issue regarding whether Raymond committed a threatened-battery assault depends on whether Dean was in actual fear of harm. Assuming that Dean's laughter and subsequent battery invitation were genuine, this element appears absent.

---

mutual consent of the engaging parties. Consent to serious bodily injury is never authorized, however.

The mere application of force incident to reckless and negligent conduct also was not sufficient to consummate battery at common law. Furthermore, battery at common law required some measure of bodily contact resulting in pain or discomfort; actual injuries, however, were not necessary to constitute the crime of battery. Figure 4-6 outlines the elements required of battery at common law. Pause for Thought 4-17 further evaluates the crime of battery with respect to mental fault.

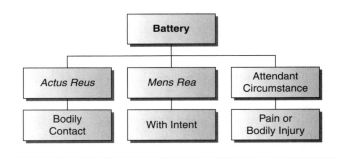

**Figure 4-6** Battery.

---

### Pause for Thought 4-17

Consider the following: Blake runs out of the office because he is late for a meeting. While exiting the office, he runs into a fellow employee, who subsequently falls and breaks his arm. Is this a criminal battery or simply an accident?

### Scenario Solution

Blake's action is considered battery in most states today because most such statutes include recklessness and negligence within their battery (or assault) statutes. Moreover, Blake's action was not an accident because he purposely set into motion a disregard for the welfare of others.

## Modern Assault Examined

Assault in most states today (and federal law) has annexed (of sorts) the common law crime known as battery. As such, it is now uncommon for a state to have a freestanding battery statute. The reason for this widespread statutory revision is grounded within court rulings in the early 1970s. At that time, most states had merged the crimes of assault and battery into what is still generically referred to as "assault and battery." The courts ultimately declared that such a merger was unconstitutional because a person could commit an assault without committing a battery. Thus, charging a person with "assault and battery" in many cases was not accurate. As a result, most jurisdictions chose to redefine their statutes by eliminating the battery terminology, opting to move common law assault and battery under the umbrella of one crime called assault.

Federal law proscribes assault as felonious in Title 18, Chapter 7 of the U.S. Criminal Code (§§111-118). Ranging from assaults on federal officers (§111) to interfering with protection functions (§118), assault federal statutes address many behaviors that aim to protect parties of federal interest. An example of particular federal interest can be found in the regulation of female genital mutilation (§116b) stating that

> whoever knowingly circumcises, excises, or infibulates the whole or any part of the labia majora or labia minora or clitoris of another person who has not attained the age of 18 years shall be fined under this title or imprisoned not more than 5 years, or both.

## Aggravated and Simple Assault Distinguished

Assault is classified along two lines of severity: simple and aggravated. **Simple assault** refers to an assault which causes, intends to cause, or threatens to cause less-than-serious bodily harm (referred to as bodily harm). Realistically, the best way to determine when an assault is simple is to understand what is not aggravated. With that understanding, most jurisdictions define **aggravated assault** as occurring under several unique circumstances. The two most common circumstances are those pertaining to serious bodily injury (or use of a deadly weapon) and mayhem. **Serious bodily injury** is defined as an action that has a "high probability of causing death." One must keep in mind that high probability of causing death is not synonymous with near certainty. Rather, an action which a reasonable person would associate with death is sufficient to infer the requisite intent to cause serious bodily injury. The use of a deadly weapon will almost always constitute such intent regardless of the true mental state of the accused. Pause for Thought 4–18 illustrates the legal interpretation for serious bodily injury.

Mayhem is a second factor that can elevate a simple assault to aggravated assault. **Mayhem** causes the dismemberment or disablement to a bodily part or organ. **Dismemberment** is the loss of some portion of a body part or organ. One of the best examples of dismemberment occurred when Lorena Bobbitt cut off the penis of her alleged abusive husband while he was sleeping. **Disablement**, on the other hand, causes loss of the use of a body part or organ. Poking someone in the eye with a pencil and causing permanent blindness is an example of disablement. Permanent disfigurement to the person of another also often constitutes mayhem. States vary in their interpretation of what constitutes permanent disfigurement, but disfigurement historically had to be visible to the public. Under this rule, disfigurement to one's private areas would be ineligible for aggravated

---

### Pause for Thought 4–18

Consider the following: Craig hates Kenneth for assorted reasons. While approaching a pedestrian crossing, Craig notices Kenneth about to cross the street and decides to scare him. As Kenneth walks across the street, Craig accelerates rapidly in his direction. Scared out of his wits, Kenneth dives out of the way, suffering cuts and bruises. Outraged at the reckless conduct, Kenneth files a criminal complaint. Is Craig guilty of an assault? If so, does the conduct rise to that of aggravated assault?

### Scenario Solution

Craig could be charged with aggravated assault even though he never truly intended to cause physical harm. Unfortunately for Craig, the actual causing or attempting to cause serious bodily injury is not the only aggravating factor by which assault can be elevated. Intent to place one in fear of serious bodily injury also constitutes aggravated assault in most jurisdictions today. In this case, it would not be difficult to presume that Craig intended to place Kenneth in fear of serious bodily harm by driving the automobile toward him.

assault consideration but could still be aggravated assault if the action causing the disfigurement has a high probability of causing death or mayhem as a product of dismemberment or disablement. Pause for Thought 4–19 illustrates the proper legal interpretation regarding the requirements for mayhem.

Many states (such as Massachusetts) categorize as aggravated assault any assault (even simple assault) committed in furtherance of a felony even though the accused did not cause (or intend or threat to cause) serious harm. Violence against children, spouses, and older persons is protected under aggravated assault statutes, once again regardless of the actual mental intent of the accused. Finally, states often elevate simple assault to aggravated assault when an accused knew the victim was acting in an official employment capacity (as specified in the statute). The most notable professions covered within such statutes are police and fire personnel.

## Modern Assault-Related Crimes

In response to the aforementioned unconstitutionality of assault and battery merger statutes, some states chose not to consolidate assault offenses under the umbrella of one statute. Some states instead created and continue to use statutes aimed at actions that fall short of physical harm. Menacing is one such statute created to singularly address serious threats that do not rise to physical action. With **menacing**, the threat must be imminent and can be directed against persons or their property. **Harassment** is another statute designed to adjudicate assaults of a minor nature, or less than bodily injury that usually occur from mere pushing or shoving. Along this same line, **jostling** has emerged in certain jurisdictions (usually large cities) to handle persons who bump or push others for the specific purpose of committing theft.

Most notable among special assault crime is domestic violence. States differ with regard to their legislative approach regarding this serious problem. Some states create specialized crimes that draw attention to the nature of the assault, whereas other states consider the relationship during the adjudication process. Regardless, **domestic violence** is nothing more than an assault that takes place within familial environments. Often referred to as interpersonal violence, these assaults ordinarily are defined as occurring between spouses, children, and other close relationships among both heterosexual and homosexual parties. Domestic Violence and Stalking, Title 18, Chapter 110A of the U.S. Criminal Code (§§2261-2266) outlines federal efforts to protect spouses and intimate partners against domestic violence and stalking. Specifically, §2266(7) defines a spouse or intimate partner as

> a spouse or former spouse of the abuser, a person who shares a child in common with the abuser, and a person who cohabits or has cohabited as a spouse with the abuser; or a person who is or has been in a social relationship of a romantic or intimate nature with the abuser, as determined by the length of the relationship, the type of relationship, and the frequency of interaction between the persons involved in the relationship. . . .

With that understanding, interstate domestic violence (§2261(1)) is regulated as

> whoever travels in interstate or foreign commerce or within the special maritime and territorial jurisdiction of the United States, or enters or leaves Indian country, with the intent to kill, injure, harass, or place under surveillance with intent to kill, injure, harass, or intimidate another person, and in the course of, or as a result of, such travel places that

### Pause for Thought 4–19

Consider the following: Boxer A removes his gum shield after the second round. During the next round, he bites off a portion of his opponent's ear. Does this action constitute aggravated assault? Does it matter that the harm occurred in a boxing match?

### Scenario Solution

This scenario may seem familiar because that is exactly what Mike Tyson did to Evander Holyfield in a 1997 heavyweight title bout in Las Vegas. Aside from Tyson's disqualification and 1-year suspension, no criminal charges were filed even though his action appeared to be one of aggravated assault. Biting off part of an ear is a classic example of mayhem when a person intends to cause dismemberment or disablement of such organ. The question, then, is did Tyson intend to cause the harm? Boxers always wear gum-shields to protect their teeth, and therefore, its removal most certainly shows premeditation. The fact that the assault occurred within a boxing ring does not excuse such criminal behavior.

person in reasonable fear of the death of, or serious bodily injury to, or causes substantial emotional distress to that person, a member of the immediate family . . . or the spouse or intimate partner of that person. . . .

Along the same line as threat-based assault, many states have created the crime of **stalking** to criminalize those who, without justification, intentionally scare (over time) another through repeated harassment, watching, and following. Repeated is the central theme, in that most states do not support a stalking conviction for an isolated episode of harassment. Moreover, as with traditional assault, victims must actually fear harm to themselves or specific others. As such, it is impossible to scare one who is unaware of stalking behavior, and thus, victim awareness is critical to a successful stalking prosecution. For example, it would not be criminal stalking if one was told of the stalking behavior without having been aware of the actions. Punishment for stalking is ordinarily increased when committed in violation of a restraining order.

Many states have yet to include the use of computers within their stalking statutes, but federal legislation does include computer use (called cyberstalking). Federal law defines (§2261A) stalking as the use of

> the mail, any interactive computer service, or any facility of interstate or foreign commerce to engage in a course of conduct that causes substantial emotional distress to that person or places that person in reasonable fear of the death of, or serious bodily injury. . . .

## Kidnapping

At common law, people often would restrain the kings' relatives for ransom. In response to these acts of political terrorism, the crime of kidnapping was created to deter removal of those "loved ones" from the king's protective jurisdiction. With that background, **kidnapping** at common law was defined as moving a person through force or deception to another country with the intent to deprive the person of freedom (confinement). Hence, movement of a substantial distance was required. In cases where a victim was not moved, the crime at common law was not kidnapping but rather **false imprisonment**. Thus, the only difference between kidnapping and false imprisonment was the act of asportation (movement). Over time, however, the crime of kidnapping evolved to protect what is now universally regarded as a citizen's **right of locomotion**, which purports that people have the right to come and go as they please without fear of unlawful seizure (or restraint).

Today, kidnapping statutes vary among the states with respect to the movement component. Some states are more liberal with their interpretation and require no actual movement to constitute the crime of kidnapping, instead requiring only evidence of the intent to move others against their will. Conversely, other states have chosen a more strict legislative strategy requiring that a victim actually be moved to some secret location. In those states, however, the movement usually cannot be merely incidental to the commission of another crime. States following this rule, for example, could not charge an accused with kidnapping (in addition to rape) if the forcible movement was only for the purpose of committing the intended rape. The rationale for this legal approach is that movement of the victim did not significantly enhance the danger to the victim compared with the harm already planned within the rape. Other states reject such an argument and allow a kidnapping charge in addition to incidental crimes. Regardless of the approach, kidnapping is widely considered a serious felonious offense punished up to life imprisonment.

Federal law proscribes kidnapping as felonious in Title 18, Chapter 55 of the U.S. Criminal Code (§§1201-1204). Federal kidnapping law addresses kidnapping in general, while also attending to more specific concerns such as ransom for money, hostage taking, and international parental kidnapping. The Federal Kidnapping Act (referred to as the Lindbergh Act of 1934) states (§1201):

> Whoever unlawfully seizes, confines, inveigles, decoys, kidnaps, abducts, or carries away and holds for ransom or reward or otherwise any person, except in the case of a minor by the parent thereof . . . shall be punished by imprisonment for any term of years or for life and, if the death of any person results, shall be punished by death or life imprisonment.

The statute also warns that "failure to release the victim within twenty-four hours . . . shall create a rebuttable presumption that such person has been transported in interstate or foreign commerce."

Federal legislation also has sought to resolve the ever-growing problem of parental kidnapping and child abduction. The most notable of such legislation pertains to child stealing by a parent (Title 28, Chapter 115 of the U.S. Criminal Code). Called the **Parental Kidnapping Prevention Act of 1980** (§1738A), the legislation eliminates jurisdiction disputes in child custody cases by usurping all state authority over such matters. In so doing, federal authorities are permitted to issue and execute warrants for parents who flee a state to avoid kidnapping charges. In essence, the legislation serves to eliminate any custodial reward a parent may derive for unlawfully transporting a child to another state. Second, the **Uniform Child Custody Jurisdiction Act of 1968** further provides that jurisdiction will always remain with the home custodial state (where the court rendered the original decision). All states participate in this legal philosophy, and thus, it does a parent little good to move a child to another state in hopes of finding a more favorable judicial ruling regarding the custodial arrangement.

## Summary

From the earliest of times, people have held sacred the importance of a habitat. This chapter sought to explain how the crimes arson and burglary aimed to protect such dwellings while also tracing their contemporary expansion to include other buildings and structures (and even the mere possession of burglar tools). Moreover, the premium placed on one's right to be free from bodily harm (or the fear of) also has been longstanding. Thus, this chapter also endeavored to provide students with a fundamental understanding of the elements associated with the crimes of robbery (armed and strong-armed), extortion, assault (aggravated and simple), and assault-related crimes (such as battery, menacing, and kidnapping). The chapter also outlined how these crimes are defined within federal law, as well as their statistical presence in the United States today.

## Practice Test

1. Common law arson and burglary were regarded as crimes against _____, which sought to deter and punish those who brazenly violated a person's home.
    a. security
    b. protection
    c. habitation
    d. public safety
    e. public order

2. _____ refers to permanent change in a material's composition.
    a. Structural degradation
    b. Material degradation
    c. Composition alteration
    d. Material metamorphosis
    e. Structural alteration

3. Common law arson was the malicious burning of _____.
    a. any building of another
    b. any dwelling of another
    c. any property of another
    d. all property
    e. any dwelling

4. _____ is defined as a primary safe haven where one habitually sleeps and eats.
    a. Home
    b. House
    c. Household
    d. Residence
    e. Dwelling

5. Common law burglary was defined as the breaking and entering of the dwelling of another in the nighttime with the intent to commit a _____ therein.
    a. crime
    b. larceny
    c. violation
    d. felony
    e. misdemeanor

6. _____ breaking occurred when one caused another to move an object that allowed entrance without physical contact of any sort.
    a. Actual
    b. Constructive
    c. Passive
    d. Active
    e. Assisted

7. _____ entry occurred when one entered without physical insertion of a body part but through the use of an instrument or tool.
    a. Actual
    b. Constructive
    c. Passive
    d. Active
    e. Assisted

8. _____ possession occurs when one causes money or property to be possessed without physical interaction.
    a. Constructive
    b. Simple
    c. Actual
    d. Physical
    e. Assisted

9. _____ possession occurs when one acquires physical custody of money or property.
    a. Constructive
    b. Simple
    c. Actual
    d. Physical
    e. Assisted

10. _____ robbery refers to robberies committed without assistance of a deadly weapon and reliance on personal weapons.
    a. Armed
    b. Simple
    c. Deadly
    d. Strong-armed
    e. Personal

11. _____ robbery occurs when a deadly weapon is used or threatened and ordinarily is punished up to life imprisonment.
    a. Armed
    b. Simple
    c. Deadly
    d. Strong-armed
    e. Personal

12. The _____ Act provides that custody will always remain with the home custodial state (where the court rendered the original decision).
    a. Parental Kidnapping Prevention
    b. Uniform Child Kidnapping
    c. Uniform Child Custody Jurisdiction
    d. Home Base
    e. Child Protection

13. Common law _____ was only committed when a public official demanded money or property that was not entitled.
    a. burglary
    b. robbery
    c. bribery
    d. extortion
    e. theft

14. Common law assault required at least a(n) _____, which required an overt act to cause harm through physical contact.
    a. simple battery
    b. simple assault
    c. attempted battery
    d. aggravated assault
    e. threatened battery

15. Most jurisdictions define _____ as serious bodily injury, mayhem, or permanent disfigurement.
    a. simple assault
    b. aggravated assault
    c. menacing
    d. simple battery
    e. aggravated battery

16. _____ refers to an assault that causes, intends to cause, or threatens to cause less than serious bodily harm (referred to as bodily harm).
    a. Simple assault
    b. Aggravated assault
    c. Minor assault
    d. Simple harassment
    e. Harassment

17. Under mayhem, _____ is the loss of some portion of a body part or organ.
    a. dismemberment
    b. amputation
    c. decapitation
    d. disablement
    e. serious bodily injury

18. _____ is a contemporary statutory creation designed to adjudicate assaults of a minor nature (less than bodily injury), which usually occur from mere pushing or shoving.
    a. Simple assault
    b. Aggravated assault
    c. Jostling
    d. Menacing
    e. Harassment

19. _____ was created in many states to address those who intentionally scare (over time) another person through repeated harassment, watching, and following.
    a. Menacing
    b. Stalking
    c. Infringement
    d. Badgering
    e. Harassment

20. Common law _____ was defined as moving a person through force or deception to another country with the intent to deprive the person of freedom.
    a. false pretenses
    b. forcible compulsion
    c. embracery
    d. kidnapping
    e. false imprisonment

## References

Federal Bureau of Investigation. (2008). *Crime in the United States, 2007: Uniform Crime Reports.* Retrieved July 14, 2009, from http://www.fbi.gov/ucr/cius2007/index.html

# CHAPTER 5

# Criminal Homicide

## Key Terms

- Accident
- Adequate provocation
- Aforethought
- Born alive standard
- Brain death
- Deadly weapon doctrine
- Deliberation
- Depraved-heart murder
- Euthanasia
- Excusable homicide
- Felony murder
- Feticide
- First-degree murder
- Genocide
- Gross negligence
- Heat of passion
- Homicide
- Imperfect self-defense
- Implied malice
- Involuntary manslaughter
- Justifiable homicide
- Law Reform (Year and a Day Rule) Act of 1996
- Malice
- Manslaughter
- Misadventure
- Misdemeanor-manslaughter rule
- Murder
- Noncriminal homicide
- Ordinary negligence
- Premeditation
- Quick fetus
- Second-degree murder
- Uniform Determination of Death Act
- Viable fetus
- Voluntary manslaughter
- Year-and-a-day rule

## Introduction

This chapter explores the legal framework for processing homicide cases in America. While progressing through its content, readers will develop an appreciation for (1) the requirements for what is considered a homicide and (2) the physical act requirements for determining which homicide cases can be criminally prosecuted. The chapter then examines an offender's state of mind to differentiate criminal homicide from noncriminal homicide, with a focus on the intent required to prosecute for criminal homicide. The chapter also chronicles the numerous standards (such as the **heat of passion** test and doctrine of transferred intent) used to formulate the standards used by courts to examine *mens rea* in homicide cases. In so doing, murder and manslaughter are distinguished at common law, along with an analysis of their modern forms within the degree system (first- and second-degree murder; voluntary and involuntary manslaughter).

## Murder in America

In 2007 (FBI, 2008), murder accounted for 1.2% of the overall number of violent crimes known to the police. Furthermore, this percentage represents a 0.6% decrease from the 2006 murder estimate. Against this backdrop, it is tempting to be unconcerned about murder as a problematic crime. Prior to planning a celebratory function, however, two additional findings should be kept in mind. First, a member of someone's family has been murdered 16,929 times. Second, the 2007 murder numbers, although lower than the previous year, actually have increased 2.4% since 2003.

## Homicide Defined

From an investigative perspective, the manner in which people ultimately die (or expire) occurs in one of four ways (referred to as NASH): natural, accident, suicide, or homicide. Homicide generates the greatest response from the criminal justice system. With that understanding, **homicide** is the killing of one human being by another. Although no portion of this definition addresses the intent of a causal actor's behavior, the mere mention of the term *homicide* causes most people to assume that some element of criminal wrongdoing exists. A homicide does not necessarily, however, constitute a crime because criminal homicide is a true crime, meaning that legal blameworthiness does not attach unless one's apparent unlawful conduct (prohibited act or omission of legal duty) is accompanied by some degree of mental fault. Deaths from natural causes, for obvious reasons, are exempt from criminal classification, but occasional prosecutions are commenced when the cause of death can be traced to some criminal conduct. For example, although heart attacks are natural malfunctions, robbers incur criminal liability for those deaths when the heart attack (myocardial infarction) results from fright in response to their use or threat of force. Known as felony murder, criminal liability attaches because the robber purposefully set into motion a dangerous chain of events, without which the heart attack and subsequent death likely would not have occurred.

Before embarking on a discussion of *mens rea* complexities, it is important to remain cognizant of the constitutional principle of state sovereignty. As long as laws are enacted in accord with constitutional provisions (state constitution and U.S. Constitution), states possess the absolute freedom to legislate in a manner consistent with their best interests. The result of such legislative autonomy, however, is a wide variation found among state homicide laws with respect to definitions, requirements, and limitations. With that in mind, a logical point to begin a discussion of homicide law and criminal culpability is with an examination of the elements required to constitute a homicide. Figure 5-1 diagrams the homicide requirements.

## Human Being

Homicide definitions require that death be of a human being. Herein lies the first of many variations associated with homicide law. What exactly constitutes a human being for criminal purposes? Common law used the **born alive standard**, requiring that a fetus be born alive and achieve independent circulation. Evidence of complete expulsion from the womb and signs of respiration (e.g., breathing, crying) satisfied the born-alive requirement, but severance of the umbilical cord was not necessary to establish independent circulation. In order to convict successfully for the death of a newborn, the prosecution had to establish beyond a reasonable doubt the existence of these two elements; however, the death of a human being could be criminally prosecuted even if the injuries were inflicted while still in the womb. Pause for Thought 5-1 illustrates the common law legalities that guide a prosecutorial inquiry into the death of an unborn child.

Critics argue that the common law definition rewards those most proficient at inflicting fatal

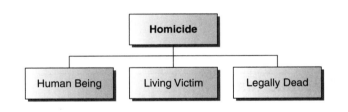

**Figure 5-1** Homicide.

---

### Pause for Thought 5-1

Consider the following: During the course of a domestic dispute, a husband intentionally kicks the stomach of his pregnant wife of 7 months. The serious injuries inflicted on the fetus require that a caesarian section be performed to remove the fetus from its mother's womb. Within a few hours of the successful procedure, the child dies from the injuries. At common law, can the husband be charged with criminal homicide (be it murder or manslaughter)?

### Scenario Solution

Even though the injuries were inflicted on a fetus, the husband would be criminally responsible for its death because the fetus was born alive, achieved an independent circulation from its mother' womb, and died as a direct result of the injuries sustained from the assault. If the fetus had been stillborn (born dead), however, the husband could not have been charged with criminal homicide because the child was not born alive.

injuries on an unborn child. In this example, for instance, the husband would still be criminally liable for assaulting his wife but would have escaped criminal homicide penalties had he continued kicking his wife in the stomach until fetal death was a certainty. In addition, the logic of withholding fetal rights in criminal cases while concurrently extending human being recognition for civil purposes is also questioned. For these and other reasons, social support for criminalizing the killing of an unborn child has increased. Despite the availability of this legal alternative, many states continue to hold to the common law definition steadfastly (Ala. §13-A-6-1; Idaho §18-4001; Ohio §2903.01; N.J. §2C: 11-2; Tex. §19.01). Even these states, however, no longer require severance of the umbilical cord for the newborn to achieve its independent circulation. Other states have expanded their homicide statutes to protect unborn children (Cal. §187; 720 Ill. §5/9-1.2; Minn. §§609.2661, 609.2662), whereas still others have created independent **feticide** statutes to deal with these complexities (Ind. §35-42-1-6; Iowa §§707.7 to 707.9). Criminal remedies for holding persons culpable for the killing of an unborn child are not, however, without their own legal dilemmas.

If a state decides to protect an unborn child in a homicide/feticide statute, at what point in neonatal development should the unborn child be regarded as a human being? As mentioned earlier, most states consider a fetus to be a human being for civil proceedings, but little consensus exists among the states concerning the developmental phase at which an unborn child becomes a human being for criminal purposes. Of the states that protect unborn children in their criminal statutes, some extend that protection beginning with the moment of conception (includes all stages of intrauterine development), whereas others implement a more strict viability standard. A **viable fetus** is defined in several ways, but states generally use one of two approaches. One option is to use a prescribed number of weeks of intrauterine development. Although some documented fetuses have survived outside the womb at 20-weeks development, medical science generally considers a fetus viable at 28 weeks after conception. Legislatures are not bound by medical definitions, however, and in an effort to protect all fetuses with a reasonable probability of life, many states elect to use a number of weeks that is substantially less than the medical standard. For example, New York protects an unborn child when the female has been pregnant for more than 24 weeks (N.Y. §125.00). A second option requires evidence of a **quick fetus**, meaning that the mother can detect fetal movement in the womb. Regardless of a state's position on the criminality of killing an unborn child, the U.S. Supreme Court's ruling in *Roe v. Wade*, decided in 1973, prohibits criminal prosecution of medical physicians who abort an embryo or nonviable fetus with the mother's consent.

## Living Victim

We have just spent a great deal of time chronicling the legal requirements for one to be considered a human being, and part of that definition often mandates that a fetus be born alive. Regardless of a state's position on when human being status attaches, however, an implicit requirement of the homicide definition is that the killing be of a living human being. Death must be the result of some cause to constitute a homicide, and although some acts or omissions are equally reprehensible to societal values, conduct designed to take the life of another is not a homicide unless the victim was alive at the time of the conduct. Pause for Thought 5-2 illustrates the legal interpretation for the living victim requirement.

---

### Pause for Thought 5-2

Consider the following: While Jane is sleeping, John enters her room with the intent to take her life. He points his pistol at her head, pulls the trigger, and fires a bullet into her skull. An autopsy reveals that Jane was not alive at the time of John's unlawful action because she had already died from a heart attack. Is Jane's death the result of homicide?

### Scenario Solution

Although John's actions have other criminal implications, it is not homicide and therefore is not subject to criminal homicide prosecution. Conversely, however, proof that the victim was alive at the time of the unlawful conduct would be sufficient to constitute homicide regardless of how near death she may have been.

## Legally Dead

The state cannot criminally convict until the victim is declared legally dead. The onset of death often is self-evident, but once again, legal definitions vary regarding when a human being technically expires. The common law standard for death required an irreversible cessation of circulatory and respiratory functions. That standard evolved because of the misguided yet persistent belief of the times that the heart was the most vital organ. Since then, however, medical science has proven that these functions are not independent indications of life. It is now possible through mechanical artifices to prolong a heart beat and respiration without indications of viable functioning in the body's real seat of life—the brain. Organ transplant technology, for example, has presented states with a multitude of legal challenges to common law death definitions. In short, the challenges are based on a general rule of law that states are bound by common law unless changed through legislative action or judicial interpretation. In the absence of such action or interpretation, an argument can be made that a state cannot justify a criminal homicide charge when a transplanted heart (functioning solely through mechanical means) continues to perform its circulatory and respiratory functions in the body of another.

In response to such challenges, many states have adopted the death standard proposed in the **Uniform Determination of Death Act** (1980). Drafted with input from the medical (American Medical Association) and legal (American Bar Association) communities, the National Conference of Commissioners on Uniform State Laws set forth a brain death standard to be used in conjunction with the common law definition. **Brain death** is defined as a complete cessation of all electrical impulses in the brain and is commonly associated with a flat electroencephalogram (EEG) reading. In practice, however, an EEG reading is but one of four criteria used by medical professionals. The other criteria focus on spinal reflexes and include (1) a lack of response to externally applied stimuli (e.g., pinching), (2) no spontaneous movements or respiration, and (3) no signs of reflex activity. Figure 5–2 diagrams the act's components.

Legislative responses have been mixed, with some states choosing continued adherence to common law and others opting for the brain death standard (*People v. Driver*, App.3d 847, 379 N.E.2d 840, 1978; *State v. Meints*, 212 Neb. 410, 322 N.W.2d 809, 1982). From a practical perspective, it really makes no difference in the wake of court rulings stipulating that states are free to use medical definitions even in the absence of formal legislative enactments. The reason for this line of thought is the prevailing notion that legislatures adopt homicide statutes so that persons causing harm will be held accountable for their conduct. In the absence of formal death definitions in statutory codes, then, legislative intent can surely be implied as voluntary deference to the prevailing medical judgment of the times.

Strict adherence to common law or modern statutes not only can create unintended legal loopholes for certain crimes, but likely would also deter medical personnel from using the most advanced technology, in some situations, to avoid possible criminal liability. For example, strict adherence to common law would profoundly influence a family's ability to remove a brain-dead loved one from life support systems because to do so would constitute for doctors an intentional taking of life, otherwise known as murder.

## Noncriminal Homicide Defined

Upon meeting the legal requirements that a homicide has been committed, it must then be determined whether its cause is criminal or noncriminal. Required of all sanctioned criminal conduct, the *actus reus* component for criminal homicide is death; however, criminal homicides are true crimes and therefore also require proof that the defendant possessed some *mens rea* before criminal punishment is authorized. The law does not treat all homicide as reprehensible and mandates that killings, even intentional ones, be classified as noncriminal homicide in the absence of *mens rea*. There are two well-delineated

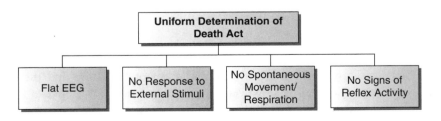

**Figure 5–2** Uniform Determination of Death Act.

exemptions from criminal culpability without which there would be no alternative to the conclusion that the homicide is of criminal origin. Although basically synonymous in modern law codes, the two exemptions are separated in the forthcoming discussion for clarification purposes. Figure 5–3 diagrams the types of **noncriminal homicide**.

## Justifiable Homicide

The first exemption from criminal culpability refers to situations in which persons intentionally cause death but do so under appropriate circumstances. With **justifiable homicide**, a person kills out of some duty or right (or necessity) and would repeat the conduct under the same circumstances. For example, as is discussed in Chapter 10, the law recognizes one's legal right to use deadly force, within legal parameters, to preserve their life (self-defense) or the life of another (defense of another) when placed in imminent danger. Also, the use of deadly force by police officers in the performance of their legal duty is justified, as are court-ordered executions pursuant to capital-crime convictions.

## Excusable Homicide

In the absence of some lawful justification, criminal culpability for homicide can still be excused. In these **excusable homicide** instances, a person acknowledges his or her role in the homicide but counters with some good reason why there should be no penalty under the extenuating circumstances. Insanity (forthcoming in Chapter 10) is one such mental capacity excuse that can result in exoneration, but is atypical of most affirmative defenses because most criminally accused persons are capable of understanding the nature of their actions. For most defendants, the first step in making a determination is to examine the incident for, at a minimum, evidence of culpable or gross negligence. Although ordinary negligence is not sufficient for criminal punishment, recklessness and gross negligence are degrees of fault in which an actor's conduct, sometimes knowingly and sometimes not, causes fatal consequences in a culpable manner. On the other hand, if a person did use a reasonable standard of care, the conduct is exempt from liability and is classified as an excusable homicide. Accidents and misfortunes are two of the more common excuses in homicide cases. One must remember, however, that popular generic definitions often differ substantially from legal definitions.

An **accident** is a lawful conduct during which one person unintentionally causes harm to another and does so with no substantial carelessness. If the driver of an automobile unlawfully speeds (with no legal excuse) through a residential neighborhood and hits and kills a child who runs into the street to retrieve a ball, the risks were foreseeable, and criminal penalties are warranted as a result of the driver's culpable negligence. Under the same circumstances, however, would it be fair to criminally punish the driver for the child's death if the driver was obeying the rules of the road? In this latter scenario, the driver was using a reasonable standard of care, and therefore the homicide would be the result of an accident. **Misadventure** (or misfortune), on the other hand, is conduct intentionally committed by a person, but where the chain of events causing the death of the unintended party was justifiably set into motion. Pause for Thought 5–3 illustrates the legal principle known as misadventure.

## Criminal Homicide Defined

It has now been well established that several legal requirements must be satisfied to proceed with a criminal homicide prosecution:
- The victim must be a human being.
- The victim must be alive at the time of the homicidal conduct.
- The victim must be legally dead.
- There must be no legal justification or excuse that would negate criminal culpability.

Upon a positive showing of the preceding elements, the conclusion that a criminal homicide has been

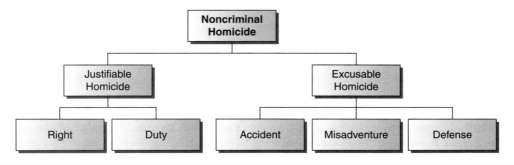

**Figure 5–3** Noncriminal homicide.

> **Pause for Thought 5–3**
>
> Consider the following: John Doe enters a convenience store, brandishes a firearm, and threatens to kill the clerk if the money in the safe and cash register is not handed over. Because the clerk's life is in imminent danger, he or she is justified in the use of deadly force directed at the source of the threat. After the clerk's discharging of the firearm, however, the bullet misses John Doe, and hits and kills an innocent bystander. Is the clerk criminally culpable for the bystander's death?
>
> **Scenario Solution**
>
> The clerk is not criminally culpable for the bystander's death because he or she was justified in using such force. Instead, the death would be classified as misadventure (or misfortune). The conduct cannot be classified as an accident, however, because the clerk purposely set into motion a chain of events for which harm was foreseeable.

committed is acceptable; however, before the specific nature of the criminal homicide, murder or manslaughter, can be debated, the prosecution must first establish the reasonableness of their accusation against the defendant through meeting the *actus reus* burden.

## *Corpus Delicti*

Assuming that the prosecution has established the existence of a homicide and that it was not committed in a justifiable or excusable manner, it must then be established (beyond a reasonable doubt and without the assistance of a defendant's uncorroborated confession) that the victim's death was caused through criminal measures. In reality, establishing the *corpus delicti* (body of the crime) in cases involving suspected foul play has become routine. The volume of direct evidence that now can be collected from a corpse is nothing short of miraculous thanks to the union of medicine and forensic science. It is often said that these technologies temporarily restore the breath of life to the dead so that they can testify from the grave, but the successful efforts of some criminals to hide victims' bodies make the *corpus delicti* requirement more problematic because of the minimization of direct evidence. How, then, can the state satisfactorily establish that a person is legally dead without the benefit of their corpse?

Even though most missing persons eventually surface as alive, it is legally permissible to presume one dead based on nothing more than circumstantial evidence. To do otherwise would serve only to reward those proficient in the disposal of cadavers. As long as circumstantial evidence independently establishes the *corpus delicti*, a jury can make a reasonable inference that a person's disappearance is the result of criminal purpose (see Chapter 1). Evidence establishing a defendant's homicidal involvement varies from case to case, but a prosecutor's most persuasive circumstantial tools usually relate to a defendant's motive, the discontinuation of a missing person's routine activities, and the absence of a reasonable explanation for the disappearance.

## Proximate Cause

It stands to reason that successful establishment of the *corpus delicti* in no way proves that the criminally accused caused the victim's death. Instead, it only clarifies that the victim is in fact dead and that someone caused the death through criminal measures; therefore, phase two in proving a defendant's *actus reus* is to establish that the defendant's conduct was the proximate cause of the victim's death. In short, evidence must exist of a direct cause-and-effect relationship between the person's conduct and the fatal outcome. Although not always the case, reliable assessments for determining the actual cause of a person's demise are now made possible through evolving medical technologies (see Chapter 1).

## Year-and-a-Day Rule

It is understandable that some perceive the moment of injury infliction as facially irrelevant to the issue of culpability, but proximate cause requirements sometimes render this information a legal necessity. At common law, an accused person could not be convicted of murder without a prosecutorial showing that his or her conduct was the proximate or direct cause of another's death. These proximate cause determinations were often complicated, however, when death occurred long after the injuries were inflicted. Understanding that many factors can intervene and contribute to death over such a long period of time and because of medicine's inability to establish with any degree of reasonable accuracy the cause of one's death, the courts began to uniformly prohibit murder trials after the passing of

1 year and a day. As a result of this practice, what became known as the **year-and-a-day rule** evolved, and even when evidence of causation could be established for deaths beyond the 1-year time restriction, that information did not negate this absolute rule; however, this common law proscription has been completely abolished in its country of origin (England) through passage of the **Law Reform (Year and a Day Rule) Act of 1996**.

Because no constitutional proscriptions regarding a passage of time exist, states are free to legislate (or not legislate) this legal matter based on the evolving standards of their communities. Some states (such as Idaho and Wisconsin) have elected to retain the year-and-a-day rule (Idaho §18-4008; *State v. Picotte*, WI 42, 2003), but states differ widely on this issue. The difference of opinion is illustrated best in a North Carolina Supreme Court ruling in which they argue that it would be folly to remain oblivious to advances in medical science when considering whether to apply this ancient rule (*State v. Hefler*, 310 S.E.2d 310, N.C. 1984). States also have chosen to revise the rule by expanding the time period to 3 years and a day or whatever restriction is considered reasonable by their legislature. Still other states have abolished the rule completely and are thus allowing criminal prosecution for any death caused by injurious conduct regardless of when the injuries were inflicted (*State v. Cross* 260 Ga. 845, 401 S.E. 2d 510, 1991; N.J. §2C:11-2.1); however, neither the absence nor presence of time limitations excuse the state from meeting its proximate cause burden.

## Murder in General

At common law, **murder** was defined as the unlawful killing of one living human being by another with malice aforethought. The courts believed that all death stemming from malice aforethought warranted the punishment of death and saw no need to grade levels of intent through the formation of a degree system. Given the finality of the ensuing punishment, however, it sought vigorously to ensure that the penalty was imposed fairly. In doing so, murder convictions were permissible only when a person's mental state possessed malice aforethought, whereas an absence of malice aforethought in unlawful killings required the criminal liability be mitigated to manslaughter in order to spare the offender's life. Modern definitions of malice aforethought do not truly encompass its original intent, however, and have since prompted most states to abandon the use of that term in favor of statutes that more specifically delineate the mental elements deserving of criminal punishment: intentionally, knowingly, recklessly, and negligently. The malice concept, albeit inconsistent, is still much the same, however, and necessitates the following examination.

### Malice Aforethought

The popular (or societal) meaning of the term **malice** has come to represent feelings restricted to those of ill will, hatred, and revenge, and although that terminology is somewhat consistent with early common law definitions, modern legal definitions do not require the accompaniment of such volatile emotions. If such negative emotions were a requirement for charging one with murder, mercy killings (**euthanasia**) would be beyond the realm of the most severe punishment because those intentional killings are accomplished with no ill will or hatred. Furthermore, the term **aforethought** mandates that the intent to kill or cause serious bodily harm be contemplated before commission of the act. Thus, as previously noted, the literal meaning of malice aforethought has in modern legal circles become so convoluted that there is little agreement among individual states and legal scholars concerning its proper interpretation. The following excerpt from the Royal Commission on Capital Punishment (1965) is commonly cited because of its clarity and insight into the malice aforethought controversy:

> It is now an arbitrary symbol. For the malice may have in it nothing really malicious: and need never really be aforethought except in the sense that every desire must necessarily come before though perhaps only an instant before the act which is desired. The word aforethought, in the definition, has thus become either false or else superfluous. The word malice is neither; but is apt to be misleading, for it is not employed in its original (and popular) meaning.

Modern malice definitions require only that an intentional act or omission be legally impermissible (e.g., unjustifiable, inexcusable, unmitigated), irrespective of a wrongdoer's intention. The range of mental states required in states today for the charge of murder includes the following:
- Intent to kill
- Intent to cause serious bodily injury
- Intent to commit dangerous felonies (felony murder)
- Conduct demonstrating extreme indifference to human life (depraved heart)

Of the aforementioned factors, the final three clearly establish that inadvertently caused deaths brought about through the intentional commission of some unlawful conduct can nonetheless be murderous. All persons are presumed to intend the natural and probable consequences of their purposeful conduct, and because death is a foreseeable risk associated with many activities not designed to kill or even seriously injure, malice can sometimes be implied when express evidence of its existence is not available. A variety of factors can be examined for implicit signs of malice, but the least debatable and most noteworthy is the perpetrator's use of a deadly weapon. Other factors for the identification of **implied malice** will be individually addressed when pertinent to upcoming murder discussions.

## Deadly Weapon Doctrine

At common law, proof of malice aforethought was required before a murder conviction was permissible, and even in our modern times of often abandoning the formal malice requirement, evidence of some required mental intent is still very much necessary. Express evidence of this mental state rarely exists, however, because most homicidal perpetrators do not verbalize their malicious intent before causing the death of another. There must exist, then, a legal mechanism whereby the judicial system can infer malice without express evidence, and one such mechanism is the **deadly weapon doctrine**. In essence, this legal doctrine allows the trier of fact (usually a jury) to infer that a perpetrator desired harm when using a deadly weapon. Remember that the law presumes that persons intend the natural and probable consequences of their conducts, and the consequences of using a deadly weapon against another are certainly foreseeable to a person of ordinary intelligence; however, because there are circumstances under which the use of a deadly weapon is not evidence of malice, the malice presumption is one that can be successfully rebutted by the defendant. After all, without this presumptive allowance, juries would be legally bound to infer malice to persons who kill with deadly weapons in legitimate self-defense. The result, then, would be the killing of one living human being by another with malice aforethought. Does that not sound like murder?

Although the deadly nature of some weapons is highly and consistently recognizable because of their potential harm, other weapons are not so easily classified. Are personal weapons (e.g., fists, hands, and feet) or umbrellas deadly weapons? Although it is possible that these items can be used to cause death, their deadliness depends exclusively on the circumstances attending their use. The size of an object, the manner in which the object is used, and the respective sizes, strengths, and ages of the involved parties are all relevant factors for the jury to consider in determining the deadly nature of an object. Pause for Thought 5–4 applies the deadly weapon doctrine to a hypothetical scenario.

## Murder Defined

Many states continue to follow common law philosophies in drafting their criminal homicide statutes but require a showing of aggravating factors to impose the death penalty. Other states, however, argue that one's intent is relevant to the fairness of criminal punishment and have abandoned common law practices in favor of a degree system, but even these states share no universal structure for degree placement. Rather, states have unbridled discretion to define statutory degrees consistent with their legislative barometer. For this reason, one should become familiar with the law of their jurisdiction before making assumptions regarding murder statutes. In states that assign degrees to the crime of murder, the most serious of the intent gradations is first-degree murder and receives

### Pause for Thought 5–4

Consider the following: Two men of similar stature become embroiled in a heated argument during which one strikes the other in the face with his fist. If the assaulted party dies as a result of complications from a broken nose, can malice aforethought be inferred via the deadly weapon doctrine?

### Scenario Solution

No, under these circumstances, the fists were not used as deadly weapons because death or serious bodily injury is not a natural and probable consequence of such an action. A charge of manslaughter may be permissible, but a murder charge would not survive legal challenge because proof of malice was absent. The same fists used against a child could easily be classified as deadly weapons, however, in that it is foreseeable that serious harm likely will result from a physical inequity of that magnitude. It also would be reasonable to consider fists as a deadly weapon when used with skillful precision and designed to injure another person seriously.

the greater punishment, whereas second-degree murder is punished to a lesser extent. All of these states agree that intent-to-kill murders deserve first-degree classification but experience more disagreement concerning the proper placement for those who unintentionally kill during the purposeful commission of unlawful conduct that do not meet the requirements for mitigation to voluntary manslaughter. Figure 5–4 diagrams the *mens rea* forms required for the crime of murder.

## Intent-to-Kill Murder

The fundamental legal requirement for what constitutes murder is that conduct must be the result of specific intent to kill. Historically, and in most states using **first-degree murder** statutes, that intent must further be accompanied with premeditation and deliberation. Although these terms are often used collectively to describe the specific intent of a person to cause the death of another, there is a legal and practical difference between their respective meanings. In fact, many students fail to comprehend the autonomy that premeditation can potentially experience apart from deliberation, and when that distinction is misinterpreted, the elemental boundaries for homicides resembling intent-to-kill murders can become blurred. Within states using the degree system, then, **second-degree murder** represents the taking of life with malice but absent premeditation and/or deliberation.

**Premeditation** refers to the mental process whereby a person, at least for a short time, thinks about some forthcoming homicidal conduct. Some planning is usually involved in premeditation, but the planning need not be extensive or elaborate. From a practical perspective, the element of premeditation can be satisfied just seconds before the conduct. **Deliberation**, on the other hand, refers to careful reflection on the wisdom of putting into action premeditated thoughts. In short, it must be established that the defendant formulated the manner in which the killing would take place and carefully weighed the consequences of such a killing. Pause for Thought 5–5 demonstrates the independent existence of premeditation and deliberation and how premeditation can exist without consummation of deliberation.

While highly controversial in contemporary American society, the requirements for intent-to-kill murder are clearly satisfied for one who actively participates in helping another human being take their own life. The law does not permit one the consensual authority to allow others to take their life; therefore, the act of euthanasia, compassionate though it may be, legally amounts to a homicide (the killing of one human being by another), committed without legally permissible justification or excuse, and with the specific intent to kill. These elements, as you are now well acquainted with, define the essence of murder and will

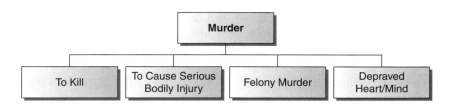

**Figure 5–4** Murder.

---

### Pause for Thought 5–5

Consider the following: John Hill leaves work early and arrives home to find his wife in bed with another man. Mentally enraged, John retrieves his gun from the closet and shoots the man three times in the head. The man then dies as a result of the gunshots. Has John committed first-degree murder?

### Scenario Solution

Assuming the jury accepts the argument that John committed the harm while in a heat of passion, it would be impossible for John to have deliberated the killing. Without this essential element, John would not be guilty of first-degree murder. He is still criminally liable for homicide, but only for a reduced charge of voluntary manslaughter because the adulterous provocation is legally adequate (a concept yet to be discussed).

continue to do so without legal reform classifying these kinds of killings as justifiable.

## Intent-to-Cause-Serious-Bodily-Injury Murder

Deaths often are the unintentional result of a desire to cause serious bodily harm, but not death. In such cases, should the state nonetheless proceed with a murder charge, or would manslaughter be more appropriate? Despite the unintended nature of the death, the state can nonetheless justify murder charges based on the principle of implied malice.

It has already been established that persons are accountable for the natural and probable consequences of their actions. The use of a deadly weapon, or sometimes words and gestures, clearly can indicate intent to seriously injure, from which death is foreseeable to persons of ordinary intelligence. Many modern states, however, have abandoned the formal use of intent-to-do-serious-bodily-injury murder, opting instead to treat those unintended deaths as originating from extreme indifference to human life (soon to be discussed depraved-heart murder). Despite a state's preference on the legislative approach to these killings, the results are the same because they both culminate with a charge of murder.

## Doctrine of Transferred Intent

Considering that malice is a requirement of murder, what happens when persons other than those for whom harm was intended are the homicide victims? Remember, successful murder prosecutions require the specific showing that a defendant possessed either the intent to kill or the intent to cause at least serious bodily injury. Does it matter that the actual intent was not harbored for the persons killed? Pause for Thought 5–6 illustrates the legal interpretation for the doctrine of transferred intent.

## Felony Murder

One of the most controversial legal principles, the felony-murder doctrine serves to punish unintended killings through the elimination of well-established *mens rea* murder requirements. In essence, **felony murder** is a legal doctrine that treats deaths caused during the perpetration of designated felonies as murder, irrespective of the felonious perpetrator(s) intent. Critics argue that the true crime of murder requires malice aforethought and that the systematic exclusion of that mental state in the felony-murder doctrine serves only to satisfy man's retributive passions, not the interests of justice.

Given the apparent exclusion of the malice aforethought requirement, how is the fairness of the felony-murder doctrine defended? The legal justification for this elemental deviation is the theory of constructive malice, whereby no actual malice exists but can be inferred (or implied) from one's conscious decision to commit a felony. The popular belief is that reasonable persons should foresee that death can be a natural and probable consequence of some felonious conduct, and therefore, those persons should be treated as though they did possess the intent to kill, especially with respect to the felonies of arson, rape, robbery, burglary, and kidnapping.

In light of the debate concerning the rule's legitimate existence, the interpretational complications and inconsistencies that exist from one state to another should not at all be surprising. The first of these inconsistencies relates to the now misleading language of the doctrine's name. Although at com-

---

### Pause for Thought 5–6

Consider the following: Roger and Laura have been involved in an extramarital affair for some time. Laura insists that she is no longer in love with her husband, Frank, but fears as credible his threats to kill her if she ever left. Roger decides that the only way he will ever have the unconditional love of Laura is to kill Frank and proceeds to place a bomb in his car. Unknowing to Laura, however, she uncharacteristically borrows Frank's car and is killed in the explosion.

### Scenario Solution

Laura's death was clearly not the intent of her lover, Roger. Should Roger's guilt, then, be mitigated to something less than murder, or should we proceed with a prosecution that treats the killing as though Laura was the intended victim? There is no legal ambiguity with regard to this dilemma. Long ago, courts established the common law doctrine of transferred intent to prevent successful defense arguments predicated on just these kinds of legal loopholes; therefore, if intent to harm or kill is present, it matters not who actually dies. The law will treat that killing as though the intent found its directed target.

mon law the felony-murder doctrine did serve to punish as murder all death resulting from the commission of any felony, states today do not strictly follow the letter of that law. The reason for the transformation rests simply in the fact that all common law felonies were dangerous to human life, but inclusion of all felonies in modern times would be unfair (and probably immoral, e.g., larceny, counterfeiting, forgery). Many states now require that the resulting harm be a foreseeable consequence of the independent underlying felony before authorizing a murder conviction, whereas others choose to allow prosecution for *malum prohibitum* related deaths but require some accompanying proof that the felonious perpetration was committed in a dangerous manner. As a result of these legal changes, the common law rule has been formally abandoned in most states through the creation of statutes that (1) list the specific felonies eligible for felony-murder treatment or (2) require that the felony be inherently dangerous to human life (*malum in se*) or committed in a dangerous manner. A second matter of importance to the rule's interpretation regards the mandate that the killing be caused within the scope of the crime. Most states define the scope of a crime to include not only a crime's actual commission but also the attempt to commit the crime and immediate flight from the crime. Three factors are used to make scope determinations, but courts differ in their determinations because of differing philosophical positions regarding the moral justness of the felony-murder doctrine. As such, the scope can be analyzed from narrow and broad extremes but nonetheless use three primary factors in their analysis. Figure 5–5 diagrams the requirements for the felony murder doctrine.

## Causal Connection

The cause of death must be connected to an underlying felony. A mere coincidence of time and place is insufficient to justify murder liability under the felony-murder doctrine. For example, a death coincidentally occurring at the same time and place as a felony commission not only is unconnected to the felony but, for that matter, likely possessed no awareness whatsoever of the felony attempt or commission. In these situations, felony murder cannot be pursued.

## Timing

A felony must be so closely aligned with time of death to allow the presumption that the death occurred within the scope of the crime. This time relationship is largely determined by the nature of the crime. For example, the scope of a felonious arson could continue until the fire is suppressed, despite successful flight from the crime, whereas the scope of a felonious burglary likely would not end until flight resulted in reaching a temporary place of safety, despite the conclusion to the crime's actual commission at some earlier moment. A precise time limitation rule is impossible to formulate, but it is safe to say it usually ranges from minutes to hours.

## Proximity

It also must be established that the place for the homicide occurred within the scope of the crime. The scope of a crime usually ends when there is breakage in the chain of events, and this normally occurs when the perpetrator reaches a place of temporary safety. Murders that occur beyond that point do not lie within the parameters of the felony-murder doctrine, but instead require proof of some intent to kill or intent to cause serious bodily injury, or conduct that demonstrated extreme indifference to human life.

Another concern pertaining to the felony-murder doctrine's proper interpretation relates to who must commit the actual homicide to constitute murder. Most states now limit this punishment to situations where active participants in the felony are the actual causal actors of death, thereby creating what is commonly known as the third-party exclusion rule. In other words, most states now will not apply the felony-murder rule to felony perpetrators when the actual killings are committed by police officers, victims, or other third parties attempting to resist or prevent the felonious commission. Other states, however, adopt the formal position that punishment is warranted for the perpetrator regardless of who commits the actual killing because the killing would never have happened without the felony initiation. It is their contention that holding felony perpetrators criminally responsible for any death is both moral and legal, regardless of who actually commits the killing, because it more greatly deters others from committing similar crimes than do the limited felony-murder rules; however, even states adhering to the third-party exclusion rule usually hold responsible felony perpetrators for indirect killings when the cause of death results from their (1) using

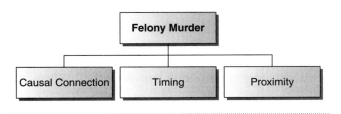

**Figure 5–5** Felony murder.

the victim as a shield or (2) placing the victim in a position of obvious danger.

## Depraved-Heart Murder

Persons who commit unjustifiable and inexcusable acts that cause another's death, but do so without any intent to kill, cause serious bodily injury, or commit a dangerous felony, are usually guilty, at most, of manslaughter because of the absence of malice. On rare occasions, however, as demonstrated with the felony-murder doctrine, it is necessary that malice be implied for the interests of justice and public welfare to be promoted. With **depraved-heart murder**, malice can be implied from a person's actions (or omissions of legal duties) when those actions clearly evince signs of an abandoned and malignant heart. In these cases, the trier of fact (usually the jury) may find that a person's mental state constitutes a murderous predisposition when their actions or omissions cause death, albeit unintended, under circumstances manifesting a depraved indifference to the value of human life.

Historically, the inherent vagueness associated with phrases like abandoned heart and malignant heart made it difficult for jurors to determine exactly what conducts rise to a level of depravity and hence murder, as opposed to normal recklessness that constitutes the crime of manslaughter. In response to these interpretational complications, the Model Penal Code abandons the common law standard in lieu of a more explicit definition based on an aggravated form of recklessness, which is more readily understood because of its current use in defining some manslaughter crimes. Recklessness, or gross negligence, refers to the creation of a high risk, but not very high risk, of death or serious injury, whereas the creation of merely an unreasonable risk, less than a high risk, is called ordinary negligence. Elevation to the more heinous conduct of depraved-heart murder, however, requires that another's death be the product of conduct so reckless that it demonstrates a total disregard for the well-being of others, meaning that the conduct must create a very high risk of death.

A common debate regarding the essence of depraved-heart murder is whether the crime requires the risk-creating party to possess an actual (or subjective) awareness of the harmful probabilities. Many jurisdictions have historically adhered to an objective standard that permitted murder convictions on a limited finding that a reasonable person would have been aware of pending dangers, regardless of whether an actual awareness existed. Many critics, however, challenged this position on the theory that persons cannot possibly possess a depraved heart if they are unaware of the potential consequences of their conduct. To clarify its position, the Model Penal Code specifically requires the showing of some subjective awareness but, in reality, admits that some highly unusual circumstance would have to exist for someone to be unaware of the perceived harms associated with the creation of such a very high degree of risk to human life. Just a few examples of deaths clearly caused through conduct demonstrating a depraved indifference to the value of human life include the following:

- Excessive speeding through a residential neighborhood where children are playing
- Firing a bullet near another with only an intent to scare
- An owner knowingly permitting his vicious dog to roam free in areas known to be frequented by persons
- Repeated and violent shaking of an infant

Depraved-heart murder is reserved for general intent homicides that are unrelated to the commission of a dangerous felony. If specific intent to kill or harm another (or others) does exist, the prosecutor cannot escape the burden of proving beyond a reasonable doubt the *mens rea* elements associated with intent murders. Pause for Thought 5–7 illustrates how the depraved heart doctrine is applied within legal circles.

## Manslaughter in General

**Manslaughter** at common law was defined as the unlawful killing of a human being without malice. In practice, the crime of manslaughter encompasses killings that are not justifiable or excusable, yet undeserving of punishments reserved for murderous (with malice) incidents. Manslaughter is ordinarily divided into two degrees, **voluntary manslaughter** and **involuntary manslaughter**, with degrees distinguished by its attendant circumstances. Many states continue to use these traditional designations, whereas others have expanded or redefined the terminology. The following discussion will follow historical tradition, but will also reference modern evolutions where appropriate. Figure 5–6 diagrams the two major forms of manslaughter.

## Voluntary Manslaughter Defined

American jurisprudence follows the principle of determinism. Persons are presumed to possess free will and therefore are capable of choosing from a host of behavioral options at any moment in time. When the selected behaviors are harmful to others, however, and committed without lawful justification or excuse, a series of appropriate punishments are reserved for

## Pause for Thought 5-7

Consider the following: Billy was recently fired by his corporate employer. Upset that his 20 years of service was so unappreciated, Billy returns to his former workplace and begins to randomly fire gunshots into the windows of the establishment. The evidence clearly showed that Billy had no intent to kill or harm any of the persons inside the building, but two people are nonetheless stricken and die from sustained injuries.

## Scenario Solution

The state could not proceed with an intent-to-kill or intent-to-cause-serious-bodily-injury murder trial because each of those prosecutions requires identification of some specific intent to harm another (or others). In this case, no specific victims were targeted, and the homicides were not the result of a felony commission gone awry; thus, the state must seek another avenue to hold Billy criminally accountable for the deaths. Depraved-heart murder would be one such avenue.

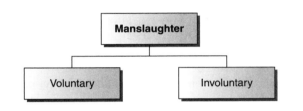

**Figure 5-6** Manslaughter.

implementation. The law does, however, recognize the emotionality of human beings and that even the most law-abiding citizen can be provoked to act in an undesirable manner in extreme circumstances. Often these homicides are retaliatory in nature and are commonly perceived by society as a moral restoration to the balance of justice. Regardless of the empathy that these killings might evoke, choosing to withhold punishment for such killing episodes would amount to societal acceptance of a right-to-kill doctrine. Given the inappropriateness of such a doctrine and societal message, how then can we fairly adjudicate killings where the victim contributed to the killer's weakened mental state? Is it moral to pursue a murder conviction (in light of the presence of malice), or should the legal system mitigate the punishment, thus partially excusing the killer's culpability for an inability to control anger and rage? The law also recognizes the imperfection of human reasoning and thus further mitigates culpability when people erroneously use self-defense to take the life of another. In such extraordinary situations, and only in those circumstances, our legal system has chosen to mitigate blameworthiness from murder to voluntary manslaughter. Figure 5-7 diagrams the two avenues through which voluntary manslaughter is a justifiable charge.

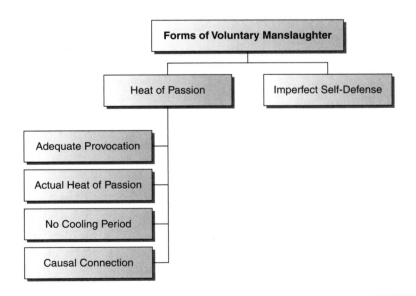

**Figure 5-7** Forms of voluntary manslaughter.

Voluntary Manslaughter Defined

## Adequate Provocation

American courts have consistently adhered to the notion that certain provocations may understandably inhibit the abilities of reasonable people to control inflamed emotions. Mere provocation, however, is not enough to qualify for voluntary manslaughter. The adequate provocation rule refuses to recognize the inability of temperamental hot heads to control rage. Instead, it mitigates only those behaviors that were not produced from a cool reflection of the consequences. Specifically, the victim's response must result from **adequate provocation**, meaning that it must have been intended and calculated to interfere with the rational-thinking skills of persons. Broadly defined, adequate provocation exists when some unlawful conduct is sufficient enough to provoke a reasonable person to kill. Adequate provocations are usually revenge motivated, but there is no legal requirement that such motivations fuel the passion. What exactly, then, are the kinds of conducts considered legally adequate to cause persons to momentarily lose control of their good judgment?

Courts have given juries much latitude in making determinations regarding adequate provocations. For example, it is commonplace for juries to accept as adequate finding one's spouse in the act of adultery. Conversely, however, the law presumes that a reasonable person would not be greatly provoked by a desire to eat a piece of chicken, and therefore, a homicide resulting from a fight over the last piece of chicken (even when it is the coveted wing) would not be recognized as adequate provocation.

Words and gestures alone, without some accompanying conduct, have historically been regarded as insufficient to constitute adequate provocation. Simply put, hurtful and emotion-laden words such as "I hate you" or "You are ugly" are not perceived adequate to enrage persons of reasonable prudence to lose control of their conduct. Regardless of their influence on certain individuals, reasonable persons would walk away or formulate a witty reply to save face; however, although insulting words are inadequate provocation, there is growing judicial support for regarding informational words as adequately provocative. For example, words spoken by a rapist to a father that his little girl was "the best he ever had" convey information of a painful nature, and even though no act accompanied the communication, these words alone are considered sufficient to stir the passions of reasonable persons.

## Heat of Passion

An adequate provocation has the potential to inflame the passions of reasonable persons, but does not always do so. It is quite possible some individuals subjected to adequate provocation may not experience any significant impairment with their ability to deliberate and reflect on pending actions. When this happens, the deliberate result amounts to a killing in cold blood, and because the rule of law for voluntary manslaughter requires that killings be committed in the **heat of passion**, these homicides correctly remain within the realm of murder.

## No Cooling Period

If a jury concludes that a killing was committed in a heat of passion stirred by some adequate provocation, a reduction from murder to manslaughter still may not be warranted unless a third component can also be satisfied. At this point, the law requires the jury to make subjective and objective evaluations regarding cooling-off periods. These evaluations are necessary because of the absence of legal guidelines regulating expected cool-down times. It is reasonable to assume that longer time periods are appropriate for greater passions than for minor passions, but beyond this basic assumption the law recognizes that people react in different ways to differing stimuli and therefore leaves these decisions to jurors.

The first evaluation, a subjective test, requires an assessment of the defendant's actual actions and a determination as to whether the defendant's passions had cooled prior to the killing. If the defendant's passions are found to have cooled, murder would be the most appropriate conviction. On the other hand, a finding that the defendant's passions did not cool would require the jury to then make a second evaluation of whether a reasonable person would likely have cooled down during the same time period and under similar circumstances. If the results of the objective evaluation are that a reasonable person would not have had an opportunity to cool down during the time interval between the provocation and the killing, then a charge reduction would be rendered subject to the satisfaction of one additional requirement (the causal connection); however, if the jury believes that a reasonable person would have cooled down, a murder conviction would be rendered even if there were no actual cooling because the defendant had the opportunity to cool off.

## Causal Connection

Establishment of the first three requirements may still result in a murder conviction if some breakage in connection can be shown to exist among the adequate provocation, the passion, and the fatal act. In short, the purpose of the voluntary manslaughter mitigation is to make allowances for intentional killings committed under some of the most difficult situations. The result of this practice, however, has been the reduction of a severe penalty for what otherwise would be murder. The final requirement, causal connection, functions to ensure that desperate defendants cannot use what otherwise could serve as legal loopholes. Simply put, the law requires each of the following connections be established or the crime remains murder:

- Adequate provocation must be the causal actor of the passion.
- Ensuing passion must be the cause of the fatal act.
- Fatality must be of the person legally responsible for the adequate provocation.

The first connection requiring that the adequate provocation be the causal actor of the passion is important because it prevents misuse of the mitigation for those who kill for some other motivation. Just as some people are unaffected by provocations that would evoke passion in other reasonable persons, there are also those who possess the passion to kill for some reason completely unrelated to any adequate provocation. The second connection requiring that the passion be the cause of the fatal act serves to eliminate premeditated and deliberate murders that occur after a sufficient cooling-down period. The purpose of voluntary manslaughter is to mitigate the blameworthiness of those who were momentarily incapable of making good decisions, not to mitigate the social contempt for murders committed with proper motivation. Finally, the third connection requires that the victim be the source of the adequate provocation. This requirement makes the voluntary manslaughter defense unavailable to those who kill, while in a heat of passion, anyone other than the provocative source, accidentally or purposefully. Pause for Thought 5–8 illustrates this point of law.

## Imperfect Self-Defense

People sometimes possess a subjective (or personal) belief that circumstances warrant the use of deadly force in self-defense, but in actuality are mistaken as to the objective circumstances. These situations are referred to as **imperfect self-defense** and present the legal system with a unique and regrettable duty to prosecute some killings that in essence possess no underlying desire to do wrong. Abused women often erroneously resort to deadly force when they mistakenly believe their domestic abusers are going to kill them. There is no question that these circumstances are undeserving of a murder charge, in that malice is absent. Still, however, despite the fact that many people disagree with these prosecutions and instead prefer that no criminal prosecution commence at all, it must by definition be manslaughter if it is neither murder or noncriminal homicide, and it cannot be noncriminal homicide without some legal justification or excuse.

## Involuntary Manslaughter Defined

Figure 5–8 diagrams the two forms of involuntary manslaughter.

## Unlawful Act Manslaughter

Because the common law felony-murder doctrine made "murderous" all death caused by felonious conduct, the law treated all other misdemeanor-caused death as involuntary manslaughter. Referred to as the **misdemeanor-manslaughter rule**, it authorized punishment without regard to an actor's awareness of pending danger to others. The rationale, similar to that for the felony-murder doctrine, was that the decision to embark purposely on unlawful conduct substituted for the *mens rea* requirement.

> ### Pause for Thought 5–8
>
> Consider the following: At the annual family reunion, John Smith is told by a close relative that his wife, Jane, and brother, Peter, have been involved in an extramarital affair for the past few months. Engulfed with rage and anger, he grabs a gun from his car's glove compartment and immediately, without any reasonable opportunity for cooling, walks to the backyard to confront Peter. Peter makes light of the whole thing and comments that "a real man's wife wouldn't need to seek satisfaction elsewhere." Further enraged and out of control, John wildly fires his pistol at Peter three times. Peter dies as a result of two gunshot wounds, and a distant cousin also is killed in the incident.
>
> ### Scenario Solution
>
> Clearly, the death of the distant cousin, even though caused from the same passionate state, is second-degree murder, not voluntary manslaughter.

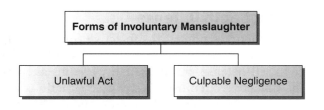

**Figure 5–8** Forms of involuntary manslaughter.

This line of thinking became the subject of much criticism, however, because unlike the commission of felonies which are dangerous to human life, the commission of most misdemeanor conduct does not generally present foreseeable dangers. Uncomfortable with a blanket crime of that magnitude, many states began to question the fairness of such automatic punishment for misdemeanants. As a result, roughly two thirds of all states abolished its use and instead opted to handle those kinds of killings in their existing criminal negligence manslaughter statutes (presented in the next section). Despite the widespread abolition, many states continue to use this common law form of involuntary manslaughter, which now also includes those felonies ineligible for the felony-murder doctrines of respective jurisdictions.

Of the states opting for its continued use, most have limited it to caused deaths during the commission of *malum in se* crimes. The mere commission of prohibited crimes (or *malum prohibitum*) resulting in another's death is not enough to warrant an involuntary manslaughter conviction. Instead, the crime must be wrong in itself. For example, assault is a *malum in se* offense for which involuntary manslaughter would be an appropriate punishment for caused deaths regardless of the person's awareness of risk. Speeding, on the other hand, is a *malum prohibitum* offense for which a caused death would require more than just evidence of the offense's commission. Furthermore, these involuntary manslaughter prosecutions, not unlike all criminal homicide prosecutions, must establish a causal connection between the unlawful act and fatality for the state to warrant a conviction.

## Culpable Negligence Manslaughter

When persons create an unreasonable risk of harm to others but are in actuality unaware of the potential harm creation, their actions are said to be inconsistent with those of a reasonable person and are referred to as ordinary or simple negligence. Even though the risk-creating party should be aware of the potential harm associated with the conduct, most states concur that this form of negligence is unworthy of manslaughter punishment given the absence of some active mental wrongdoing; despite the decriminalization of such negligence in many states, numerous other states do punish the conduct but as a lesser misdemeanor crime, commonly called criminally negligent homicide. The states do agree, however, that when a party chooses to disregard a risk of which he or she was aware and an unintended death results, the **gross** (or culpable) **negligence** is paramount to recklessness and is worthy of the more serious and usually felonious punishment of involuntary manslaughter.

Although most people believe that moral virtues should guide conduct, the law requires affirmative actions of people in only the most limited of circumstances. In most situations, the law permits freedom of choice but requires that due care accompany the voluntary commission of those conducts. For example, there is no legal requirement that we drive automobiles, but if we choose to do so, there is a legal requirement that the person drive with care so to avoid injuring others. Additionally, criminal liability could be found to exist for failure to help another when the danger posed to one's self is minimal, and the omission of some affirmative act of assistance is certain to result in death. As citizens, we all possess duties to some extent. In order to treat conduct (acts or omissions) as grossly negligent, however, the following elements must be shown to exist:

- There must be a duty to act.
- There must be evidence of an unjustifiable and unreasonable breach of duty.
- The breach of duty and resulting injury must be causally related.
- There must be an actual injury, namely death in the case of involuntary manslaughter.

As mentioned in the unlawful act manslaughter section, some affirmative duties are required of citizens, but these duties are normally limited to situations involving familial (parent/child, spouse, etc.) and contractual (lifeguard, babysitter, etc.) relations where persons have accepted responsibilities to ensure the well being of others. Good Samaritan laws requiring that citizens offer some limited assistance in emergency situations are also becoming more common in populous cities. Pause for Thought 5–9 illustrates how to legally interpret the issue of negligence within homicide law.

## Homicide and Genocide in Federal Law

Federal law proscribes varying forms of criminal homicide in Title 18, Chapter 51 of the U.S. Criminal Code (§§1111-1122). Ranging from general definitions of murder (§1111) and manslaughter (§1112) to more unique statutes offering protection against the human immunodeficiency virus (§1122), homi-

> **Pause for Thought 5–9**
>
> Consider the following: Distracted with drug and alcohol addictions, Mary Smith has not been providing her infant child with food of the appropriate volume or nutrition. At 6 months of age, the child dies and an autopsy reveals that the child died from malnutrition.
>
> **Scenario Solution**
>
> Parents have a legal obligation (or duty) to care for the well-being of their children; therefore, the mother clearly possessed a legal duty that she breached and in so doing caused the death of her infant child. Without some legally permissible justification, there would clearly be, at a minimum, some degree of negligence, and because drug and alcohol addictions are inappropriate legal justifications, the question of criminal liability revolves around the extent of negligence. In most states, **ordinary negligence** is not criminal, whereas gross negligence is sufficient to incur criminal liability. Is the mother's omission criminal? If the omission amounts to gross negligence or recklessness (gross deviation from the conduct of a reasonable person), the answer is yes. In this case, then, the mother can definitely be charged with involuntary manslaughter.

cide law within the federal criminal code focuses on the prohibition of uncommon homicide traditionally regulated by the states.

The crime of murder at federal law (§1111a) is defined as

> the unlawful killing of a human being with malice aforethought. Every murder perpetrated by poison, lying in wait, or any other kind of willful, deliberate, malicious, and premeditated killing; or committed in the perpetration of, or attempt to perpetrate, any arson, escape, murder, kidnapping, treason, espionage, sabotage, aggravated sexual abuse or sexual abuse, child abuse, burglary, or robbery; or perpetrated as part of a pattern or practice of assault or torture against a child or children; or perpetrated from a premeditated design unlawfully and maliciously to effect the death of any human being other than him who is killed, is murder in the first degree. Any other murder is murder in the second degree.

Conversely, the crime of manslaughter at federal law (§1112a) is defined as "the unlawful killing of a human being without malice." It goes on to explain that there are two kinds of manslaughter: (1) voluntary manslaughter occurs "upon a sudden quarrel or heat of passion," and (2) involuntary manslaughter occurs when "in the commission of an unlawful act not amounting to a felony, or in the commission in an unlawful manner, or without due caution and circumspection, of a lawful act which might produce death."

Federal law also regulates the commission of acts of genocide in Title 18, Chapter 50A (§§1091-1093). In short, **genocide** refers to any action(s) possessing the "specific intent to destroy, in whole or in substantial part, a national, ethnic, racial, or religious group" (§1091). It is important to keep in mind that the prohibitions embedded within the statute are not limited to death. Rather, the statute also regulates (1) causing serious bodily injury; (2) causing permanent impairment of mental faculties through drugs, torture, or similar techniques; (3) subjecting the group to conditions of life that are intended to cause the physical destruction of the group in whole or in part; (4) imposing measures intended to prevent births within the group; and (5) transferring by force children of the group to another group or attempting to do so.

## Summary

Death occurs through one of four means (or manners): natural, accident, suicide, and homicide. This chapter sought to explain the legal requirements that must be established to constitute a homicide and then aimed to guide students through the legal complexities for determining when an accused is not culpable for a homicide (justifiable homicide, excusable homicide). When established that no valid justification or excuse exists, the chapter then focused on the *actus reus* requirements for establishing a criminal homicide charge: *corpus delicti*, proximate cause, time limitations (year-and-a-day rule). With regard to *mens rea*, the crimes murder and manslaughter were then differentiated through the defining of malice aforethought. The modern degree system also was examined to separate the severity within murder (premeditation, deliberation) and manslaughter (heat of passion, imperfect self defense) offenses. Federal homicide (and genocide) law and the extent to which murder is problematic in America (UCR data) also were examined.

## Practice Test

1. Common law used the _____ standard regarding the protection of a fetus within criminal homicide codes.
   a. human being
   b. born alive
   c. living victim
   d. viable
   e. quick

2. _____ is defined as complete cessation of electrical impulses in the brain and is associated with a flat electroencephalogram (EEG) reading.
   a. Brain arrest
   b. Cranial herniation
   c. Catatonic hematosis
   d. Brain death
   e. Brain expiration

3. The _____ Act represents legislative and medical efforts to create a formal death standard acceptable to most states.
   a. Universal Death
   b. Death Standard Reformation
   c. Uniform Determination of Death
   d. Death Declaration
   e. Death Revision

4. _____ homicide is defined as a person who kills based on duty or right.
   a. Excusable
   b. Misdemeanor
   c. Culpable
   d. Criminal
   e. Justifiable

5. _____ is a lawful conduct during which one person unintentionally causes the death of another and does so with no degree of carelessness.
   a. Accident
   b. Adequate provocation
   c. Involuntary manslaughter
   d. Voluntary manslaughter
   e. Unintentional homicide

6. Killings can be classified as noncriminal homicide in the absence of _____.
   a. *actus reus*
   b. excuse
   c. *mens rea*
   d. *mala in se*
   e. justification

7. _____ conduct intentionally committed by a person but where the chain of events causing the death of an unintended party was justifiably set into motion.
   a. Misadventure
   b. Recklessness
   c. Gross negligence
   d. Criminal negligence
   e. Criminal mischief

8. The _____ is the common law standard which dictated that a person must die within a proscribed time period to become eligible for a charge of murder.
   a. felony-murder doctrine
   b. elapsed-time principle
   c. uniform homicide code
   d. heat of passion test
   e. year-and-a-day rule

9. _____ can often be the result of a "mercy killing."
   a. Voluntary manslaughter
   b. Misery murder
   c. Feticide
   d. Euthanasia
   e. Good Samaritan homicide

10. At common law, _____ was defined as the unlawful killing of a human being with malice aforethought.
    a. homicide
    b. murder
    c. manslaughter
    d. first degree murder
    e. second degree murder

11. _____ has come to represent feelings of ill will, hatred, or revenge.
    a. Cruelty
    b. Bad intentions
    c. Spite
    d. Malice
    e. *Mens rea*

12. A(n) _____ fetus is an unborn child from whom movement has been detected.
    a. intrauterine
    b. viable
    c. contraceptive
    d. quick
    e. conceptual

13. _____ refers to the mental process whereby a person, at least for a short time, thinks about some forthcoming conduct.
    a. Premeditation
    b. Recklessness
    c. Deliberation
    d. Malice
    e. Negligence

14. _____ refers to a killing agent's careful reflection on the wisdom of putting into action planned thoughts.
    a. Premeditation
    b. Reflection
    c. Meditation
    d. Malice
    e. Deliberation

**15.** _____ aims to punish, as murder, unintended killings of others even though traditional *mens rea* murder requirements are not present.
   a. Feticide
   b. Perfect self defense
   c. Uniform Determination of Death Act
   d. Depraved-heart murder
   e. Malice

**16.** _____ encompasses killings with no legal justification or excuse, yet perceived undeserving of punishments reserved for malice-based incidents.
   a. Euthanasia
   b. Genocide
   c. Murder
   d. Noncriminal homicide
   e. Manslaughter

**17.** _____ exists when conduct is sufficient to provoke a reasonable person to kill.
   a. Justifiable homicide
   b. Murder
   c. Excusable homicide
   d. Adequate provocation
   e. Reasonable provocation

**18.** The use of necessary and reasonable deadly force in the defense of self or another or to prevent the commission of a violent felony is called _____ self-defense.
   a. preventive
   b. perfect
   c. reasonable
   d. necessary
   e. imperfect

**19.** The _____ rule authorizes punishment without regard to an actor's awareness of pending danger to others.
   a. felony murder
   b. heat of passion
   c. misdemeanor-manslaughter
   d. *corpus delicti*
   e. self-defense

**20.** _____ negligence is paramount to recklessness and worthy of the more serious and usually felonious punishment of involuntary manslaughter.
   a. Gross
   b. Mere
   c. Ordinary
   d. Contributory
   e. Unintentional

## References

Federal Bureau of Investigation. (2008). *Crime in the United States, 2007: Uniform Crime Reports.* Retrieved July 14, 2009, from http://www.fbi.gov/ucr/cius2007/index.html

Royal Commission on Capital Punishment. (1965). 1949–1953 Report. London: Her Majesty's Stationery Office.

# Sex Offenses

**CHAPTER 6**

## Key Terms

| | | |
|---|---|---|
| Affinity | Cunnilingus | Rape by instrumentation |
| Age of consent | Fellatio | Rape shield laws |
| Bestiality | Fondling | Secondary traumatization |
| Buggery | Forcible rape | Seduction |
| Carnal knowledge | Incest | Sexual battery |
| Chaste character | Marital rape exemption | Single legal entity theory |
| Chattel | Megan's Law | Sodomy |
| Child exploitation | Necrophilia | Statutory rape |
| Consanguinity | Pederasty | Voyeurism |
| Contractual theory | Penetration | |

## Introduction

Sex offenses can be especially traumatic for victims. In many cases, sexual assault victims are left to cope with years of psychological and emotional trauma in addition to the stigma of being violated. Victims report difficulty functioning in social situations and romantic relationships. As such, the consequences of sexual crimes are enduring, and for many victims an appropriate support system for recovery does not exist.

The legal system and society have struggled with awareness of sexual crimes. In many instances, victims describe a criminal justice system wherein protection not only appears absent, but is actually perceived as enhancing the traumatic experience. **Secondary traumatization** occurs when interaction between victims and the criminal justice system is harmful; examples of such traumatic experiences include delays in court proceedings, repeated interviews, stigmatization, "blame the victim" trial tactics, and failure to offer counseling and/or support services. Although many strides have been made to alleviate the secondary trauma of rape victims, much work remains to be done.

In general, there are two overarching categories of sexual offenses: sexual assault and sexual exploitation. Sexual assault offenses specifically address physical contact between the victim and perpetrator. Sexual exploitation offenses address cases in which the victim is manipulated or tricked into sexual behavior or pornography.

## Forcible Rape in America

In 2007, 90,427 forcible rapes, accounting for 6.4% of the overall number of violent crimes known to the police, were committed in the United States. This percentage represents decreases of 2.5% over the previous year (2006) and 2.9% over the past decade (1998). When compared with the even greater 3.7% decline over the last 5 years (2003), the reduction associated with forcible rape appears to be one with staying power (FBI, 2008).

## Forcible Rape Defined

At common law, **forcible rape** was defined as the carnal knowledge of a female against her will and through the use or threat of force or violence. Modern rape statutes, however, have far outgrown the scope of their statutory predecessors and as such now universally include elements outside the force requirement.

Thus, for the purposes of our discussion, we use the common law elements as a basis for evaluating modern forcible rape offenses and discuss those legislative changes where appropriate. Figure 6–1 diagrams modern forcible rape elements.

## Carnal Knowledge and Gender of Participants

In states with a common law rape definition, carnal knowledge is a required burden of proof. **Carnal knowledge** refers to penile-vaginal intercourse, meaning proof must be established that the male sex organ penetrated the female sex organ. **Penetration**, however, occurs at the moment the penis passes into the outer genital lips of a female. Thus, the penis need not even enter the vagina to constitute penetration for legal purposes.

At common law, males under the age of 14 were deemed legally incapable of committing the crime of rape. In states with traditional statutory elements, if intercourse cannot be established, but rather the parties engaged in some other form of sexual behavior, the offense of rape cannot be proven. Moreover, some jurisdictions maintain that if genital organs of a victim are torn or lacerated, penetration need not be proven by the state. Modern statutes may also include penetration by an object inserted into the genitals, anus, or perineum of the victim, referred to as **rape by instrumentation**.

In many jurisdictions, the definition of intercourse has been modified to include penetration of the sexual organs of a male or female victim by a male or a female perpetrator, thus making the statutes *gender neutral*. The move to make sexual offenses gender neutral has two underlying purposes: (1) to protect male victims and (2) to address cases where the perpetrator is a female; however, change has come slowly to some degree because the public has been resistant to the premise that a male can be raped by a female. Then again, this social disregard should not be too surprising given that men are widely perceived as sexually-charged beings. For example, the Uniform Crime Reports confirm that men represent the overwhelming majority (more than 90%) of sexual predators (FBI, 2008).

In states retaining the traditional rape definition or with sexual attacks involving same-gender participants or in cases in which the sexual assault does not include sexual intercourse as defined by the statute, prosecutors must evaluate the possibility of charging a perpetrator with an offense other than rape. Before the evolution of sex offense laws and the addition of gender-neutral offenses, prosecutors were often left with few alternatives. In certain cases, assault charges were filed, but unless the assault involved a deadly weapon, it was only chargeable as a misdemeanor and therefore provided little punishment. Moreover, in such cases, the perpetrator was not required to register as a sex offender.

## Force

Force or threat of force was required to prove the crime of rape at common law. In such cases, the prosecution had to establish the victim's will was overcome by the perpetrator. In most jurisdictions, a lack of physical resistance by a victim does not constitute consent to the act. Courts have more recently added, too, that in certain cases physical force need not be proven if evidence shows the victim chose not to resist because of a reasonable fear of great bodily harm. Thus, a state need only establish the victim resisted to the extent permitted under the circumstances.

Many rape cases today arise from situations in which the perpetrator has administered to a victim some substance (such as drugs or alcohol), rendering them incapable of resisting an attack because of extreme intoxication or unconsciousness. In most jurisdictions, legislatures or courts have dealt with such situations, concluding that it is nonetheless forcible rape when a substance is administered to induce or render the victim unable to resist or in cases where the victim is in an incapacitated state (such as a coma or drunken stupor). As such, consent has become the main ingredient for proving

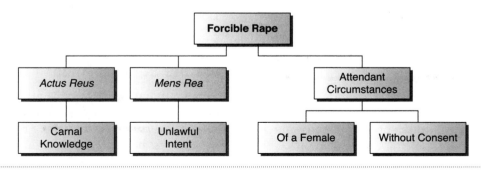

**Figure 6–1** Forcible rape.

the crime of rape, replacing the previous requirement that it be against the will of a victim. Pause for Thought 6–1 illustrates the force principle examined within rape statutes.

## Consent

A lack of consent by the victim must also be established in rape cases. In cases in which force is proven, a lack of consent may be presumed. For example, if the victim is raped with a knife to her throat, a lack of consent may be presumed from the circumstances. In cases involving adult victims where no physical force is established, the issue of consent may be more complex.

Evidence of consent may include an invitation into the victim's home or a prior relationship between the perpetrator and the victim. Ultimately, consent is an issue for the jury. Consent is not established just because the victim requests the perpetrator wear a condom. In a Texas case, the victim, in an attempt to protect herself from sexually transmitted diseases, asked her attacker to wear a condom. He did and later argued during his trial that her request constituted consent to the sexual behavior. The jury disagreed and convicted the defendant of rape.

If the victim is unable to consent to sexual behavior, factual consent is no defense. Such cases typically include victims under the age of consent or who are mentally defective. Thus, if a child under the age of 13 consents to sexual intercourse out of fear of physical harm, the perpetrator may be charged with rape. As discussed in the section on statutory rape, factual consent by a female under a certain age is not the equivalent of legal consent and is invalid in the eyes of the law.

## Marital Rape

At common law, a man could not legally rape his wife. Thus, sexual assaults within the marital home and between spouses were not punishable as a crime. The mere fact that parties were married was a valid defense to prosecution for rape. Three justifications were endorsed by the courts of England to sustain the **marital rape exemption**. The property theory suggested that women were the **chattel** (or property) of their husbands. As such, a husband could do whatever he wanted with his property as he had ownership. The **single legal entity** (or unity in marriage) **theory** suggested that at the time of marriage, two become one. Thus, husbands and wives cease to be separate individuals in the eyes of the law. Under this theory, at the time of marriage, a married woman no longer exists. Thus, there is no longer a perpetrator and a victim. The final theory is perhaps the justification that is cited most often. Lord Hale, in 1 Hale P.C. 629, suggested that by consenting and entering into a marital contract, the wife consents to all forms of sexual intercourse, consensual and forcible. This theory, better known as **contractual theory**, uses principles of implied consent to resolve the issue of sexual assaults during the marriage.

In most states today, the marital rape exemption has been repealed by legislatures or overturned by appellate courts; however, remnants of this exemption may be found in other sexual offense statutes. For example, many statutes do not apply to parties who

> ### Pause for Thought 6–1
>
> Consider the following: Sandy and Gary dated for 2 years in high school. Once while in college, the two went separate ways and began dating others. After a homecoming game, Sandy and Gary saw each other at a campus party. Each had been drinking. They decided to leave together and go to Sandy's apartment to reminisce about old times. Once there, they continued drinking and had sexual intercourse. Sandy awoke the next morning and could not remember much about homecoming night except that she repeatedly told Gary that she did not want to have sex and tried to get away from him. Despite her objections, Gary continued and scolded her for leading him on. Sandy notices bruises on her upper thighs and arms and tells her best friend what happened. Her friend tells her she must file a police report because she was raped. Was Sandy raped by Gary?
>
> ### Scenario Solution
>
> In most American jurisdictions, Gary's actions would be considered rape. This is the classic date-rape scenario. Despite their prior relationship, Sandy repeatedly told Gary "no" and tried to get away. As such, the sexual intercourse was committed without her consent. The bruises are evidence of force; however, rape trials are not easy and are especially tough on victims. At trial, the prosecution will face challenges by the defense because both parties were drinking and had a previous relationship.

are married and living together. Thus, unless the couple is separated, the perpetrator may not be charged with a crime. As such, although the common-law version of the marital rape exemption has been eliminated, it is imperative to examine other sexual offense statutes to determine whether the offense applies to married individuals who are living together. In such jurisdictions, the marital rape exemption appears to be alive and well.

## Penalties

In the majority of states, the crime of rape is a felony. States vary on the potential length of sentences, which can range from no imprisonment to life in prison. All states require individuals found guilty of rape to register as sex offenders. Capital rape, however, has traditionally been punishable by life in prison or death. Capital rape is a distinct offense requiring proof that a victim was under a certain age. Few states, however, continue to classify rape as a capital offense. Until recently, Louisiana was the only state actively seeking the death penalty for child rapists. After the ruling of the U.S. Supreme Court in *Coker v. Georgia* (1977), the majority of states with capital rape statutes eliminated the death penalty as a possible sentence; Louisiana did not. In 2008, however, the U.S. Supreme Court held that the use of the death penalty in rape cases violated the Eighth Amendment's prohibition against cruel and unusual punishment. As a result, states may not use the death penalty as punishment in rape cases. Pause for Thought 6–2 illustrates the case decision addressing this legal issue.

**Rape shield laws** are a modern strategy of most states to minimize trauma to rape victims. Generally speaking, these laws prohibit introduction of evidence during a criminal trial regarding the victim's past sexual behavior and reputation. By excluding such evidence, the laws protect rape victims from being on trial; however, rape shield laws do not prohibit introduction of such evidence in all circumstances. Previous sexual conduct may be admitted as evidence (1) when necessary to rebut evidence offered by the prosecution; (2) if the previous conduct occurred between the victim and the perpetrator and is relevant to the issue of consent; (3) if the conduct occurred between the victim and others but is necessary to rebut scientific or medical evidence, to prove the source of semen, injury, disease, or knowledge of sexual matters; or (4) to prove a similar pattern of sexual behavior to establish consent by the victim.

## Statutory Rape Defined

At common law, wives and children were regarded as chattel (or personal property) of the husband. As such, many sexual offenses that developed during this period embody attempts to protect the property interest in wives and children. Such statutes also convey an interest in shielding young females from the consequences of pregnancy, disease, and trauma. Statutory rape laws were also a mechanism through which bloodlines were maintained for inheritance purposes. The history of statutory rape reveals, from ancient times, that the law has created special protections for those regarded as too young to understand or appreciate the consequences of their actions. The offense of statutory rape was originally created to provide strict accountability for adult males who engaged in sexual behavior with young virgin females. Early versions of **statutory rape** laws required proof of the following elements: carnal knowledge, between a male over a certain age and a female under a certain age, and where the female is of chaste character and not the spouse of the adult male. Thus, for the purposes of our discussion, we use the common law elements, once again, as the basis for our discussion and discuss

---

### Pause for Thought 6–2

Consider the following: In a tragic case involving the rape of an 8-year-old child, the prosecutor elected to seek the death penalty. The jury agreed and sentenced the offender to death. The Louisiana Supreme Court affirmed the verdict of guilt as well as the penalty. The Court concluded that rape cases involving child victims were unique and in light of the extremely traumatic consequences for child rape victims the penalty for such crimes could be death. The Court dismissed the constitutional challenge by the defendant that a death sentence in such cases was cruel and unusual and violated the 8th Amendment.

### Case Decision

In June, 2008, the U.S. Supreme Court by a vote of 5 to 4, in *Kennedy v. Louisiana*, held that use of the death penalty in rape cases violated the 8th Amendment's prohibition against cruel and unusual punishment.

those areas in which legislative changes have occurred. Figure 6–2 diagrams the statutory rape elements.

## Carnal Knowledge and Gender of Participants

Under traditional statutory rape laws, the prosecution is required to establish carnal knowledge (or sexual intercourse) between the male and female. Proof of penetration of the female sex organ by the penis of the male is required; however, in some states, proof of penetration is not required if the genitals of the child are torn or lacerated as a result of an attempt to have sexual intercourse. In states with traditional statutory elements, if intercourse cannot be established but rather the parties engaged in some other form of sexual behavior, the offense of statutory rape cannot be proven.

Many jurisdictions have modified their intercourse definition to include penetration of the sexual organs of a male or female, thus making the statutes *gender neutral*. The move to make sexual offenses gender neutral, as was the case with forcible rape statutes, is an attempt to protect male victims and to address cases in which the perpetrator is a female. Modern statutes also may include sexual penetration by the penis of a male or an object that is inserted into the genitals, anus, or perineum of the victim. Although many states have modified statutory rape laws and now have gender neutral provisions, the U.S. Constitution did not require such changes. Pause for Thought 6–3 illustrates the case decision addressing this legal issue.

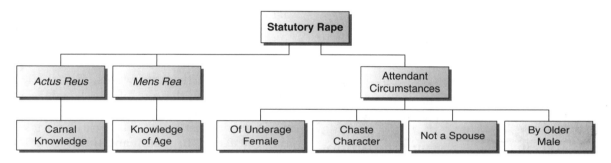

**Figure 6–2** Statutory rape.

### Pause for Thought 6–3

Consider the following: In *Michael M. v. Superior Court of Sonoma County* (1981), the Court addressed an equal protection challenge raised by an adult male convicted of statutory rape in California. In this case, Michael M. argued that such statutes violate the Equal Protection Clause of the 14th Amendment to the U.S. Constitution because the offense embodied a gender-based classification that only allowed prosecution of males for sexual intercourse with minors.

### Case Decision

The Court concluded that gender-based classifications may be upheld as long as such classifications are intended to serve a legitimate state interest. The Court held that the Equal Protection Clause does not require that a statute necessarily apply equally to all persons or require things that are different in fact . . . to be treated in law as though they were the same. Thus, a gender-based classification will be upheld where the law "realistically reflects the fact that the sexes are not similarly situated in certain circumstances." The Court recognized that the prevention of illegitimate teen pregnancies was a legitimate state interest and therefore allowed California to treat males and females differently for purposes of statutory rape. Furthermore, the Court observed that the statute protected young females from sexual intercourse and pregnancy at an age when the physical, emotional, and psychological consequences are particularly severe. In conclusion, the Court acknowledged that

> virtually all of the significant harmful and identifiable consequences of teenage pregnancy fall on the female, a legislature acts well within its authority when it elects to punish only the participant who, by nature, suffers few of the consequences of his conduct.

## Age

Age is the defining characteristic of statutory rape. Statutory rape requires proof beyond a reasonable doubt that participants are of the ages established by statute. Thus, the prosecution must establish that the male is at or above the age and that the female is at or below the age identified in the statute. At common law, females under the age of 10 years were legally incapable of consenting to sexual acts. Today, however, ages vary among states, with a majority of jurisdictions requiring males be at least 17 years of age and females be under the age of 14 or 16 for purposes of statutory rape. Thus, in most modern jurisdictions, females cannot consent to sexual behavior until attaining the legal **age of consent**. In most states, a birth certificate is not required to establish age at a criminal trial; rather, age can be proven through testimony of a witness. However, birth certificates or other documentary evidence would certainly enhance such proof at trial.

## Chaste Character

Chaste character (as defined within Black's Law Dictionary) was "never having had sexual intercourse" (Garner, 2009). In other words, the prosecution must establish that the female was a virgin before the sexual intercourse at issue. The fact that the victim was not of previous chaste character could be raised as a defense and result in acquittal of the adult male. In many cases, trials quickly became centered on the sexual history of the victim rather than the alleged conduct of the perpetrator. Defense attorneys quickly developed trial strategies that involved summoning witnesses to testify regarding their sexual exploits with the alleged victim. Such situations often resulted in trauma and humiliation for the victim. In many cases, the potential for this occurrence deterred victims and/or families from reporting the offense to authorities. Modern efforts to protect rape victims from the admission of evidence regarding sexual history have resulted in significant changes, including elimination of chaste character as an element of statutory rape.

## Consent

Unlike forcible rape, force and lack of consent are not elements for the crime of statutory rape. Rather, statutory rape crimes may occur in cases in which the female factually consents to the sexual behavior; however, factual consent by a female under a certain age is not the equivalent of legal consent and is therefore invalid in the eyes of the law. Thus, legally speaking, females under a certain age are unable to consent to sexual behavior. Statutory rape laws in most states now clearly reveal an attempt to protect young females from the consequences of sexual behavior such as pregnancy, disease, and physical and emotional trauma.

## Intent

Noticeably absent from statutory rape laws in the majority of American jurisdictions is the intent element required of most crimes. Historically, statutory rape has been a strict liability offense. Strict liability crimes do not require the state to establish *mens rea* as an element of the crime. Rather, the defendant's evil intent is presumed from the act in question. In statutory rape cases, courts have held that the intent of the defendant to commit statutory rape can be derived from his intent to commit the act of fornication.

The lack of an intent requirement in statutory rape law raises an important question. Does a mistake or lack of information regarding the age of the victim matter? In other words, if the defendant is mistaken about the age of the female, can he be acquitted of statutory rape? A defendant may try to argue at trial that he was unaware or mistaken about the true age of the victim and therefore is not guilty of statutory rape. For example, may a defendant argue at trial that he mistakenly believed the victim was 18 rather than 14? Whether mistake of age can be raised as a defense to statutory rape during a criminal trial has been raised in many state appellate courts. Generally speaking, those states that view statutory rape as a strict liability offense have declined to allow defendants to raise the mistake of age defense at trial. If no intent has to be proven by the state, then the state of mind or belief of the defendant is not a matter for the jury. Defendants are held strictly liable for their actions, regardless of their mistaken belief about the age of the victim. Such an approach requires adults to exercise caution and judgment when engaging in sexual relations with others. Pause for Thought 6–4 illustrates how to properly interpret the consent issue within statutory rape criminal guidelines.

## Penalties

In the majority of states today, statutory rape is a felony ranging in punishment from no imprisonment at all to several years in prison. Individuals found guilty of statutory rape are required to register as a sex offender in all states.

> **Pause for Thought 6-4**
>
> Consider the following: Jamie and Suzie grew up together. Their families lived in the same neighborhood, and the siblings in each family attended the same schools and played together. Jamie was 5 years older than Suzie. Jamie went off to college when Suzie was in the 8th grade. Suzie blossomed into a beautiful young woman who was 5'10" and looked much older than her chronological age. Jamie returned for summer break and was instantly smitten. Unbeknownst to others, the two began seeing each other and eventually had consensual sexual intercourse. Jamie was 19 and Suzie was 14. The age of consent in the state where Jamie and Suzie reside is 16. Is Jamie guilty of statutory rape?
>
> **Scenario Solution**
>
> Yes, Jamie could be prosecuted for statutory rape. Jamie and Suzie believe they engaged in consensual sexual intercourse; however, because the age of consent is 16, Suzie is unable to give consent. It would be difficult for Jamie to argue that he was reasonably mistaken about Suzie's age because they grew up together.

## Sexual Battery Defined

The need for sexual battery statutes is twofold. First, such laws provide a statutory alternative to common law rape statutes that failed to address same-gender or female on male sexual attacks. Second, the offense of sexual battery criminalizes attacks that involve sexual penetration other than traditional intercourse. As such, for prosecutors, sexual battery laws have provided an extremely useful alternative in sexual offense cases. The essence of the crime of **sexual battery** is sexual penetration of another without their consent. Figure 6–3 diagrams the elements for the crime of sexual battery.

## Penetration

The crime of sexual battery requires proof of penetration. Penetration for purposes of sexual battery is defined in much broader terms than in forcible or statutory rape offenses. Specifically, the manner and means of perpetration in sexual battery statutes is markedly different in that many states include genital, anal, and oral penetration of victims. Thus, sexual battery statutes can apply to cases involving acts of **cunnilingus** (oral stimulation of the female sex organ), **fellatio** (oral stimulation of the male sex organ), **buggery** (anal intercourse), or **pederasty** (unnatural intercourse between a man and boy), most of which may not have provided a basis for prosecution under more traditional rape statutes. Sexual battery statutes also are broader in terms of the instruments of penetration, with statutes usually including penetration by a body part or an object.

## Of Another

As stated earlier, sexual battery statutes are gender neutral and are therefore not limited to cases involving male attackers and female victims. Such a shift reflects the reality of sexual assault: that it is not about physical or lustful attraction, but that such crimes are committed by perpetrators who desire power and control over their victims. Thus, although

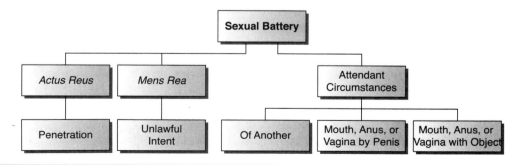

**Figure 6–3** Sexual battery.

male on female attacks may be prosecuted under sexual battery statutes, the parties involved may also include same-gender attacks or female on male.

## Consent

Sexual battery requires proof that the penetration occurred without the consent of the victim. Unlike rape statutes, force is, generally speaking, not an element of most sexual battery statutes. Thus, sexual battery cases may include victims who are too young to give legal consent to sexual acts; cases in which victims are intoxicated or drugged; or cases involving victims with mental, developmental, or physical limitations. In each of these situations, the victim is unable to give legal consent to sexual acts. In several states, sexual battery offenses apply to cases involving perpetrators who occupy a position of trust or authority over a child victim. These offenders use their position of trust or authority to coerce the child into sexual behavior. Despite factual consent or acquiescence by the child, legal consent is impossible, and thus, the acts are without consent in the eyes of the law. Such cases may include perpetrators such as teachers, counselors, physicians, clergy, coaches, parents, or other relatives. Pause for Thought 6–5 illustrates how persons in positions of trust or authority are culpable within sexual battery statutes even when relations were consensual.

## Penalties

In the majority of states today, sexual battery is a felony that is punishable up to life in prison. Individuals found guilty of sexual battery are required to register as a sex offender in all states.

## Sodomy Defined

At common law, the crime of sodomy—involving certain sexual acts traditionally viewed as unnatural—was a felony punishable by death. In fact, many states, in lieu of the sodomy label, labeled their statutes as unnatural sexual intercourse. Generally speaking, **sodomy** refers to oral or anal intercourse between human beings. Essentially, it targeted acts including penetration of the anal cavity of another human or animal. Modern statutes, however, have limited the crime of sodomy to living human beings, meaning that **necrophilia** (sex with a human corpse) and **bestiality** (sex with an animal) are now treated as separate offenses. As with other sexual offenses, the state is required to establish penetration of the victim in sodomy cases. The most common ways to establish such penetration is through the testimony of the victim or proof of injury. The state must also establish that the victim was penetrated by the penis of the perpetrator. Thus, if the jurisdiction has maintained a traditional sodomy statute, the perpetrator must be a male; however, sodomy victims may be male or female. Figure 6–4 diagrams the elements for the crime of sodomy.

## Force and Consent

Forcible sodomy or penetration without consent of the victim is a crime. Historically, most sodomy

---

### Pause for Thought 6–5

Consider the following: Christy is a high school math teacher who offers tutoring sessions to assist students. Matt, who is struggling, has been attending these sessions regularly and has developed a crush on Christy. Christy is 23 and Matt is 14. They begin meeting at locations away from school and soon become friends. Their friendship eventually turns into romance. The rumors are rampant around school, and Christy's principal eventually calls her in for a conference. During that meeting, Christy admits that the two have been intimate and have had oral sex. She insists that the couple has not had sexual intercourse. The age of consent in this state is 16. Can Christy be charged with sexual battery?

### Scenario Solution

Yes, Christy may be charged with sexual battery. Matt is an impressionable young man under her supervision and is subject to her trust and authority. Despite the lack of traditional sexual intercourse, the parties have engaged in oral sex, which is sufficient under sexual battery statutes. Because Matt is under the legal age of consent, he is unable to consent to sexual behavior. Also, because sexual battery statutes are gender neutral, the fact that Christy is a female perpetrator does not prohibit a charge of sexual battery.

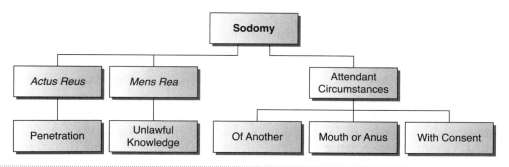

**Figure 6–4** Sodomy.

statutes have not required elements of force or lack of consent. Sodomy was traditionally criminalized because the acts themselves were considered unnatural, or an abomination against nature. As such, many characterize consensual sodomy as a crime against morality as opposed to a crime against a person; however, despite the lack of consent or force elements, consensual sexual acts between adults have been criminalized under some state sodomy statutes. This has given rise to much debate in the last few years and attracted the attention of the U.S. Supreme Court on two separate occasions. Pause for Thoughts 6–6 and 6–7 provide the details for those cases.

## Child Molestation Defined

Sexual abuse of children is an enduring social issue with widespread implications. The numbers continue to escalate, with most child sexual abuse occurring within the home or by perpetrators known to the child. The vulnerability of children in their

### Pause for Thought 6–6

Consider the following: In *Bowers v. Hardwick* (1986), law enforcement officers encountered a homosexual couple in the act of sodomy up on entering a bedroom in a home. The law enforcement officers were in the process of serving a warrant. On appeal, the defendant challenged the criminalization of private and consensual sexual acts between adults. The defendant specifically argued that Georgia's statute, as applied in this case, violated the right to privacy.

### Case Decision

In *Bowers*, the High Court (in a controversial 5–4 count) upheld Georgia's sodomy statute, concluding that the right to privacy did not include the right to engage in unnatural sexual acts, consensual or not.

### Pause for Thought 6–7

Consider the following: Like its predecessor (*Bowers v. Hardwick*), *Lawrence v. Texas* (2003) involved law enforcement officers executing a warrant and encountering a homosexual couple involved in certain sexual behavior. John Lawrence was charged with sodomy and was prosecuted and convicted. Despite recent precedent as established in *Bowers*, the Court granted the petition for writ of certiorari.

### Case Decision

The Court overruled (in a 6–3 vote) the holding in *Bowers* and concluded that private sexual conduct between consenting adults should not be criminalized. After *Lawrence*, states may not criminalize consensual oral or anal sexual penetration between adults. The holding, however, does not in any way impact the right of a state to criminalize forcible or nonconsensual sodomy.

own homes has increased because of alarming divorce rates and increased numbers of single parents. Each situation allows opportunities for access to children by stepparents, paramours (or lovers), and others.

Although the majority of child sexual abuse occurs in the home, children in today's society face many threats to their welfare and are extremely vulnerable to exploitation and victimization because modern technology allows crimes such as Internet stalking and child pornography to flourish. In order to combat significant increases in child sexual abuse, states have strengthened penalties for existing offenses and have enacted new laws to address modern problems. This section will survey many criminal offenses that may apply when the victim is a child.

Fondling is one of the more common charges brought against perpetrators in child sexual abuse cases. In general, **fondling** is defined as the handling, touching, or rubbing of a child under a certain age by an offender over a certain age. In addition to child victims, fondling statutes may also address such behavior with victims who have mental, physical, or developmental limitations.

The prohibition against fondling reflects an attempt to allow individuals to be secure in their persons, while also demonstrating an appreciation for trauma inflicted on children when subjected to forms of sexual abuse that do not involve force or sexual penetration. This is extremely important as pedophiles and sexual predators often groom their victims with forms of abuse that are perceived as "milder" and less intrusive. As a result, many victims may begin to question their perception of the events, or other adults to dismiss the behavior as a misunderstanding. Strict penalties for fondling, therefore, may interrupt and impede the ability of sexual predators to escalate their offending. The criminalization of such behavior also serves to dispel the myth that such conduct is not traumatic or harmful to the victim. Figure 6–5 diagrams the elements for fondling.

## Touching, Rubbing, or Handling with Body Part or Member

In fondling cases, the prosecution must establish that the perpetrator touched, rubbed, or otherwise handled the body of a child or vulnerable adult. In most cases, this element is established through the testimony of the victim or witnesses. Fondling cases often present unique challenges not present in other types of sexual molestation cases. For example, in fondling cases, physical evidence is rarely present. Unless the physical acts have involved a significant amount of force (such as extreme rubbing, which may result in chafing or other skin injuries), no physical damage will be present on the body of the victim. As such, photographs or medical testimony may not be available. The lack of physical injury is often a defense strategy at trial. The defense commonly attempts to challenge victim accounts on cross-examination by arguing that the lack of visible physical injury (bruising, tearing, chafing, etc.) indicates that no crime occurred.

Another common issue in fondling cases is the subtlety of the acts. In many cases, the defendant may claim that his or her body accidentally touched, rubbed, or handled the child. Sexual predators are experts at manipulating others and are often able to convince children or other witnesses that the contact was accidental or negligent. For example, consider a case in which an individual attempts to assist a 5-year-old female child when the child gets off playground equipment. The adult touches the chest area of the young girl. The child feels uncomfortable and wriggles away from the adult and runs to her mother who is reading a book on a nearby bench. The child is crying and believes that she was touched in a bad way. The mother is confused about the account;

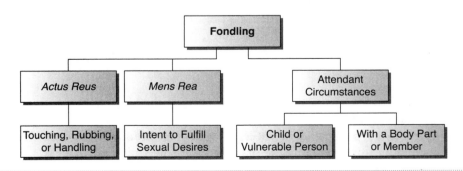

**Figure 6–5** Fondling.

however, the adult walks over and explains that he accidentally touched the child as he was assisting her. Obviously, cases involving touching, rubbing, or handling with a sexual organ are less subtle and do not pose such challenges. Finally, the act of handling, touching, or rubbing may be accomplished with the use of a body part or any member thereof. Thus, an individual may be guilty of fondling if he or she uses hands, fingers, sexual organs, or any other body part during the course of the acts.

### Children and Vulnerable Individuals

Significant variety exists among state fondling statutes. In many states, fondling statutes require the victim be a child under a certain age or an individual with physical, mental, or developmental limitations. In others, statutory penalties may be enhanced if the perpetrator occupies a position of trust or authority over the child. For example, if the perpetrator is a parent, teacher, physician, or clergy member, potential penalties may be greater. Such practice reflects a strong public policy directed toward deterring individuals from exploiting unique relationships with children subject to the influence and control of persons in positions of trust or authority.

### Intent

In the majority of states, fondling statutes specifically require proof that the perpetrator engaged in the act for the purpose of gratifying his or her sexual desires. The *mens rea* element in fondling offenses can present a challenge for prosecutors. Again, the state must overcome common defense explanations of accident or mistaken perception by the victim. Most jurisdictions designate fondling as a felony punishable by imprisonment, fines, or both. Possible sentences range from 1 to 20 years in the penitentiary. In many states, subsequent convictions for fondling result in enhanced penalties. As with other sexual offenses, individuals convicted of fondling are required to register as sex offenders.

## Other Sexual Offenses

The following section provides readers with an overview of other forms of sexual offenses that may not involve sexual assault, but nonetheless constitute an intimate violation. Offenses such as child pornography, exploitation, incest, seduction, and voyeurism are included in this section. The evolution of technology has especially enhanced the level of difficulty for prosecutors in child pornography, exploitation, and voyeurism crimes. As such, these cases involve significant and complex legal issues for federal, state, and international authorities.

## Child Pornography and Exploitation

Federal and state laws exist to address the increasing problem of child pornography and the exploitation of children. Child pornography is more fully explored in Chapter 7. Statutes that address child pornography typically prohibit all forms: visual or computer depictions of actual or simulated sexual activity, sketches, drawings, and photography. Furthermore, possession, transmission, or receipt of child pornography is criminalized.

Federal law also criminalizes child pornography and specifically addresses cases that involve the interstate transportation of such materials (see The Federal Protection of Children Against Sexual Exploitation Act of 1977 as codified in 18 U.S.C.A., Section 2252). **Child exploitation** refers to a variety of acts calculated to derive financial, sexual, or other benefits from the manipulation of children. Exploitation as a criminal offense typically applies to cases in which a child is used in pornography or prostitution; however, exploitation is a separate offense and may be charged in addition to either pornography or prostitution. Pause for Thought 6–8 illustrates the proper legal interpretation regarding child pornography and exploitation.

## Incest

The crime of **incest** refers to intercourse or marriage between individuals too closely related by blood or marriage. If two individuals are closely related because of marriage, the acts are referred to as incest due to **affinity**. If two individuals are closely related by blood, the acts are referred to as incest due to **consanguinity**. At common law, incest was a felony, and thus offenders could be sentenced to death. The strong prohibition against incest is derived from three sources. Ecclesiastical or religious law historically condemned sexual intercourse between such individuals. Moreover, many cite biological or genetic reasons for prohibiting such relationships. Thus, the fear that genetic abnormalities, deformities, or defects will plague the offspring of these unions is a common reason for prohibiting the relationships. Finally, others cite disruption to marital harmony as a reason for prohibiting such conduct.

All states criminalize incest to some degree, but states differ with regard to the manner in which incest is defined. Some states criminalize only sexual intercourse between individuals too closely related by blood or marriage, whereas others also criminalize marriage between such individuals. The Model Penal Code defines incest as sexual intercourse or marriage between individuals too closely related. Each state identifies which individuals are too closely related by blood or marriage to either

### Pause for Thought 6–8

Consider the following: Bob Stone is currently living with his girlfriend (Holly) and her 4-year-old daughter (Jesse). Things go well at first, but Bob eventually begins viewing and becomes addicted to online pornography. A few months later, Bob approaches Holly about posing with her daughter for some photographs. Bob argues they can make thousands of dollars for photographs of Jesse. Holly initially resists but eventually complies with the request. Holly is uncomfortable with the photographs but makes Jesse pose nude, promising it will be only this once. Bob takes the photographs, which eventually end up on a European child pornography website. Bob receives $2,000 for the pictures (which he splits with Holly). Of what offenses are Bob and Holly guilty?

### Scenario Solution

Bob and Holly may be charged with child pornography and/or exploitation. Recall that these are distinct offenses. Bob took nude photographs of a 4-year-old child, which is sufficient for a child pornography charge; however, Bob later sold these photographs for a profit and therefore could also be charged with exploitation. Holly also participated in the child pornography. She assisted Bob and was a principal player in these events. As such, she too can be charged with child pornography. Also, Holly derived a financial benefit from the pictures and therefore could be charged with exploitation.

---

marry or engage in sexual relations. Thus, it is necessary to review relevant state statutes to determine prohibited relationships. For example, most states prohibit relationships between mothers, fathers, children, siblings, grandparents, aunts, and uncles; however, states differ with respect to relationships among cousins, step-relatives, and half-blood relationships. Some states allow first cousins to marry, whereas others do not. Incest is also grounds for an annulment or divorce in most states. In many states, children who are born into a marriage that is incestuous are considered illegitimate. This consequence reflects an attempt to stigmatize incestuous relationships in hopes of deterring them. Pause for Thought 6–9 explain the legal process for resolving incestuous accusations.

## Seduction

Many jurisdictions have retained **seduction** as a crime, occurring when an adult male entices an unmarried young woman of previous chaste character to engage in sexual intercourse, but also requires that the male entice the female with a false promise of marriage. Like statutory rape, seduction aims to protect young chaste females from sexual advances of older males. Pause for Thought 6–10 illustrates how to legally combat the crime of seduction.

## Voyeurism

*Voyeurs* are more commonly known as *Peeping Toms*. The crime of **voyeurism** is committed when an individual views or attempts to view others in private set-

### Pause for Thought 6–9

Consider the following: Bob and Betty have decided to get married. Their families are extremely happy and wish them well on their new life. After the nuptials, Betty is unpacking some family keepsakes that Bob has in his home. Betty notices a family Bible with a genealogy chart and recognizes several names. After contacting her mother, Betty realizes that she and Bob are third cousins. Is this an incestuous relationship?

### Scenario Solution

In most states, this would not be considered an incestuous relationship. States typically prohibit relationships between first or second cousins, but rarely is marriage or a sexual relationship prohibited between third cousins; however, if in doubt, it is always best to consult with an attorney to determine whether the relationship falls within those prohibited by law.

### Pause for Thought 6–10

Consider the following: A young girl, Laura, lives with her mother and works part-time at a fast food restaurant. A delivery man named John often visits to the restaurant. John and Laura strike up a friendship and eventually begin dating. Laura is 15 and John is 42. John convinced Laura to move in with him and promises to get married. Unbeknownst to Laura, John is already married to another woman. Laura is still in high school and initially does not want to leave her mother; however, John promises that she can finish high school and even attend college. During the middle of the night, Laura leaves to meet John at the restaurant. They live together for 2 months, but John no longer wants to marry Laura. Is John guilty of any criminal offense?

### Scenario Solution

First, John promises to marry Laura but does not. Moreover, he is unable to fulfill the promise because he is married to another. If the prosecution can establish that John enticed Laura to move in and become sexually involved with him, with a false promise of marriage, then he may be convicted of seduction.

---

tings. The most common scenario arises when an individual places himself or herself in a position on the property of another to view activities within the privacy of their home; however, voyeurism statutes are not limited to perpetrators who attempt to view others in their homes. Some jurisdictions include other settings in which there is an expectation of privacy such as hotel rooms, public restrooms, or dressing rooms. Pause for Thought 6–11 provides a hypothetical scenario regarding the crime of voyeurism.

### Sexual Offenders and Megan's Law

**Megan's Law** allows states to provide information about registered sex offenders to the public. Information is available and may be accessed by the public. Named after Megan Kanka, every state now has a version of Megan's Law. Megan was a 7-year-old girl who was raped and murdered by a sexual offender who had moved, unknowingly to the family, into a home across the street from their home. Megan's Laws reflect an attempt to allow the public to protect children from sexual offenders and increase awareness about the location of sexual offenders.

### Federal Law

Title 18, Chapter 109A of the United States Criminal Code (§§2241-2248) delineates a multitude of sexual offenses as crimes pursuant to federal law. Merely entitled Sexual Abuse, this criminal chapter comprises most all forms of sexual abuse (as described throughout this chapter). One example is that of aggravated sexual abuse (§2241)

> Whoever, in the special maritime and territorial jurisdiction of the United States or in a Federal prison, or in any prison, institution, or facility in which persons are held in custody by direction of or pursuant to a contract or agreement with the Attorney General, knowingly (1) causes another person to engage in a sexual act by threatening or placing that other person in fear (other than by threatening or placing that other person in fear that any person will be subjected to death, serious bodily injury, or kidnapping); or (2) engages in a sexual act with another person if that other person is—(A) incapable of appraising the nature of the conduct; or (B) physically incapable of declining participation

### Pause for Thought 6–11

Consider the following: Peeping Pete owns a photography studio in a small town outside of Atlanta. While women change clothes for photography sessions, Pete secretly films the women. He does not sell these films or videos but, rather, keeps them for his own pleasure. Is Pete guilty of voyeurism?

### Scenario Solution

Pete may be charged with voyeurism. The women have an expectation of privacy in the dressing rooms. Whether Pete sells or transfers the films is irrelevant under a voyeurism statute.

in, or communicating unwillingness to engage in, that sexual act; or attempts to do so, shall be fined under this title and imprisoned for any term of years or for life.

## Summary

The number of sexual offenses has been significantly expanded to cope with changing notions of morality, culture, and the status of women and children. Reference to common law offenses is important to understand the foundation of contemporary sex crimes and to understand their evolution. Modern sex offenses tend to be gender neutral and carry greater penalties than their common law counterparts. Moreover, the increasing social problems associated with pornography have led to significant expansion of sexual exploitation laws.

## Practice Test

1. At common law, wives and children were regarded as _____, defined as the personal property of the husband.
   a. pecuniary
   b. tangible assets
   c. bond
   d. inferior collateral
   e. chattel

2. The forcible insertion of the penis or a foreign object into the vagina, mouth, or anus of a female constitutes the crime of _____.
   a. seduction
   b. sexual battery
   c. forcible rape
   d. statutory rape
   e. capital rape

3. Modern rape statutes often include penetration with an object into the genitals, anus, or perineum of the victim, referred to as _____.
   a. affinity
   b. fondling
   c. rape by instrumentation
   d. voyeurism
   e. consanguinity

4. At common law, _____ was defined as penile-vaginal intercourse.
   a. cunnilingus
   b. sodomy
   c. consanguinity
   d. carnal knowledge
   e. rape

5. _____ refers to sexual intercourse with a human corpse.
   a. necrophilia
   b. sadism
   c. flagellation
   d. buggery
   e. infibulation

6. At common law, males under the age of _____ were deemed legally incapable of committing the crime of rape.
   a. 13
   b. 14
   c. 16
   d. 17
   e. 18

7. Many jurisdictions have modified the definition of intercourse to include penetration of both male and female sexual organs by both male and female offenders, thus creating what is legally referred to as a(n) _____ statute.
   a. gender-neutral
   b. unbiased
   c. equal protection
   d. unconstitutional
   e. ex post facto

8. _____ was NOT required to prove the crime of forcible rape at common law.
   a. Carnal knowledge
   b. Actual bodily harm
   c. Use or threat of force
   d. Intent
   e. Penetration

9. _____ refers to a variety of acts calculated to derive financial, sexual, or other benefits from the manipulation of children.
   a. Neglect
   b. Pederasty
   c. Abuse
   d. Statutory rape
   e. Exploitation

10. _____ is the defining characteristic of statutory rape.
    a. Intent
    b. Age
    c. Consent
    d. Identity
    e. Personality

11. _____ is defined as "never having had sexual intercourse."
    a. Virginity
    b. Purity
    c. Virtuous
    d. Novice
    e. Chaste

12. Strict liability offenses (such as statutory rape) do not require the state to establish the _____ of a defendant.
    a. identity
    b. personality
    c. age
    d. intent
    e. consent

13. _____ is defined as oral stimulation of the male sexual organ.
    a. Necrophilia
    b. Cunnilingus
    c. Pederasty
    d. Buggery
    e. Fellatio

14. In _____, the U.S. Supreme Court upheld a constitutional challenge to Georgia's sodomy statute.
    a. *Bowers v. Hardwick*
    b. *Lawrence v. Texas*
    c. *Roe v. Wade*
    d. *Johnson v. Georgia*
    e. *Coker v. Georgia*

15. _____ refers to incestuous relationships between blood relatives.
    a. Affinity
    b. Lineage
    c. Consanguinity
    d. Sodomy
    e. Contractual kin

16. The U.S. Supreme Court, in _____, held that the death penalty for rapists violated the 8th Amendment's prohibition against cruel and unusual punishment.
    a. *Coker v. Georgia*
    b. *Bowers v. Hardwick*
    c. *Kennedy v. Louisiana*
    d. *Lawrence v. Texas*
    e. *Furman v. Georgia*

17. _____ occurs when an adult male entices an unmarried woman of chaste character to engage in sexual intercourse based on a fraudulent promise to marry.
    a. Seduction
    b. Perversion
    c. Enticement
    d. Molestation
    e. Predation

18. The majority of child sexual abuse occurs at _____.
    a. church
    b. school
    c. playgrounds
    d. home
    e. neighbors' homes

19. _____ are commonly known as "Peeping Toms."
    a. Perverts
    b. Voyeurs
    c. Molesters
    d. Stalkers
    e. Eavesdroppers

20. With _____ cases, the prosecution must establish that a perpetrator touched, rubbed, or otherwise handled the body of a child or vulnerable adult.
    a. obscenity
    b. child pornography
    c. fondling
    d. bestiality
    e. incest

## References

Federal Bureau of Investigation. (2008). *Crime in the United States, 2007: Uniform Crime Reports*. Retrieved July 14, 2009, from http://www.fbi.gov/ucr/cius2007/index.html

Garner, B. A. (Ed.). (2009). *Black's law dictionary* (9th ed.). Eagan, MN: West Group.

# Crimes Against Moral Values

**CHAPTER 7**

## Key Terms

Adultery
Bigamy
Blood alcohol level
Breathalyzer
Drug possession
Field sobriety test
Fornication
Gambling
Gaming
Hate Crimes Statistics Act of 1990
Implied consent statutes
Indecent exposure
Interstate Wire Act of 1961
Johns
Mann Act
Nystagmus gaze
Polygamy
Possession with intent to distribute
Precursors
Prostitution
Protection of Children from Child Exploitation Act of 1977
Prurient interest
Public intoxication
Purported marriage
Uniform Controlled Substances Act of 1970
Violent Crime Control and Law Enforcement Act of 1994
Volstead Act of 1919

## Introduction

Crimes against public morals reflect an attempt to protect values important to a civilized society. These crimes usually prohibit acts in conflict with traditional conservative values. Opponents challenge the continued enforcement of moral offenses on the basis that morality should not be legislated; rather, it is a matter of personal freedom. Opponents frequently contend that these offenses, with some exceptions, target consensual behaviors. Despite infrequent enforcement of these offenses there appears to be no general consensus to abolish such conduct.

## Morality Legislation in America

The origins of American law are deeply rooted in English common law but were informed significantly by Ecclesiastical offenses, which use biblical sins as a basis for defining culpability; such offenses were adjudicated by the Ecclesiastical courts governed by the Church of England. Despite the passage of much time, biblical sins, traditional values, and a sense of right and wrong continue to inform the development of American law. For example, crimes that are *malum in se* violate the collective moral standards of society and as such are considered wrong in and of themselves, not simply because the legislature has deemed them to be so. The potential for social harm is great, and there is usually an identifiable victim. Examples of crimes classified as *malum in se* include rape, robbery, murder, and assault. These classifications demonstrate that morality continues to inform the development of American criminal law. Moreover, American jurisprudence has retained many of the historical offenses adjudicated by the Ecclesiastical courts. The modern versions of these offenses are discussed in this chapter to illustrate the enduring attempts to ensure the sanctity of traditional values in American society.

## Sex Offenses

Sex offenses were thoroughly discussed in Chapter 6, but some sexual crimes (such as bigamy, polygamy, prostitution, fornication, adultery, and sodomy) are more appropriately discussed in this chapter to illustrate efforts at curtailing immoral sex acts.

## Bigamy and Polygamy

**Bigamy** occurs when an individual enters into a purported marriage while legally wed to another. In all states, a second or subsequent union is referred to as a **purported marriage** to indicate that the marriage is not legitimate. The offense of bigamy exists to protect the sanctity of marriage. Penalties for bigamy aim to deter individuals from entering into further marital unions without having dissolved the legally recognized marriage. In most states, the crime of bigamy is a felony and often serves as a basis for divorce or annulment. States differ with regard to the status of children born during a bigamous union. Some states recognize such births while other states refuse to recognize children born of these secondary marriages as legitimate through the eyes of the law.

Although not recognized by any state today, the crime of **polygamy** was part of common law and as such requires some degree of attention in this section. Whereas bigamy addressed one additional unlawful marriage, polygamy sought to deter multiple (two or more) extramarital unions. It is essential to understand, however, that the separate and distinct crime of polygamy existed to distinguish its less severe encroachment on the sanctity of marriage. At first glance, it may not seem rationale to suggest that one extra wife is a more serious breach of trust than the taking of multiple extra wives. When considering the influence of natural law (see Chapter 1), however, the evolution of polygamy as distinct from bigamy makes perfect sense. In biblical times (and even within several fundamentalist sects of the Mormon church), the taking of multiple wives was a common practice with many women consenting to such communal arrangements—a harem of sorts. In this light, one should be able to see that the breach of trust is absent in that the wife voluntarily consented to such a lifestyle. Although the act of polygamy is still practiced in some religions, American law has removed all distinction between polygamous practices and bigamous acts. In so doing, only the crime of bigamy now exists within legal statutes even when the act causing the bigamous charge was based on a consensual polygamist arrangement.

Recently, the nation watched as a religious sect in Texas was raided by law enforcement and child protective services following reports of arranged underage marriages and child abuse. Although most children were eventually returned to their parents, these events focused the nation's attention on the religious group's practices regarding sexual conduct and marriage. The leader of the sect, Warren Jeffs, was prosecuted for promoting and arranging plural marriages with underage females. Arrested in Nevada, Jeffs was convicted in Utah as an accomplice for two counts of rape. Jeffs' prosecution suggests that prosecutors remain vigilant about enforcement of sexual offenses within religious practices that may constitute a crime. Thus, while the crime of polygamy may not be used in modern jurisdictions, those who engage in its practice risk prosecution. Pause for Thought 7–1 compares and contrasts bigamy and polygamy.

## Prostitution

Often referred to as the world's oldest profession, prostitution is a crime in virtually all states. Prostitution was not a crime at common law but, with the exception of Nevada, is now illegal throughout the United States. Even in Nevada, however, prostitution is mostly sanctioned by local ordinances and is therefore still highly regulated. **Prostitution** is defined as the act of engaging in sexual favors for hire. Payment may take the form of money or goods and in many cases drugs. The essence of a prostitution offense is the business transaction. Exchange of money or other forms of compensation for sexual services is the central focus of the offense.

Prostitution is criminalized for many different reasons; the two predominant reasons relate to public health concerns and the victimization of prostitutes.

### Pause for Thought 7–1

Consider the following: Ken so enjoys being married that he receives permission from his wife to wed four other women, all of whom are knowledgeable about the lifestyle. He then formalizes the marriages through proper regulatory channels. Is Ken guilty of polygamy or bigamy?

### Scenario Solution

At common law, the preceding polygamist acts would have been charged as one crime of polygamy because the first (or lawful) wife and all subsequent purported wives were consensual participants in the unlawful marital scheme. Considering that the crime of polygamy no longer is recognized within modern legal codes, however, it is clear that the preceding scenario could result in four separate bigamous charges.

Sexually transmitted diseases are of paramount concern in this day and age. Public health officials have worked for decades to decrease the prevalence of sexually transmitted diseases such as AIDS, gonorrhea, syphilis, and others.

The circumstances that may attend prostitution certainly raise the level of concern regarding the spread of sexually transmitted diseases. Although prostitution is often characterized as a victimless crime, prostitutes are often victimized during these transactions. Many are raped, beaten, and robbed. Moreover, children and young adults are often recruited or otherwise used in the sex trade. As such, most agree that the high risk of victimization warrants the criminalization of these acts.

Prostitution exists on many different levels. Although most think of the lone prostitute strolling along an urban street, many prostitution rings are highly organized. In the latter situations, prostitutes are often referred to as "call girls" or "escorts" and are screened and supervised thoroughly by their employers. Supporters of prostitution as a legitimate commercial enterprise advocate for its legalization. Advocates claim legalization would transform a black market business into a regulated industry, which would concomitantly serve to increase the safety of its participants while also providing additional tax revenue for state coffers. Despite the arguments, advocates have not met with much success.

Where criminalized, prostitution is typically a misdemeanor and thus carries minimal penalties. Arrest for prostitution typically causes only minor inconvenience for the prostitute. The Model Penal Code also criminalizes prostitution in section 251.2 but makes prostitution either a misdemeanor or felony. The Model Penal Code also includes the offense of Promoting Prostitution, which addresses behaviors commonly associated with pimps and makes such conduct either a misdemeanor or felony. **Johns**, as the customers in the transaction, are typically charged with solicitation of prostitution. Under some statutes, the crime of solicitation (discussed more fully in Chapter 9) is charged when an individual offers payment for sex. Like prostitution, solicitation is usually a misdemeanor and, other than embarrassment and humiliation, results in minimal fines or penalties.

Federal law also addresses prostitution. The **Mann Act** was originally designed to prohibit interstate transportation of women and young girls for the purpose of prostitution or other immoral behavior. Later amendments to the act included gender-neutral provisions and now prohibit the interstate or foreign transportation of a male or female for the purpose of engaging in prostitution. Violations of The Mann Act are felonious and punishable by fines and imprisonment for a term up to 10 years.

## Fornication and Adultery

At common law, fornication and adultery were, in part, criminalized to minimize the risk of illegitimate births. An additional motivation for regulating adultery was to protect the sanctity of marriage; however, criminalization of adulterous behavior has been traced even farther back to primitive civilizations. In the Bible, for example, an account is given regarding the attempted stoning of a woman accused of adultery.

**Fornication** is defined as consensual sexual intercourse by an unmarried individual, whereas **adultery** is sexual intercourse with someone other than a lawful spouse. Although these laws are still on the books in many states, neither fornication nor adultery is actively (if at all) enforced and prosecuted. The sheer number of these cases alone would likely overwhelm any prosecutor's office. Generally speaking, fornication and adultery have been repealed from statutory codes or decriminalized because of an increasing social agenda against the regulation of morality. Other states, however, wish not to repeal the crime of adultery for fear it would send a social message that their state permits (perhaps even condones) adulterous relationships. In such states, there has been a compromise, of sorts, regarding what elements will continue to encompass the crime of adultery, and these definitions do not always resemble their historical predecessor originating in the ecclesiastical (or religious) courts. Alabama, for example, stipulates that a person has not committed the crime of adultery unless living in adultery. The Alabama adultery statute (13A-13-2) specifically states that "a person commits adultery when he engages in sexual intercourse with another person who is not his spouse and lives in cohabitation with that other person when he or that other person is married."

## Sodomy and Homosexuality

Sodomy was discussed in the previous chapter on sex offenses; however, because sodomy laws originally evolved from the belief that the act was unnatural or an abomination against nature, sodomy may also be characterized as a crime against morality. Historically, consensual sodomy, despite the lack of force, was criminalized; however, after the holding in *Lawrence v. Texas*, states may not criminalize consensual oral or anal sexual penetration between adults. The holding, however, does not in any way impact the right of a state to criminalize forcible or nonconsensual sodomy.

The historical characterization of sodomy as an unnatural act is consistent with the views on homosexuality; however, it is important to understand that acts that constitute sodomy may occur between homosexual or heterosexual couples. As such, the criminalization of sodomy is not limited to acts committed by homosexuals. The holding in *Lawrence v. Texas* coupled with changing social attitudes toward homosexuality has resulted in more moderate treatment of homosexuals in American culture. This trend can be seen in the increasing numbers of states permitting marriage or civil unions for homosexual couples.

## Indecent Exposure

Indecent exposure was a crime at common law mostly known as *lewdness*, which in turn is often titled as *lewd and lascivious behavior*. Most modern statutes and the Model Penal Code now refer to this offense as indecent exposure. The purpose of such statutes is to shield individuals from lewd and indecent displays by others.

**Indecent exposure** requires proof an individual intentionally exposed private parts in a manner that others are likely to view and for the purpose of gratifying licentious desire. At common law, the exposure was required to occur in a public place; however, if the statute limits exposure to that which occurs in public, cases involving individuals who expose themselves to others in private homes or through windows could not be convicted. As such, many states have modified their statutes and require only that the exposure occur in a manner likely to cause alarm. Indecent exposure is a misdemeanor offense in most jurisdictions, but approximately 10 states do regulate it as a felony. In such states, however, there is often a lesser included offense designated as a misdemeanor for cases where the exposure was the result of reckless disregard or negligence (such as urinating in public but with no specific intent to cause alarm).

Figure 7-1 diagrams the essential elements for the crime of indecent exposure.

Challenges to indecent exposure statutes typically are grounded in the First Amendment's freedom of expression. Although individuals possess the freedom to express themselves, there are limits. Cases involving exotic or nude dancing illustrate the tension between the constitutional freedom of expression and indecent exposure. Dancing is considered a form of expression entitled to First Amendment protection, but many communities nonetheless choose to prohibit nude dancing. Thus far, appellate courts have upheld reasonable restrictions on nude dancing. In most cases, bans on totally nude dancing have been deemed constitutional. For example, in *Barnes v. Glen Theatre* (1991), the U.S. Supreme Court upheld a ban on nude dancing, concluding that the requirement that dancers wear G-strings and pasties was not unreasonable.

## Obscenity

The U.S. Supreme Court has held that obscene materials or expressions are not protected by the First Amendment. At common law, obscenity was prohibited. Federal and state laws, as well as local ordinances, also prohibit obscenity, including the transmission, mailing, or sale of obscene materials. The Model Penal Code in Section 251.4 addresses obscenity. Although there is general consensus that obscene materials should not enjoy constitutional protection, determining what materials are obscene is a more difficult task, as evidenced by the often-cited Justice Stewart "I know it when I see it" (*Jacobellis v. Ohio*, 1964).

In *Roth v. United States* (1957), the U.S. Supreme Court upheld a conviction for sending obscene materials through the mail, holding that such materials were not protected by the First Amendment. Moreover, the Court announced that alleged obscene materials were legally obscene if "to the average person applying contemporary community standards, the dominant theme of the material taken as a whole,

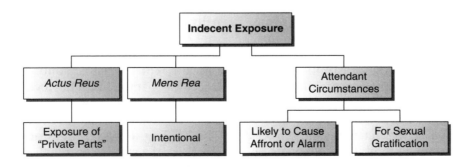

**Figure 7-1** Indecent exposure.

appeals to the prurient interest." **Prurient interest** (as defined by *Black's Law Dictionary*) is "a shameful or morbid interest in nudity, sex or excretion" (Garner, 2009). Despite the *Roth* decision, the Court's standard was a difficult one to apply. Moreover, there was continuing debate regarding whether the obscenity standard should be applied nationally or just locally. Clearly, the diverse nature of individual communities could lead to very different interpretations of obscenity. As such, obscenity cases continued to trigger litigation in the federal and state systems.

In *Miller v. California* (1973), the U.S. Supreme Court revisited the issues associated with obscenity. The Court acknowledged the difficulties inherent in the application of the *Roth* standard and seized the opportunity to develop new standards by which obscenity should be evaluated. These standards have since become known as the *Miller* test and use the following three prongs to determine whether materials are obscene: (1) whether the average person applying contemporary community standards would conclude that the dominant theme of the material as a whole appeals to prurient interest; (2) whether the work depicts or describes, in a patently offensive way, sexual conduct specifically defined by applicable state law; and (3) whether the work, taken as a whole, lacks serious literary, artistic, political, or scientific value.

Later, in *Pope v. Illinois* (1987), the U.S. Supreme Court held that although local or community standards could be used when applying the first two prongs of the *Miller* test, application of the third prong must be evaluated from the perspective of a reasonable person. Thus, lower courts must assess whether a reasonable person would conclude that the "work, taken as a whole, lacks serious literary, artistic, political or scientific value." Although the standards used to determine whether materials are obscene have been clearly set forth by the High Court, there remains great difficulty in the practical application of these guidelines.

Many obscenity statutes and ordinances have been successfully challenged on the ground that they are void for vagueness. In order to withstand constitutional challenges, statutes must be sufficiently specific to place individuals on notice as to what behavior, conduct, or expression is prohibited. If the statute fails to place a reasonable person on notice, then the statute may be void for vagueness, and thus violates the fundamental notions of due process under the Fifth and Fourteenth Amendments. The original purpose of such requirements was to limit the discretion of the police and prosecutors who may interpret vague statutes differently for different cases. As such, statutes should be specific so that little room is left for individual interpretation or guesswork.

Obscenity within federal law is addressed in Title 18, Chapter 71 of the United States Criminal Code. Several sections (§§1460-1470) define the nature and scope of obscenity protection at federal law. A few examples of such federal efforts include mailing indecent matter on wrappers or envelopes (§1463), broadcasting obscene language (§1464), and obscene visual representations of the sexual abuse of children (§1466A).

## Pornography

To conclude that materials are pornographic does not necessarily mean they are obscene. As such, pornographic materials that do not rise to the level of obscenity are protected by the First Amendment; however, child pornography is different. Congress and the U.S. Supreme Court have declared possession, transmission, or sale of child pornography as illegal. Federal law details well the litany of prohibitions related to the Sexual Exploitation and Other Abuse of Children in Title 18, Chapter 110 of the United States Criminal code (§§2251-2260). Most importantly, Congress passed the **Protection of Children from Child Exploitation Act of 1977**. This legislation specifically applied to parents or custodians of children and prohibited the use of children under the age of 16 in sexually explicit materials. In order to be prohibited under this Act, the materials must have been (or would be) transported in interstate commerce. Violation of this law was a felony punishable up to 10 years and/or a $10,000 fine. The act was later modified in 1984, in which the age of the child was increased to 18 years to provide greater protection. The act also eliminated the requirement that materials be transported in interstate commerce and that they be obscene. Furthermore, in *New York v. Ferber* (1982), the U.S. Supreme Court held that child pornography, regardless of whether it appealed to the prurient interest of the viewer, was not protected by the First Amendment. In the *Ferber* case, the Court acknowledged that the government has a compelling interest in protecting the welfare of children.

Pornographic materials come in all forms: books, magazines, movies, and electronic visual depictions. The pornography industry is a multibillion dollar international business, and as such, enforcement or regulation of its practices is quite difficult. Congress did, however, enact legislation to restrict transmission of pornography through the Internet. The Communications Decency Act of 1996, however, was later ruled unconstitutional on the grounds that it violated the First Amendment (*Reno v. American Civil Liberties Union*, 1997). In response, Congress enacted the Child Online Protection Act of 1998. Criminal penalties are available to punish such violations, but the real deterrent value of the act appears to be the fines that can be meted out—a whopping $50,000 per day. Despite these efforts to remedy the constitutional issues presented by earlier legislation, the U.S. Supreme Court has held that the Child Online Protection Act, too, would not survive a First Amendment challenge. Despite the continuing legal issues associated with finding the appropriate balance between protection of First Amendment guarantees and shielding children and the public from obscene materials, federal and state laws continue to prohibit the sale, transmission, and possession of obscene materials.

## Gambling

Characterization of gambling as immoral originated with natural law (the Bible, to be exact), but perceptions of gambling as a crime against the moral fabric of our nation are becoming increasingly uncommon. Rather, the legality of gambling appears simply to reside with whether a state wishes to permit and regulate such an industry. With that said, we nonetheless continue the tradition, although somewhat reluctantly, of placing its discussion squarely within the confines of a moral offense.

At common law, gambling was not a crime. In America, however, gambling has been historically prohibited among states and the federal government. As such, gambling was an industry ripe for organized crime. **Gambling** refers to an activity in which one risks something of value to win something of greater value. The essence of gambling then is the risk involved in the transaction: the gambler may win something of value or possibly lose everything. As such, gambling activities are referred to as games of chance. The terms *gaming* and *gambling* are used interchangeably. In most cases, they are used in a generic sense to describe the same risk-taking behavior. In some circles, gambling refers specifically to illegal participation in games of chance, with **gaming** then reflecting legal participation in said games of chance.

Numbers rackets and other forms of gambling were operated by various crime families and organizations in the early years. In 1931, Nevada became the first state to allow organized gaming. In more recent years, however, other states (except Utah and Hawaii) have followed suit in legalizing some form of gambling, and thus, its allure has become a viable source of revenue for states that have convinced their citizens to move past the moral implications of gambling. State lotteries represent one popular form of legalized gambling; other common forms include bingo, pari-mutuel betting, and jai alai. Once approved, however, gambling becomes highly regulated. For example, some states have approved casinos but then restrict their operations to certain locations (such as riverboats or otherwise adjacent to water). Additionally, Native Americans have embraced gambling as a viable industry, resulting in many reservations devising a significant source of revenue from casinos operated by the Indian Nation.

Although many forms of gambling have been legalized, federal and state statutes continue to prohibit others. Most gambling laws are fairly simple to construct and enforce, but Internet gambling, with its off-shore casinos outside the jurisdictional control of the federal government, has posed a significant problem for American regulation. Internet casinos emerged in the mid-1990s and over the course of a decade experienced unparalleled growth topping more than 2,000 worldwide. With more than half of its wagering business in the United States (approximately $15 billion annually), it simply was too much for the federal government to ignore. In general, federal law proscribes gambling violations in Title 18, Chapter 50 of the United States Criminal Code (§§1081-1084). At first, federal authorities relied on the **Interstate Wire Act of 1961** (§1084) to regulate online gambling as illegal, but the Fifth Circuit Court of Appeals (*In Re MasterCard International Inc.*, 313 F.3d 257, 5th Cir. 2002), held that the Wire Act was restricted to sports wagering, thereby excluding Internet-based casinos. Several strategies have since been employed to restrain the operation of internet gambling (referred to as i-gaming), but it remains true that placing a bet with an Internet casino is not illegal under federal law. As such, states have become more active in exercising their constitutional powers to regulate the industry. The State of Washington, for example, considers online wagering (in 2006) as a felony, which is punishable up to 5 years in prison and $10,000 in fines (Wash. Rev. Code §9.46.240, 2006).

## Alcohol and Drugs

Modern criminal offenses also address issues generated through the misuse of alcohol and drugs. The evolution of American crime policy on alcohol and drugs is an interesting one and serves as a useful backdrop for a discussion of modern offenses. During the 1960s, drug use and abuse reached epidemic proportions. The collateral social issues resulting from the influx of illegal drugs necessitated a national response to the drug problem. After a decade of discussion and debate regarding the use and abuse of narcotics, Congress passed the Drug Abuse Prevention and Controlled Substances Act of 1970. In 1972, Congress then passed the **Uniform Controlled Substances Act of 1970** wherein it sought to assist federal and state government agencies with the regulation and control of controlled substances. According to Congress, "A main objective of the Uniform Act is to create a coordinated and codified system of drug control, similar to that utilized at the federal level, which classifies all narcotics, marijuana, and dangerous drugs subject to control into five schedules, with each schedule having its own criteria for drug placement."

The five schedules created by Congress are as follows:

1. Opiates, high potential for abuse, no acceptable medical use or unsafe under medical supervision
2. High potential for abuse, acceptable for medical use with restrictions, may lead to psychological or physical dependence
3. Lower potential for abuse, but may lead to psychological or physical dependence
4. May lead to limited physical or psychological dependence
5. Least potential for physical or psychological dependence

Classification of drugs into any particular schedule under this act is determined by the application of eight criteria as follows:

1. Actual or relative potential for abuse
2. Scientific evidence of the drugs pharmacological effect, if known
3. State of current scientific knowledge regarding the substance
4. History and current pattern of abuse
5. Scope, duration, and significance of abuse
6. Risk to the public health
7. Potential of the substance to produce psychic or physiological dependence
8. Status of the substance as an immediate precursor of a substance already controlled.

The act also schedules and classifies controlled substances:

1. Uniform definitions for manufacturing, preparation, propagation, compounding, conversion, and processing
2. Criteria and qualifications for individuals and/or corporations seeking licenses to dispensing controlled substances
3. Open records and premises provisions for drug manufacturers

4. Criminal penalties for violations of the act
5. Provisions that facilitate the sharing of information among federal, state, and local law enforcement agencies

## Drug Offenses

The mere use of illegal substances is not illegal under federal and state laws, nor is being a drug addict. Rather, drug offenses, on federal and state levels, primarily exist in three different forms: possession, manufacture, and delivery or sale. Although other types of statutes exist, these classifications make up the general framework of statutory drug offenses.

The crime of **drug possession** occurs when an individual knowingly has dominion and/or control over illegal drugs. Whether an individual "possesses" a drug can be legally complex. Although most criminal statutes address actual possession of illegal drugs, many allow prosecution when individuals have constructive possession. The doctrines of actual and constructive possession are not only relevant to a discussion of drug offenses, but may apply in a variety of other situations where possession is an issue. For example, arguments regarding possession may occur in cases involving pornography, weapons, alcohol, stolen goods, and drug paraphernalia. Figure 7–2 diagrams the essential elements for the crime of drug possession.

As addressed in Chapter 4 (robbery), actual possession occurs when the prohibited item is on the person of the defendant. Conversely, constructive possession occurs when the prohibited items are in the immediate vicinity of the individual. In constructive possession cases, prosecutors must establish that items were accessible and subject to his or her control. Proximity also is an important issue. In other words, the farther away items are from the person, the less likely it is that he or she has accessibility, control, or dominion over them. Pause for Thought 7–2 examines a hypothetical scenario regarding the crime of drug possession.

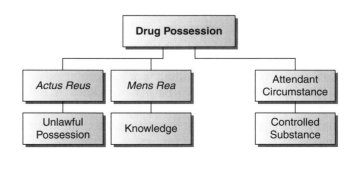

**Figure 7–2** Drug possession.

> **Pause for Thought 7–2**
>
> Consider the following: Danny Drug Dealer is walking down a street when stopped by undercover drug agents. In his pockets, the officers find two vials containing what appears to be crack cocaine. Danny is in actual possession of the drugs as they are on his person; however, assume for one moment that Danny Drug Dealer is sitting on a park bench. Undercover drug agents then approach Danny but find nothing on his person. There is, however, a brown bag about 2 feet away from the bench. In the brown bag are two vials containing what appears to be crack cocaine. Can Danny be prosecuted for possession?
>
> **Scenario Solution**
>
> Danny Drug Dealer may be charged with possession. The prosecutor would likely argue that Danny is in constructive possession of the vials. In order to secure a conviction, the prosecutor would have to convince the jury that the vials are within Danny's immediate span of control. A savvy defense attorney will, however, argue that Danny was not in control or possession of the vials and therefore not guilty of possession. The defense will argue that Danny was not sufficiently close to the vials to establish ownership or possession. The defense will argue that the vials could have belonged to anybody.

Possession statutes also require proof that the defendant knowingly possessed the illegal drug. Thus, the *mens rea* requirement is one of knowledge. In such cases, knowledge exists along two dimensions. First, the prosecution must establish the defendant knew he or she was in possession of the drug. Second, the prosecution must establish that the defendant knew the possessed drug was illegal.

Possession offenses may be a felony or a misdemeanor. The classification of the charge is typically determined by one of two criteria: (1) type of drug or (2) amount of drug. Thus, if the drug is of a certain type or amount, a possession charge may be a felony. The *mens rea* in possession cases may also transform a charge into a more serious matter. For example, **possession with intent to distribute** (deliver or sell) is usually classified as a felony by state and federal statutes. The backbone for these cases is the ability of a prosecutor to establish the intent of the defendant, that of specific proof the defendant intended to distribute illegal drugs. Intent to distribute is often proven by circumstantial evidence, which may include the amount of the illegal drug, packaging products at or near the drugs, mea-suring equipment, or customer lists. In possession with intent cases and where the prosecution uses the amount of drug to establish the intent to distribute, the amount should be one in excess of what could be consumed by the individual. For example, possession of 10 pounds of cocaine may arguably be more than one individual could consume.

Possession of precursors is also a criminal offense in many jurisdictions. **Precursors** are ingredients used to manufacture illegal drugs. In many states, mere possession of these ingredients, alone or in combination, is a crime. For example, precursors for the manufacture of methamphetamine commonly include over-the-counter medications, which contain ephedrine or pseudo ephedrine, hydrochloric acid, cleaning products, battery acid, and antifreeze.

## Drug Manufacture

The manufacture and/or production of illegal drugs are separate and distinct criminal offenses in most jurisdictions. Absent a license to produce certain controlled substances for use in medical or scientific fields, manufacture is illegal. Drug manufacturing has been a significant issue for many decades. Marijuana, cocaine, methamphetamine, and ecstasy are all examples of drugs illegally manufactured or produced in this country. Evidence of drug manufacturing may include (1) equipment used to distill, cook, or otherwise produce illegal drugs, (2) ingredients for the production of the substance, or (3) packaging materials. Drug agents are commonly called on to investigate and seize clandestine laboratories in which illegal drugs are produced. Given the nature of the ingredients utilized to produce many of these drugs, this can be a very dangerous task.

## Delivery or Sale of a Controlled Substance

Delivery or sale of a controlled substance is a serious felony in all jurisdictions. Often, delivery and sale of controlled substances are separate statutes. Delivery refers to the unlawful transfer of a controlled substance, whereas sale requires proof of the unlawful transfer as well as compensation for the product. In contrast to possession with the intent to deliver or sell cases, delivery or sale requires proof of the completed transaction or transfer. Despite the amount of substance delivered or sold, policy makers have specifically targeted drug dealers, even low-level street dealers, with severe penalties. This is a clear attempt to control the supply of illegal drugs. Figure 7–3 diagrams the essential elements for the crime of delivery or sale of a controlled substance.

## Alcohol Offenses

After years of advocacy directed at the evils of alcohol, Congress approved the Eighteenth Amendment (in 1917), officially ushering in Prohibition by restricting the production, sale, and transport of alcohol. After ratification of the Eighteenth Amendment, the **Volstead Act of 1919** was enacted as a necessary corollary to the Eighteenth Amendment, as it provided the means through which prohibitions against alcohol could be enforced. Despite valiant efforts, attempts to criminalize alcohol were largely unsuccessful. Alcohol use and production simply moved underground so to speak. In 1933, during the Great Depression, Congress approved the Twenty-First Amendment, which repealed Prohibition. Like illegal drugs, the impact of alcohol on society is significant. Although alcohol is not an illegal substance in and of itself, several criminal offenses exist to control its harmful impact on society.

## Public Intoxication

**Public intoxication** (or drunkenness) statutes reflect an attempt to prohibit the debauchery that may arise when consuming alcohol to excess in public places. Public intoxication requires proof of the following: (1) intoxication or drunkenness, (2) in a public place, and (3) resulting in the inability to care or control for oneself. Figure 7–4 diagrams the essential elements for the crime of public intoxication (or drunkenness).

## Driving Under the Influence

The impact of alcohol on traffic accidents and fatalities is an enduring issue. According to the U.S. Department of Transportation, nearly 13,000 people were killed by intoxicated drivers in 2007. In an effort to reduce the number of fatalities and injuries related to alcohol, every state has enacted a statute that criminalizes the operation of a motor vehicle while intoxicated. What constitutes operation of a motor vehicle, however, can be more complex than first appears. For example, defendants have been convicted of driving under the influence (DUI) in cases where they were intoxicated and sitting or sleeping in a running, parked vehicle. Technically, because the vehicle is running and therefore operational, the individual could be charged with DUI. Moreover, the term *vehicle* may include boats, all-terrain vehicles, motorcycles, or other forms of transportation.

DUI statutes are not necessarily limited to intoxication resulting from alcohol. Rather, it is common for statutes to apply to intoxication resulting from

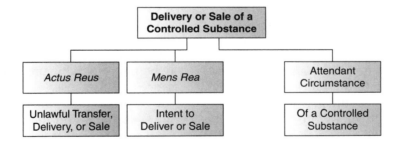

**Figure 7–3** Delivery or sale of a controlled substance.

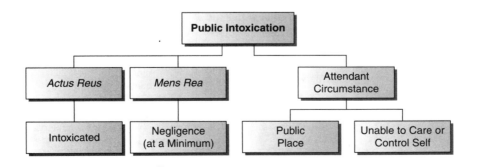

**Figure 7–4** Public intoxication.

drugs, too. Whether the drug is legal or illegal is irrelevant; the key is the intoxicating effect of the substance on the driver. Acceptable blood alcohol levels (BACs) are determined by each state for purposes of DUI. **Blood alcohol level** refers to the number of milligrams of alcohol per milliliter of blood. In most states the level is .08. States are fairly uniform because of conditions placed on receipt of federal highway funds. Drivers must register below a certain level or be charged with DUI. BAC levels may be lower for juveniles, individuals with previous DUI convictions, and commercial drivers.

A variety of methods are used to determine whether an individual is impaired. Drivers may refuse to participate in the tests, but there are consequences associated with such refusals (discussed more fully later). If stopped for a DUI, the driver will usually be required to participate in a **field sobriety test**—designed to detect impairment of drivers. These tests include walking heel to toe in a line, verbal exercises, and the ever-popular Horizontal Gaze Nystagmus test. Commonly known as the **nystagmus gaze**, this test monitors the automatic tracking mechanisms of the eyes in response to moving objects. In short, alcohol slows the eyes' ability to track objects rapidly, thereby causing one's eyes to oscillate (or jerk) long before they normally would in a sober person. The test proposes to gauge intoxication by measuring involuntary eye oscillations, but other valid reasons can also explain one's nystagmus.

Officers also are trained to observe the physical appearance of drivers, as well as the presence or smell of alcohol. One final test measures BAC level through the use of breath, blood, and urine samples. Alcohol content of the breath may be estimated with the use of a **breathalyzer**, a portable unit to obtain an initial BAC reading while the officer and driver are in the field. Upon arrival at the police station, the individual will be asked to blow into a stationary breathalyzer. Urine or blood tests may also be used when impairment is suspected. These tests measure the volume of alcohol in urine or the blood. Individuals are not required to submit to these tests, but refusal to submit can itself result in the loss of one's driver's license. This is possible because of the existence of **implied consent statutes**. In order to obtain a driver's license in most states, drivers give their consent to submit to field sobriety and breathalyzer tests. Failure to do so violates the implied consent law. Figure 7–5 diagrams the essential elements for the crime of driving under the influence.

## Hate Crimes

Congress passed the **Hate Crimes Statistics Act of 1990** (28 U.S.C. § 534), requiring the Attorney General to collect and publish data regarding the extent to which hate crimes occur in America. Based on the most recent 2007 Uniform Crime Report estimates (FBI, 2008), 7,722 incidents (comprising 9,080 offenses) were reported by participating agencies in 2006. The findings indicate that the majority (61%) of hate crimes were committed against the "person" of the victim, with intimidation (46%) and simple assault (32%) accounting for most harmful acts (three people were murdered). With regard to property crime, the overwhelming majority (81%) were acts of vandalism or destruction. More importantly, more than half of all hate crimes (in 2006) were motivated by race (52%), primarily against African Americans. To a lesser extent, though equally important, hate crimes were also frequently committed against Jewish persons (religion), homosexuals (sexual orientation), and Hispanics (ethnicity). Essentially, then, a hate crime refers to criminal conduct perpetrated against a member of a certain group and purely motivated by prejudice. Thus, bias is the essence of the offense.

In response to alarming increases in hate-based crimes, 45 states and the District of Columbia have passed some sort of legislation to deter crimes motivated by hate. Nationally, hate crime statutes vary

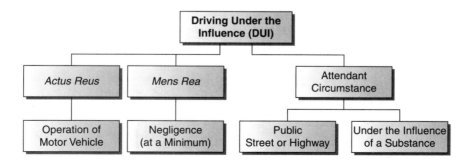

**Figure 7–5** Driving under the influence (DUI).

greatly regarding what groups are protected, but all states include, at a minimum, race, ethnicity, and religion. A majority of states have also included sexual orientation, gender, and disabled persons within their protective codes. On the federal level, Congress passed the **Violent Crime Control and Law Enforcement Act of 1994** (28 U.S.C. § 994), which mandated that penalties be enhanced for federal crimes committed on the basis of race, color, religion, national origin, ethnicity, gender, disability, and sexual orientation.

There are three hate crime categories. The first category prohibits the intimidation of a particular person or group. The second category includes statutes that criminalize offenses where the underlying intent is to harm a particular type of victim. Meanwhile, the third category includes enhancement statutes which affect the sentence a defendant will receive. Thus, if a defendant committed murder against a victim solely based on race, the defendant may receive a harsher penalty. Although intended to protect, not discriminate, hate crime statutes are not without their opponents. Some people contend that enhancing penalties for crimes committed against some and not others sends a message of inequality. What do you think?

Hate crime statutes have been challenged on several grounds. The First Amendment, with its guarantees of freedom of expression and speech, has provided the most significant challenge to hate crime laws. In 1992, in *R.A.V. v. City of St. Paul*, the U.S. Supreme Court struck down a city ordinance that prohibited burning crosses or placing Nazi swastikas or graffiti on private or public property for the purpose of arousing anger, fear, or alarm in others because of their race, color, creed, religion, or gender. The statute classified such conduct as a misdemeanor. After review, the Court held the statute violated the First Amendment in stating, "The First Amendment does not permit St. Paul to impose special prohibitions on those speakers who express views on disfavored subjects."

In another case, *Wisconsin v. Mitchell* (1993), the U.S. Supreme Court reviewed the constitutionality of a Wisconsin statute that enhanced penalties in cases involving crimes motivated by certain forms of bias. In this case, Mitchell, a black teenager, along with a group of others, chose the victim, a 14-year-old white male, because of his race. Upon conviction, the sentencing enhancement allowed the defendant to receive a term of 7, as opposed to 2, years for the crime of aggravated battery. Mitchell was sentenced to 4-years imprisonment. On appeal, Mitchell challenged the statute on First Amendment grounds. After review, the Court upheld the sentencing enhancement provision. The Court reasoned that a state has a compelling interest in preventing hate crimes where the victim is targeted due to race, gender, religion, color, disability, sexual orientation, national origin, or ancestry. Moreover, unlike in *R.A.V.*, this statute was directed at conduct, not speech, and as such did not suffer from the same deficiencies under the First Amendment.

## Summary

Criminal offenses described in this chapter evolved from American notions of morality. As a result, many of the offenses target conduct traditionally considered sinful in nature. Moral offenses range from adultery to gambling. Other offenses included in this chapter (such as alcohol-related offenses) were originally considered moral offenses but in many states may be viewed as regulatory in nature.

## Practice Test

1. _____ occurs when one enters into a second marriage while legally married.
    a. Fornication
    b. Prostitution
    c. Adultery
    d. Polygamy
    e. Bigamy

2. _____ (or plural marriage) occurs when a man takes multiple wives.
    a. Fornication
    b. Prostitution
    c. Adultery
    d. Polygamy
    e. Bigamy

3. _____ is consensual sexual intercourse between individuals not married.
    a. Fornication
    b. Prostitution
    c. Adultery
    d. Polygamy
    e. Bigamy

4. _____ is sexual intercourse with someone other than a spouse while married.
    a. Fornication
    b. Prostitution
    c. Adultery
    d. Polygamy
    e. Bigamy

5. Indecent exposure requires proof that an individual possessed a _____ desire.
    a. provocative
    b. licentious
    c. acrimonious
    d. credulous
    e. limonious

6. _____ is defined as a shameful or morbid interest in nudity, sex, or excretion.
   a. Prurient interest
   b. Perversion
   c. Sodomy
   d. Sadism
   e. Masochism

7. The _____ applied to parents or custodians and prohibited the use of children under the age of 16 years in sexually explicit materials.
   a. Child Pornography Act
   b. Exploitation of Children Statute
   c. Protection of Minors Act
   d. Guardian Guidelines
   e. Protection of Children from Child Exploitation Act

8. The U.S. Supreme Court held in _____ (1982) that child pornography is not protected by the First Amendment even when it does not appeal to prurient interest or is patently offensive to a viewer.
   a. *Williams v. Stevenson*
   b. *Kollath v. Pagan*
   c. *Jennings v. Massachusetts*
   d. *New York v. Ferber*
   e. *Arizona v. Johnson*

9. _____ refers to activity where one risks something of value to win something of greater value.
   a. Betting
   b. Gambling
   c. Game of chance
   d. Risk taking
   e. Chancing

10. In 1972, the _____ Act was passed, wherein Congress sought to assist federal and state governments with the regulation of controlled substances.
    a. Uniform Controlled Substances
    b. Drug Abuse Prevention
    c. Controlled Substances
    d. Federal Drug Regulation
    e. Congressional Substance

11. The _____ test monitors the automatic tracking mechanisms of the eyes in response to moving objects.
    a. Retinal motion
    b. Intoxilyzer
    c. Breathalyzer
    d. Optometric alert
    e. Horizontal Gaze Nystagmus

12. A(n) _____ marriage refers to a marriage not recognized as legitimate because the person is already married.
    a. constructive
    b. dissolved
    c. purported
    d. unsanctioned
    e. perpetual

13. _____ possession occurs when prohibited items are in the immediate vicinity.
    a. Constructive
    b. First-degree
    c. Actual
    d. Superior right of
    e. Drug

14. _____ are ingredients used to manufacture illegal drugs.
    a. Primaries
    b. Nuggets
    c. Precursors
    d. Luminaries
    e. Contaminants

15. The _____ Act provided the means through which prohibitions against alcohol could be enforced.
    a. Homestead
    b. Volstead
    c. Prohibition
    d. Wilkinson
    e. Alcohol

16. The _____ Amendment officially ushered in Prohibition by prohibiting the production, sale, and transport of alcohol.
    a. 8th
    b. 11th
    c. 16th
    d. 18th
    e. 21st

17. _____ statutes reflect an attempt to prohibit the debauchery that may arise when consuming alcohol to excess in public places.
    a. Public consumption
    b. Excess consumption
    c. Alcohol prohibition
    d. Excess alcohol
    e. Public intoxication

18. _____ refers to the number of milligrams of alcohol per milliliter of blood.
    a. ACL
    b. AAA
    c. BAC
    d. MCL
    e. ABC

19. The _____ Act was originally designed to prohibit the interstate transportation of women and young girls for the purpose of prostitution or other immoral behavior.
    a. Immoral Transportation Prohibition
    b. Unlawful Transportation of Children
    c. Protection Against Sexual Exploitation
    d. Mann
    e. Anti-Johns Commerce

20. The _____ Act requires the Attorney General to collect and publish data regarding the extent to which crimes are grounded in prejudicial attitudes.
    a. Violent Crime Control and Law Enforcement
    b. Statistics on Racial Equality
    c. Interstate Wire
    d. Hate Crimes Statistics
    e. Uniform Controlled Substances

## References

Federal Bureau of Investigation. (2008). *Crime in the United States, 2007: Uniform Crime Reports.* Retrieved July 14, 2009, from http://www.fbi.gov/ucr/cius2007/index.html

Garner, B. A. (Ed.). (2009). *Black's law dictionary* (9th ed.). Eagan, MN: West Group.

# Crimes Against the Administration of Justice and Public Order

**CHAPTER 8**

## Key Terms

Abatement
Blackmail
Breach of the peace
Bribery
Civil contempt
Commercial bribery
Compounding
Consideration
Constructive contempt
Courtroom decorum
Criminal contempt
Direct contempt
Disorderly conduct
Embracery
Escape
Federal Anti-Riot Act of 1968
Fighting words
Incitement of a riot
Loitering
Malicious mischief
Materiality
Nuisance
Obstruction of justice
Perjury
Prison break
Reasonable-person standard
Rescue
Resisting arrest
Riot
Rout
Sports bribery
Subornation of perjury
Trespass
Tumultuous
Two-witness rule
Uniform Vehicle Code
Unlawful assembly
Unlawful fleeing
Vagrancy
Vandalism
Victim and Witness Protection Act of 1982
Witness tampering

## Introduction

The integrity of the judicial process is of paramount importance to the public, professionals working within the system, and individuals subject to its jurisdiction. Certain offenses have been created to protect the integrity of the judicial process and instill confidence in its operation. These offenses also existed at common law but have been modified to meet the needs of modern society. Although all legal systems have flaws, it is imperative that governments do their best to ensure the system works without bias, prejudice, or unlawful interference. This chapter examines the various offenses that strive to protect the sanctity of the legal process. The chapter also examines certain offenses designed to protect public order and safety.

## Crimes Affecting the Integrity of the Judicial Process

The following section provides readers with an overview of crimes for acts that compromise the integrity of the judicial process. Crimes such as obstruction of justice, resisting arrest, and unlawful flight undermine the efficient and effective operation of the legal system. These acts are prohibited by federal and state authorities to ensure a fair judicial system that will sustain the public trust.

### Obstruction of Justice

At common law, **obstruction of justice** encompassed many forms of behavior that impeded or hindered the administration of justice, including witness and jury tampering, failing to disclose evidence, suppressing or destroying evidence, intentionally making a false statement during an investigation, interfering with the duties of a law enforcement officer, resisting arrest, or otherwise interfering with the lawful functions of the judicial process. Thus, this misdemeanor offense at common law was overarching in prohibiting many different acts.

Over time, obstruction of justice statutes have changed in two significant ways. First, modern statutes tend to isolate acts originally regulated at common law, preferring now to separate them as distinct offenses (to be addressed in forthcoming sections). Modern obstruction statutes tend to address cooperation and assistance with law enforcement, compliance with court orders, and the destruction of evidence. Moreover, most jurisdictions (including the federal government) have statutes that impose a legal duty on individuals to assist law enforcement officers when called on to do so. The second change pertains to the classification of the offense. Federal statutes (and many states) now classify obstruction of justice as a felony offense. This classification change has significantly increased the potential penalties for an offense that aims to deter those who contemplate hindering investigations or prosecutions.

Federal law classifies obstruction of justice as a felony in Title 18, Chapter 73 of the U.S. Criminal Code (§§1501-1520). Ranging from assaults on process servers to destruction of corporate audit records, these federal statutes address behaviors that threaten the integrity of the criminal justice process. Punishable by up to 5 years in prison, obstruction of criminal investigations (§1510a) criminalizes one who "willfully endeavors by means of bribery to obstruct, delay, or prevent the communication of information relating to a violation of any criminal statute of the United States by any person to a criminal investigator."

## Resisting Arrest

Federal law relies on two criminal chapters to prosecute those who resist arrest. First, obstruction of justice often is used for dealing with such actions. One specific example within the obstruction of justice chapter is that of resisting an extradition agent (§1502). Additionally, resisting can also be prosecuted under the prohibition found in Title 18, Chapter 7 (§111), which prohibits assaulting, resisting, or impeding officers.

The crime of **resisting arrest** occurs when one subject to lawful arrest attempts to thwart or avoid being taken into custody by law enforcement officers. Resistance most often takes the form of physical efforts (such as fighting) or leaving the scene. The prosecution must establish, however, that the individual knew the officers were attempting to make an arrest. A defense to a charge of resisting arrest could arise if the individual believes he or she is being kidnapped and therefore unaware of the true identity of the law enforcement officer. As such, officers are trained to inform individuals of their status before making an arrest. Furthermore, the Model Penal Code (and some state statutes) does not allow suspects to use any degree of force or resistance when officers are attempting to make an arrest, regardless of the legality of the arrest. The time to challenge such legality occurs later during pretrial proceedings. This approach seeks to ensure that the safety of officers and suspects is not compromised during an arrest.

Some states have also enacted **unlawful fleeing** statutes to deal with individuals who flee law enforcement by vehicle. Many such statutes were enacted after tragic accidents occurred during car chases between law enforcement and suspects. Under these statutes, law enforcement is not required to be in the process of making an arrest. Rather, the act of fleeing from a traffic stop or other intervention by law enforcement is sufficient.

## Perjury and Subornation of Perjury

Ensuring that witnesses present truthful testimony during judicial proceedings is essential to maintaining the integrity of the justice system. As such, the offense of perjury was created to deter untruthful testimony. Common law defined **perjury** as a false yet material statement offered during a judicial proceeding and made under oath or affirmation and without belief in the truth of the statement. Perjury existed as a separate misdemeanor offense at common law, but most modern perjury statutes now classify the offense as a felony. The essence of perjury, however, has not changed significantly since common law.

At common law, the prosecution was required to present two witnesses to establish the crime of perjury. Known as the **two-witness rule**, most states today have abandoned this approach in favor of permitting just one witness in perjury trials. In such cases, however, the prosecution must have additional proof (such as documents or previous statements) to establish the statement was false in order to corroborate the witness' account. Perjury also requires a prosecutor to establish that the declarant knew the statement was false. Thus, the jury must assess whether the defendant intentionally made a false statement. It is not sufficient that the witness was merely mistaken about the facts; rather, the declarant must have knowingly and intentionally given false information. In order to determine the intent of the defendant, the **reasonable-person standard** is commonly used. Referred to as the objective standard, it requires jurors to put themselves in the place of a hypothetical reasonable person. In doing so, the jury must assess whether a reasonable person would have known his or her statement was false. In contrast, a subjective approach would require the jury to place themselves in the position of the defendant and determine whether the defendant knew the statement to be false.

Another important issue in perjury prosecutions is **materiality**, meaning the prosecution must establish that the facts contained in the false statement were important and relevant to the proceeding, and would have affected the outcome of the proceedings. Statutory requirements may vary with regard to the nature of the proceeding during which the statement must be made. At common law, perjury occurred when a false statement was made during a judicial proceeding. Individuals who made false statements in situations that required an oath but not occurring within judicial proceedings were not considered perjury. The deception could, however, be prosecuted under the common law offense of false swearing. Modern statutes (including the federal government and Model Penal Code) have addressed this issue by simply expanding perjury statutes. Under modern perjury statutes, false statements knowingly made under oath and that occur in conjunction with administrative proceedings, depositions, or applications for government benefits or licenses may also constitute perjury.

Federal law classifies perjury as a felony in Title 18, Chapter 79 of the United States Criminal Code (§§1621-1623). In short, the laws aim to protect the judicial process from deceitful communication. Perjury generally (§1621) can be punished up to 5 years in prison and is defined as "having taken an oath before a competent tribunal, officer, or person, in any case in which a law of the United States authorizes an oath to be administered, that he will testify, declare, depose, or certify truly, or that any written testimony, declaration, deposition, or certificate by him subscribed, is true, willfully and contrary to such oath states or subscribes any material matter which he does not believe to be true."

**Subornation of perjury** is a related yet separate offense referring to the willful and corrupt procurement of false testimony from another. In such cases, the prosecution must be able to prove both the defendant and testifying individual knew the testimony would be false. Keep in mind, however, that a conviction for subornation of perjury requires the requested person actually provide the false testimony. Like perjury, subornation of perjury was a misdemeanor at common law that is now classified as a felony under most modern statutes and federal law (§1622). Pause for Thought 8–1 illustrates the crime called subornation of perjury.

## Embracery and Witness Tampering

Typically referred to as jury tampering, most jurisdictions now choose to include the common law act of embracery within obstruction of justice statutes, including the federal statute discussed above (Title 18, Chapter 73, §§1503-1504). **Embracery** refers to the unlawful attempt to influence a jury or juror. The right to a trial by jury is one of the most important features of the American justice system. This right was so central to the notions of due process that the drafters of the U.S. Constitution included the right to trial by jury in the Sixth Amendment. Given the important functions of juries, laws and procedures are designed to ensure the sanctity of the jury composition and its deliberations. Jury deliberations often involve stressful situations in which jurors attempt to persuade the votes of others. Embracery is

---

### Pause for Thought 8–1

Consider the following: Danny Drug Dealer requests that his friend Joe provide him with an alibi in an upcoming court proceeding. During the trial, Joe gives false testimony that Danny was in another state during the time the crime occurred. What crime (if any) has Danny Drug Dealer committed?

### Scenario Solution

Given that Joe actually provided false testimony on behalf of Danny Drug Dealer, Danny is clearly guilty of the crime subornation of perjury. If Joe had refused the invite, the actual subornation charge would be without legal merit.

not designed to chill the ability of a jury to deliberate, but rather seeks to protect jurors from unlawful influences. Although most such influences were external, fellow jurors could be prosecuted under embracery codes if they attempted to illegally influence a fellow juror. Pause for Thought 8–2 illustrates how the crime of embracery is legally interpreted.

Crimes that prohibit witness tampering also reflect the attempt to ensure the sanctity of the criminal justice process and protect it from unlawful influences. **Witness tampering** refers to the unlawful attempt by another to influence, delay, or prevent witness testimony or the production of evidence. Witness tampering can occur in many different ways; bribery, harassment, intimidation, threats, and coercion of witnesses are examples of means through which witness tampering may occur. Many of these could also be charged as other offenses.

Witness tampering provisions (as with jury tampering) also are typically included in obstruction of justice statutes. Federal (Title 18, Chapter 73, §1512) and state statutes both address witness tampering. The **Victim and Witness Protection Act of 1982** is an example of a federal statute that seeks to protect witnesses throughout the course of judicial proceedings. This act seeks to minimize the opportunities for witness tampering as well as ensure the safety of victims and witnesses.

## Contempt of Court

The availability of contempt of court as a crime reflects an attempt to allow judges to control decorum in the courtroom and ensure compliance with court orders, decrees, and judgments. **Courtroom decorum** refers to the orderly and professional demeanor and atmo-sphere required in courtrooms. Judges, attorneys, jurors, court staff, witnesses, parties, and spectators are expected to behave in a courteous manner and demonstrate respect for the court.

Contempt of court may be civil or criminal in nature. The distinction between the two lies with the underlying purpose for the charge. **Civil contempt** of court is purely coercive in nature, reflecting an attempt by the court to obtain compliance with a judicial order, decree, or judgment. Once compliance occurs, the defendant is absolved of contempt. For example, a witness is ordered to provide documents that will be used as evidence in a criminal trial. The witness refuses to produce the documents and is found to be in civil contempt of court. The court orders the witness to remain incarcerated until the documents are produced. After the documents are submitted to the court, the witness may be released. Thus, through the use of its civil contempt powers, the court was able to successfully enforce its order.

**Criminal contempt** is different in that its underlying purpose is to punish the offender. Unlike civil contempt, the punishment does not end merely because the offender complies with the directive of the court. Punishments for criminal contempt typically include fines or imprisonment. If the potential punishment for criminal contempt is 6 months or more, however, the defendant is entitled to due process including a jury trial (*Bloom vs. Illinois*, 1968; *Baldwin v. New York*, 1970). Federal law regulates criminal contempt through Title 18, Chapter 21 (§§401-403). Pause for Thought 8–3 illustrates the crime of criminal contempt.

Contempt can be direct or constructive. **Direct contempt** refers to behavior that occurs in the presence of the court, as with the example described previously. **Constructive contempt** is indirect and refers to behavior that does not occur in the presence of the court, but by its nature is disrespectful or negatively impacts court proceedings. For example, constructive contempt may occur if an attorney is displeased with a verdict and when leaving the courthouse shouts profanities to describe the proceedings,

### Pause for Thought 8–2

Consider the following: A defendant is on trial for capital murder. The case has been submitted to a jury, which has been sequestered for 5 days. Primarily arguing over the credibility of eyewitness testimony, jury deliberations have been ongoing for 8 hours. An initial vote (11 to 1) indicates the jury favors conviction, and two such jurors attempt to persuade the one juror favoring acquittal. A heated argument ensues, resulting in the holdout juror conceding to vote with the others. Are the two jurors guilty of embracery?

### Scenario Solution

A charge of embracery would not withstand legal challenge. If the jurors had used threats, coercion, or bribery to persuade the juror to change the vote, they would be guilty of embracery. In this case, however, the juror's change of vote did not result from such persuasive means.

> **Pause for Thought 8–3**
>
> Consider the following: An attorney is frustrated with court proceedings. Several motions made by the attorney have been overruled by the court. The attorney begins to use profane language in open court. He has been warned repeatedly by the judge regarding his contemptuous conduct but continues to use profanity. Can the attorney be charged with a crime?
>
> **Scenario Solution**
>
> Because of the nature of his behavior, the attorney is guilty of criminal contempt. Whether the attorney apologizes for the behavior is irrelevant.

the judge, and the jury. The judge and jury are unaware of this event until the evening news airs. Unfortunately, a local news station filmed the tirade.

## Misprision of Felony and Compounding Crime

Misprision of felony occurs when a person fails to report and conceals a felony committed by another. Federal and state statutes both include the felonious crime of misprision. The underlying purpose of such offenses is to encourage reporting and to deter concealment of crimes. In order to avoid a charge of misprision of felony, one with knowledge of a crime must notify authorities. Misprision of felony is prohibited at federal law in Title 18, Chapter 1 (§4) as one who has "knowledge of the actual commission of a felony cognizable by a court of the United States, conceals and does not as soon as possible make known the same to some judge or other person in civil or military authority under the United States, shall be fined under this title or imprisoned not more than three years, or both."

One notable example of this crime occurred in the State of Mississippi. Dickie Scruggs was a billionaire lawyer widely known for his successful advocacy against tobacco companies, asbestos manufacturers, and insurance companies. His son, Zach Scruggs, was charged with misprision of felony in federal court for his failure to report his father and other lawyers for attempting to bribe a circuit court judge. Although Zach was not involved in the attempted bribery, he possessed knowledge about the activities of others—which alone was sufficient.

**Compounding** occurs when an individual accepts money, property, or something of value in exchange for failing to report a crime. Recall that violations of criminal laws are considered offenses against the peace and dignity of the state or federal government. As such, crime is a transgression that causes social, as well as individual harm. After a crime or attempt occurs, discretion regarding whether to prosecute remains with the government, not private citizens. In most, if not all, jurisdictions, citizens have a legal duty to report criminal offenses. The failure to do so may result in criminal charges, even against victims. Failure to report crime is thought to be in direct conflict with the orderly investigation and prosecution of crime and the administration of justice. Acceptance of rewards (financial or otherwise) for failing to report the crime is thought to increase or compound the harm caused by the initial wrongful act.

## Escape

The criminal offense of escape exists to maintain the authority of law enforcement and corrections officials over the care, custody, and control of those charged with or convicted of law violations. A separate criminal offense usually classified as a felony, **escape** occurs when a lawfully detained or incarcerated individual leaves or fails to return without official permission. At common law, escape was one of three offenses that could be invoked in situations in which individuals unlawfully leave lawful custody or detention. **Prison break** was a common law offense that occurred when an individual used force to leave lawful custody or detention. Finally, individuals who assisted prisoners with escape or prison break could be charged with the crime of **rescue**.

Courts have addressed the issue of defenses to the crime of escape. For example, in *United States v. Bailey* (1980), the U.S. Supreme Court addressed necessity as a defense to escape. In this case, four defendants were charged with violation of the federal escape statute. After review, the Court held that the defense of necessity required a defendant to establish the following: (1) the prisoner faced imminent threat of harm, (2) escape was the only reasonable alternative, and (3) the prisoner made a *bona fide* attempt to return to custody after the threat of harm subsided. The defense of necessity often has been raised in cases in which prisoners faced the threat of physical and sexual attacks.

Federal law proscribes escape and rescue in Title 18, Chapter 35 of the United States Criminal Code (§§751-758). Ranging from immigration checkpoints to prisoners of war, escape and rescue legislation seeks to deter behaviors that compromise custodial necessities. Punishable up to 5 years in prison, federal law regulates as criminal an officer permitting escape (§755) when while "in his custody any prisoner by virtue of process issued under the laws of the United States by any court, judge, or magistrate judge, voluntarily suffers such prisoner to escape . . . or if he negligently suffers such person to escape."

## Corruption of the Judicial Process by Public Officials

The following section provides readers with an examination of crimes wherein public officials engage in acts of corruption. Bribery is the central focus, as it is the most common form of corruption by public officials. However, opportunities for bribery exist in many settings, and as such other forms of bribery also are discussed.

## Bribery

At common law, **bribery** was a misdemeanor offense defined as the agreement to do or refrain from doing an act required of a public official in exchange for money or property. Bribery also includes the individual who offers the money or property to the public official. The essence of common law bribery was the agreement between the parties, rather than the actual transfer of money or property that constituted the offense. After the agreement was complete, each party could be charged with bribery. If the offer was made but no agreement was secured, then the individual could not be charged with bribery, but rather only attempted bribery. Also, the prosecution need not prove that the public official actually carried out the terms of the agreement to prosecute successfully for bribery. Again, after the agreement is made, the crime of bribery is complete regardless of whether the objectives are achieved.

Modern bribery statutes are similar to those developed at common law. The essence of the offense continues to be the agreement between parties. Thus, the *actus reus* is the agreement and the *mens rea* is false or corrupt intent of the actors. Many states now require proof of the actual exchange of money or property, meaning the crime is not complete until the money or property is transferred to the public official. It is also common for modern bribery statutes to more broadly define the type of consideration for the agreement. **Consideration**, as defined within *Black's Law Dictionary*, refers to the requirement that something of value be exchanged or proposed for exchange (Garner, 2009). Although common law specified the exchange of money or property, modern statutes contemplate the exchange of various types of consideration from which the public official may derive benefit. For example, cases involving sexual favors, use of vacation homes, contractual services, and other forms of value have been successfully prosecuted.

Another modification of common law bribery is the expansion of the actors that may be bribed. At common law, the offense was limited to the influence of public officials or government actors, whereas modern statutes tend to have additional provisions that address the illegal influence of sports officials, athletes, jurors, witnesses, public servants, party officers, and others. Party officers are individuals who hold office in political parties. This expansion has resulted in additional categories of bribery in some jurisdictions. Thus, the inclusion of actors other than public officials may be addressed by the creation of a separate statute. For example, many states have enacted **commercial bribery** statutes, which typically address the illegal influence of those engaged in business transactions and who are requested to violate their duty to clients, partners, or employees in exchange for money or other forms of value. **Sports bribery** statutes are also increasingly common. These statutes address the illegal influence of sports officials and athletes. Examples of sports bribery may include the attempt to influence athletes to lose a game voluntarily by shaving points or for sports officials to make inappropriate calls. Federal law addresses bribery in Title 18, Chapter 11 of the United States Criminal Code (§§201-226).

## Extortion and Blackmail

At common law, extortion prohibited demands for things of value (primarily money) by public officials in exchange for the performance of official duties. The crime of extortion (also discussed in Chapter 4) was developed to prohibit public officials from collecting fees for services already provided by the government. At common law, extortion occurred when public officials, acting under color of office, collected a fee when none was due or in excess of that lawfully required to be paid or collected prior to the same being due.

The offense of extortion under modern statutes has changed significantly and now more closely resembles theft. Although common law restricted extortion to the unlawful collection of fees by public officials or servants, modern extortion statutes are much broader. Modern extortion is now generally defined as the taking of property illegally from another by (1) threat of violence, (2) exposure of secrets or damaging information, (3) coercion, or (4) in exchange for the taking or withholding of official action. The linchpin of extortion is the threat of future harm in order to obtain property of another; this is also the primary distinction between robbery and extortion. Robbery requires a threat to do immediate harm, whereas extortion requires the threat of future harm. Moreover, in extortion cases, the prosecution must be able to establish that the victim parted with his or her property because of the threats of the extortionist.

Threats to do harm to the victim may include economic harm, exposure of secrets or damaging information, or refusal to perform official actions. The Model Penal Code (Section 223.4) specifically identifies the types of threats that constitute the basis of extortion, and includes the following threats: to inflict bodily injury or commit a criminal act; to accuse another of a criminal offense; to expose secrets which may subject the victim to hatred, contempt, or ridicule, or impair credit or business; to take or withhold official action; to cause a strike or boycott; to testify or refuse to provide testimony or information in a legal matter; or to inflict any other harm that would not benefit the actor.

Federal law proscribes extortionist conduct as criminal in Title 18, Chapter 41 of the U.S. Criminal Code (§§871-880). Ranging from threats against the President to receiving proceeds from extortionist practices, the federal statute titled Extortion and Threats seeks to eliminate threats of harm used as tools to minimize free will. Most would likely agree that blackmail is the most known form of extortion. **Blackmail** constitutes the threat to expose secrets or other damaging information and is complete after the threat is made. The completed offense is not contingent on receipt of money or property. Blackmail is also addressed by the Model Penal Code as a form of theft. Punishable up to 1 year in prison, federal law defines the crime of blackmail (§873) within its extortion code as "whoever, under a threat of informing, or as a consideration for not informing, against any violation of any law of the United States, demands or receives any money or other valuable thing."

## Ethical Violations

In every state, ethical codes govern the conduct of public officials and servants. These standards were developed to supplement the criminal offenses that address official misconduct. Certain conduct may violate both the criminal law and ethical standards; however, in other cases, conduct may not rise to the level of criminal offense. Moreover, if the conduct constitutes both a crime and an ethical violation, the official may be sanctioned for each. Penalties for the violation of ethical standards may include the removal from office, monetary fines, public or private reprimands, or the forfeiture of professional licenses.

## Crimes Against Public Order and Safety

Public order offenses reflect attempts to control behavior which creates a public disturbance, such as fighting, excessive noise, use of profane language or gestures, interruption of travel on public roadways, and interruption of lawful assemblies. Generally speaking, public order offenses attempt to deter behaviors that interrupt the general peace and dignity of a civilized society; however, statutes and ordinances that define public order offenses must be extremely careful not to interfere unreasonably with the right to free expression as set forth in the First Amendment.

## Unlawful Assembly, Rout, and Riot

The First Amendment of the U.S. Constitution guarantees individuals the right to "peaceably assemble." The majority of state and local governments have enacted statutes and ordinances designed to proscribe assembly that is not peaceful in nature. Generally, **unlawful assembly** provisions restrict the ability of several individuals (usually more than three) to gather to either (1) commit an unlawful act or (2) commit a lawful act in an unlawful manner. A common element in unlawful assembly, rout, and riot statutes is the requirement that the conduct be **tumultuous**, meaning behavior that poses a significant risk of personal injury or damage to property.

**Rout** is a common law offense that refers to an intermediate stage of conduct occurring between unlawful assembly and a riot. For example, if a group gathers and decides to engage in a riot but is thwarted by the presence of law enforcement, the individuals in the group have engaged in rout. The group gathered for an unlawful purpose and possessed the specific intent to riot. It is necessary to establish the specific intent to riot in rout cases; however, rout does not usually stand alone as a distinct crime. Rather, rout is typically addressed by unlawful assembly statutes.

**Riot** refers to the unlawful gathering of a group of individuals with the intent to create a public disturbance, which poses a significant risk of personal injury or property damage. Congress enacted the **Federal Anti-Riot Act of 1968** criminalizing riots involving interstate travel or communication. Found in Title 18, Chapter 102 of the United States Criminal Code (§§2101-2102), the act prohibits the **incitement of a riot**, referring to individuals who "organize, promote, encourage, participate in, or carry on a riot." Congress limited the application of the statute, however, so as not to offend the provisions of the First Amendment. Specifically, Congress included statutory language that expressly states that "the advocacy of ideas or expression of belief without the advocacy of violence" does not constitute inciting a riot.

## Fighting

Many jurisdictions have statutes or municipal ordinances that specifically address fighting. These provisions make it illegal for individuals to engage in physical combat with each other; however, it is just as common for jurisdictions to prosecute such behavior with the use of disorderly conduct statutes. At common law, public fights were prosecuted under affray statutes, which is a misdemeanor offense defined as a physical altercation between two or more individuals in a public place. To be charged with an affray, the parties must be mutually at fault. If one or more parties physically attacked another without consent, the conduct would be chargeable as an assault or battery.

## Disturbing the Peace and Disorderly Conduct

Disturbing the peace statutes evolved from the common law offense known as **breach of the peace**, a misdemeanor that constituted a disturbance of the peace and tranquility of society. Examples of conduct that amounted to a breach of the peace included excessive noise, fighting, offensive or profane gestures or language, or engaging in dangerous activity that posed a risk to the public. These behaviors are also addressed by disturbing the peace statutes, which, like their common law counterparts, are also classified as misdemeanors. Disorderly conduct, unlike disturbing the peace, did not have a common law counterpart. **Disorderly conduct** statutes also prohibit conduct that constitutes a public disturbance or is otherwise threatening or menacing. In a majority of American jurisdictions, however, this misdemeanor offense often is indistinguishable from disturbing the peace.

Disturbing the peace and disorderly conduct statutes have been the subject of numerous legal challenges, which typically allege violations of the First Amendment guarantee of free speech. In *Chaplinsky v. New Hampshire* (1942), the U.S. Supreme Court attempted to identify certain conduct that could be criminalized under breach of the peace or disorderly conduct statutes. There, the Court stated that speech that was lewd and obscene, profane, libelous, insulting, or fighting words could be a basis on which to charge an individual with a breach of the peace. **Fighting words** are those that inflict injury, create a breach of the peace, and are not a central party of the exposition of ideas. The primary characteristic of such forms of speech is that "by their very utterance inflict or tend to incite an immediate breach of the peace"; however, the Court has clearly indicated that speech that is merely unpopular, annoying, or against the views of the majority is not an appropriate basis on which to charge one with breach of the peace or disorderly conduct. Thirty years later, in *Colten v. Commonwealth* (1972), the Supreme Court reiterated its holding that the mere expression of unpopular or annoying ideas is not, in and of itself, sufficient to uphold a conviction for breach of the peace or disorderly conduct. The Court held that breach of peace or disorderly conduct come into play "only when the individual's interest in expression, judged in the light of all relevant factors, is 'miniscule' compared to a particular public interest in preventing that expression or conduct at that time and place." Thus, state and local ordinances that define disorderly conduct and breach of the peace must always be mindful of these constitutional limitations.

## Nuisance

Nuisances may be public or private. **Nuisance** provisions generally address conditions such as excessive noise, offensive conditions, and interference with the lawful use of property that result in annoying or harmful effects. If an individual is causing a nuisance, most statutes or ordinances use the process of abatement. **Abatement** refers to an order to cease or eliminate the condition or behavior that results in the nuisance. If individuals fail to comply with an order to abate, he or she may be charged with creating a nuisance. Clearly, the conditions commonly addressed by nuisance provisions may overlap with those of other offenses discussed in this section.

## Trespass

The offense of trespass was fully discussed in Chapter 4; however, a brief discussion here also is warranted. **Trespass** occurs when there is an unlawful interference with the person or property of another. Trespass may constitute a civil or a criminal offense with attendant remedies for each. Civil trespass constitutes a tort (civil wrong) in most jurisdictions. Liability for civil trespass may result in the assessment of monetary damages or equitable remedies such an injunctions or restraining orders. Civil or criminal, trespass reflects an attempt to protect the personal and property rights of individuals from unlawful interference by others.

## Vandalism and Malicious Mischief

**Vandalism** refers to the willful or negligent physical damage to the property of another. In order to secure a conviction, the prosecution must establish that the offender knew or should have known that his or her actions would cause damage to the property. Statutes and ordinances vary with regard to the extent of damage required; however, most statutes contemplate that the damage should be of such an amount as to require repair or replacement.

In most jurisdictions, there is little distinction between vandalism and malicious mischief. **Malicious mischief**, like vandalism, was a misdemeanor at common law that prohibited the willful damage or destruction of the property of another; however, to be chargeable as malicious mischief, the offender must have acted willfully and intentionally, whereas vandalism may occur with as little as gross negligence. Federal law addresses malicious mischief in Title 18, Chapter 65 of the United States Criminal Code (§§1361-1369). Malicious mischief statutes protect, among other things, against tampering with consumer products, destruction of veterans' memorials and energy facilities, and government property in general. One specific example pertains to those who harm animals (dogs and horses) used in law enforcement. The statute (§1368a) criminalizes an act that "willfully and maliciously harms any police animal, or attempts or conspires to do so." Under this statute, an offender can be punished by up to 1 year in prison; however, the penalty can be enhanced to up to 10 years in prison if the offense causes permanent harm to the animal (disability, disfigurement, serious injury, or death).

## Vagrancy and Loitering

**Vagrancy** statutes or ordinances address a variety of conduct. Generally, these statutes are directed at idleness whereby individuals wander or loiter about with no visible means of support. Vagrancy was a crime at common law and is most often a misdemeanor in modern jurisdictions which have retained such laws. Modern provisions are usually modified to address disorderly conduct, begging, and loitering. Historically, vagrancy laws were enacted to control the behavior of able-bodied individuals who were not deemed productive and wandered about without an identifiable or lawful purpose. The underlying purpose of common law vagrancy provisions was to deter individuals from unproductive lifestyles and encourage a strong work ethic. Although early vagrancy laws were directed at idleness, policy makers also viewed them as crime-prevention tools.

Regardless of their underlying purpose, vagrancy provisions were intentionally broad and granted law enforcement significant discretion with regard to their application. For these reasons, their statutes have been subject to numerous legal challenges. Most critics contend that vagrancy laws are so vague and overbroad that a reasonable person of ordinary intelligence cannot discern what is unlawful; others, meanwhile, challenge their selective and arbitrary use by police officers.

Generally speaking, appellate courts have not been favorable toward vagrancy laws. In *Papachristou et al. v. City of Jacksonville* (1972), the U.S. Supreme Court addressed cases involving convictions under a Jacksonville, Florida, vagrancy ordinance as well as a Florida state statute. After review, the ordinance was deemed void for vagueness and the convictions of those prosecuted for violations of the state law were overturned. The ordinance and the statute in question contained language similar to that in most vagrancy provisions. More recently, the Supreme Court in *Chicago v. Morales* (1999) addressed the constitutionality of an Illinois loitering statute. Similar to vagrancy, **loitering** refers to wandering about with no apparent lawful purpose. Loitering statutes too have been plagued with legal challenges similar to those raised against vagrancy provisions. In *Morales*, the Court reviewed the constitutionality of a Gang Congregation Ordinance enacted (by the Chicago City Council) to assist law enforcement with ongoing problems associated with criminal gang activity and intimidation by gang members. The ordinance allowed law enforcement officers to order individuals to disperse and leave the area when those persons are perceived to be criminal street gang members loitering in a public place. Individuals who refused to leave would be charged with a violation of the ordinance.

On appeal, the U.S. Supreme Court concluded that the ordinance was unconstitutional under the Fourteenth Amendment for three reasons. First, the ordinance could reasonably apply to lawful conduct as well as unlawful conduct. Second, the ordinance failed to provide ordinary citizens with adequate notice of what behavior is forbidden. In short, the Court concluded that the ordinance did not adequately define loitering. Finally, the Court held that the ordinance did not sufficiently establish guidelines to govern the use of discretion by law enforcement. The Court noted that the ordinance placed complete discretion in law enforcement to determine who was loitering, whether the individuals were or were not in the company of a criminal street gang member, and whether the individuals had an apparent purpose to be where they were.

## Traffic Offenses

Traffic offenses exist in every jurisdiction because they are necessary to maintain the safety and integrity of public roadways. Traffic offenses include many different violations, which range from speeding or failure to yield or stop to careless or reckless driving and driving under the influence (DUI). Traffic offenses have become fairly standard throughout the country largely due to the development of the **Uniform Vehicle Code**, which contains the "Rules of the Road." The majority of American jurisdictions proscribe traffic offenses as criminal, not civil. As such, violators may be arrested; however, given the frequency of traffic violations and their usual misdemeanor status, law enforcement officers prefer to use citations in lieu of formal arrest. Modern DUI statutes tend to be a misdemeanor, but often increase to a felony for subsequent DUI offenses. Finally, traffic offenses are strict liability offenses in that they require no particular proof of intent. As such, an individual may be found guilty of a traffic violation whether or not they intended to commit the offense. Pause for Thought 8–4 illustrates the strict liability component for traffic offenses.

## Summary

One of the primary goals of the American judicial system is to provide a fair and just forum to determine the guilt or innocence of individuals charged with crimes. As such, the integrity of this process is critical for a myriad of reasons. Offenses described in this section were created to protect the justice system from illegal influences, such as jury or witness tampering, perjury, and bribery. Shielding the justice system from such influences is one example of the efforts to maintain public order and to sustain public confidence in the system. Public order offenses were also addressed in this chapter. These offenses represent an effort to prevent certain behaviors that pose a danger to public order and public safety.

### Pause for Thought 8–4

Consider the following: Vanessa is on her way home from college. She is surprised to see blue lights in her rearview mirror. Once stopped, the officer advises that she has been pulled over for speeding. Her speedometer is broken, and she therefore was unable to determine her actual speed. She believed that if she kept up with the flow of traffic, she would be close to the posted speed limit. She explains the situation to the police officer but is unsuccessful. The officer is not sympathetic and hands her a traffic citation. Can she be charged with speeding?

### Scenario Solution

Vanessa can be charged with speeding. Her intent (or reasonable belief) is irrelevant, as traffic offenses are strict liability offenses. The officer need only observe her operating a motor vehicle at a speed that violates the posted speed limit. Traffic offenses have no *mens rea* requirement.

# Practice Test

1. _____ is a false statement during a judicial proceeding made without belief in the truth of the statement.
   a. Abatement
   b. Obstruction of justice
   c. Subornation of perjury
   d. Perjury
   e. Compounding

2. _____ is commonly used to determine the intent of a defendant.
   a. Materiality
   b. Reasonable person standard
   c. Probable cause
   d. Two-witness rule
   e. Tumult

3. Embracery refers to the process of _____.
   a. resisting arrest
   b. criminal contempt
   c. witness tampering
   d. malicious mischief
   e. jury tampering

4. _____ is the unlawful attempt to influence, delay, or prevent testimony or production of evidence in court.
   a. Witness tampering
   b. Misprision of felony
   c. Nuisance
   d. Embracery
   e. Obstruction of justice

5. _____ refers to the orderly and professional demeanor and atmosphere required in courts of law.
   a. Rout
   b. *En banc*
   c. Decorum
   d. Structuring
   e. Abatement

6. _____ is an agreement to do or refrain from doing acts in exchange for money or property.
   a. Blackmail
   b. Bribery
   c. Extortion
   d. Consideration
   e. Vandalism

7. _____ prohibits the demand for money or other things of value by public officials for performing official duties.
   a. Blackmail
   b. Bribery
   c. Extortion
   d. Consideration
   e. Tumultuous

8. A threat to expose secrets or other damaging information in exchange for something of value is the crime of _____.
   a. blackmail
   b. compounding
   c. sabotage
   d. subornation of perjury
   e. extortion

9. _____ is the unlawful gathering of a group of persons with the intent to create a public disturbance that poses a significant risk of personal injury or property damage.
   a. Malicious mischief
   b. Nuisance
   c. Unlawful assembly
   d. Riot
   e. Rout

10. _____ was defined at common law as a consensual physical altercation between two or more parties in a public place.
    a. Incitement of a riot
    b. Disorderly conduct
    c. Assault
    d. Battery
    e. Affray

11. _____ prohibits conduct which constitutes a public disturbance or is otherwise threatening or menacing.
    a. Nuisance
    b. Vandalism
    c. Disorderly conduct
    d. Loitering
    e. Vagrancy

12. _____ addresses conditions such as excessive noise, offensive conditions, or interference with lawful use of property resulting in annoying or harmful effects.
    a. Nuisance
    b. Vandalism
    c. Disorderly conduct
    d. Loitering
    e. Vagrancy

13. _____ is unlawfully interfering with the person or property of another.
    a. Affray
    b. Vandalism
    c. Malicious mischief
    d. Contempt
    e. Trespass

**14.** A civil wrong is referred to as a _____.
   a. misdemeanor
   b. tort
   c. felony
   d. crime
   e. violation

**15.** _____ is the willful and intentional damage to the property of another.
   a. Recklessness
   b. Culpable negligence
   c. Malicious mischief
   d. Embracery
   e. Vandalism

**16.** _____ refers to the willful or negligent physical damage to another's property.
   a. Recklessness
   b. Culpable negligence
   c. Malicious mischief
   d. Embracery
   e. Vandalism

**17.** Generally speaking, vagrancy statutes are directed at _____.
   a. Boredom
   b. Idleness
   c. Mischief
   d. Assembling
   e. Congregating

**18.** _____ refers to wandering about with no apparent lawful purpose.
   a. Vagrancy
   b. Assembling
   c. Tumultuous
   d. Loitering
   e. Affray

**19.** _____ orders one to cease a condition or behavior resulting in nuisance.
   a. Abatement
   b. Criminal contempt
   c. Civil contempt
   d. Misprision of felony
   e. Compounding

**20.** Traffic offenses are fairly uniform throughout the country in large part because of the development of the _____, which contains the "Rules of the Road."
   a. National Road Ordinance
   b. Standardized Vehicle Law
   c. Uniform Vehicle Code
   d. Uniform Motor Code
   e. Uniform Automobile Code

## References

Garner, B. A. (Ed.). (2009). *Black's law dictionary* (9th ed.). Eagan, MN: West Group.

# Inchoate Offenses and Party Liability

## CHAPTER 9

## Key Terms

Accessory after the fact
Accessory before the fact
Accomplice
Attempt
Bilateral theory
Conspiracy
Conspirators
Dangerous proximity test
Equivocality test
Factual impossibility
Hearsay
Inchoate
Indispensable element test
Intermediary
Involuntary renunciation
Last act test
Legal impossibility
Overt act
Physical proximity test
Principal at the fact
Probable desistance test
Renunciation
Rule of Consistency
Solicitation
Substantial step test
Unilateral theory
Wharton's Rule

## Introduction

**Inchoate**, as defined within *Black's Law Dictionary*, is "an incipient crime which generally leads to another crime" (Garner, 2009), or what is often referred to as an incomplete crime. The offenses of solicitation, conspiracy, and attempt are considered "inchoate" given that these offenses involve acts committed in furtherance of a substantive offense. Each of these offenses is separate and distinct and therefore not dependent on the completion of other crimes. For example, one may be charged with conspiracy to commit murder despite the fact that the killing did not occur.

## Inchoate Offenses at Common Law

At common law, the offenses attempt, conspiracy, and solicitation were misdemeanors developed to penalize acts in furtherance of other crimes. The underlying logic for these offenses reflects a strong public policy to deter individuals from engaging in acts of preparation for more serious crimes. By penalizing preparatory acts, it is more likely that intended crimes will be prevented. In contrast to their status at common law, the majority of modern American jurisdictions now classify these offenses as felonies punishable by fines or substantial terms of incarceration. Figure 9–1 outlines the three major forms of inchoate offenses.

## Attempt

Attempt statutes are varied among the states while federal law does not specifically define attempt. In general, the crime of **attempt** requires the prosecution to establish the intent of the actor and the nature of steps taken in furtherance of the intended crime. Thus, the *actus reus* of attempt is an act in furtherance of the intended crime. The *mens rea* of attempt is proof that the actor specifically intended to commit the substantive crime. Given the failure to consummate the commission of the resulting crime, prosecution of attempt can be complex. The

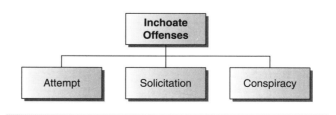

**Figure 9–1** Inchoate offenses.

elements required for the crime of attempt are outlined in Figure 9–2.

## What Constitutes an Act?

The offense of attempt requires proof of an act beyond mere preparation. Essentially, there is a spectrum of activity (ranging from mere preparation to the actual completion of a crime) that must be distinguished to properly understand the act requirement in attempt cases. Although mere preparation is insufficient for an attempt charge, proof of the last proximate act is not required. Thus, attempt lies beyond acts of mere preparation but prior to the last proximate act (or crime itself). Numerous tests are used by courts to determine whether acts in a given case constitute mere preparation or something more. The two most common tests adopted by American courts are the physical proximity test and substantial step test. All available tests are discussed, however, and are outlined in Figure 9–3.

### Proximity

Two tests focus on how close an offender is to completing a crime. The tests—physical proximity and dangerous proximity—share qualities, but differ as a matter of degree. Let us first examine the physical proximity test. Often referred to as the traditional test, the **physical proximity test** focuses on what is left for the defendant to do in order to commit the intended crime. Thus, how close was the defendant to accomplishing the crime? This test requires actions of the defendant be evaluated in light of the likelihood of committing the intended crime. Meanwhile, the **dangerous proximity test** (endorsed in several states) instead examines whether the defendant was dangerously close to committing the intended crime through an analysis of several factors. Factors may include whether the defendant had approached the victim, whether all instrumentalities had been obtained, or whether the defendant had arrived at the crime scene. Pause for Thought 9–1 illustrates how to interpret these proximity tests.

### Substantial Step

The **substantial step test**, endorsed by the Model Penal Code, seeks to determine whether the defendant has taken a substantial step toward the commission of the intended crime. The focus of the substantial step test is the nature of what actions the defendant has taken in furtherance of the intended crime rather than what remains to be done to carry out the intent of the defendant. In order to assist courts with the application of the substantial step test, drafters of the Model Penal Code specifically identify facts and circumstances that may be considered a substantial step toward commission of an offense.

### Last Act

The **last act test** is one of the earliest tests identified to evaluate whether actions of the defendant constitutes attempt; however, the test is no longer used because of the strict requirement that the defendant must have engaged in the last proximate act necessary to commit the intended crime. This approach minimizes the distinction between attempt and the completed crime.

## Intent

Attempt is a specific intent crime. The crime of attempt requires proof that the defendant specifically intended to commit a crime. Thus, if the defendant is charged with attempted murder, the prosecution must establish that the defendant specifically intended to kill another human being. Proof of general intent to engage in some crime is insufficient in attempt cases. Two major tests are used to determine whether the requisite intent was present for the crime of attempt. First, the **probable desistance test** specifically evaluates whether it is likely—in light of the facts and circumstances—that the defendant will desist from commission of the criminal act. Scholars and legal experts have criticized this test and discount its utility with violent crimes. Second, the **equivocality test** seeks to determine whether the actions of the defendant are indicative of his or her intent to commit the crime.

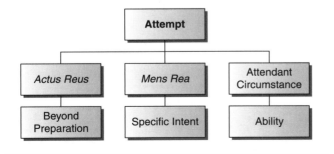

Figure 9–2  Attempt.

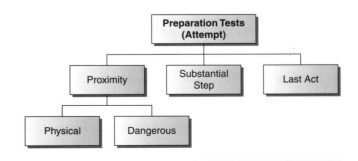

Figure 9–3  Preparation tests (attempt).

> **Pause for Thought 9–1**
>
> Consider the following: Larry and Moe decide to rob a bank to obtain money for a new car. They arrive at the bank's parking lot wearing masks and hiding guns under their jackets. Upon arrival, they walk to the front door and draw their weapons as they enter the bank but have not uttered any words to announce their intent. Just as Larry is about to shout "get your hands up, this is a robbery," the fire alarm goes off and all the tellers run out a back door. Unbeknownst to Larry and Moe, the bank was conducting a fire drill—their efforts are foiled, and they return home. Are Larry and Moe guilty of attempted robbery?
>
> ### Scenario Solution
>
> If the physical proximity test is used, the focus is on what remained to be done to rob the bank. Larry and Moe had not yet demanded money but were in the building with the requisite intent. Many jurisdictions would regard them guilty of attempted robbery. Were they dangerously close to committing the crime, however? Using the dangerous proximity test, does your conclusion change? Because they were on the premises, with weapons, and possessed the intent to rob, your outcome likely does not change. Under this test, one may reasonably conclude Larry and Moe were dangerously close to robbing the bank.

## Indispensable Element Test

The **indispensable element test** focuses on the ability of the defendant to carry out the criminal act. The central question within this test is whether the defendant has control or possession of the instrumentality to carry out the intended crime? If the defendant has acquired the instrumentality, or means to carry out the crime, then he or she can be found to have satisfied the act requirement of attempt.

## Defenses to Attempt

The following section provides readers with an examination of defenses to the crime of attempt. This examination primarily includes the defenses of impossibility and renunciation. The distinction between factual and legal impossibility, as well as between voluntary and involuntary renunciation, also are explored.

## Impossibility

The ability of the defendant to accomplish the intended crime is relevant to determine whether the defense of impossibility exists. The defense of impossibility may be raised by a defendant where commission of the crime is legally impossible; however, factual impossibility is not currently a defense to attempt nor was factual impossibility a defense at common law.

**Factual impossibility** refers to an inability to commit the intended crime because of the existence of some fact unknown to a defendant or beyond the control of the defendant. Thus, if the defendant makes a mistake about some fact, yet still possesses the specific intent to commit the crime and commits acts beyond mere preparation, he or she is guilty of attempt. Pause for Thought 9–2 illustrates the application for the factual impossibility test.

**Legal impossibility** exists when the defendant is mistaken about the law or is charged with the wrong crime. Thus, despite the actions of the defendant coupled with the specific intent to commit the crime, the acts do not constitute a crime. Pause for Thought 9–3 applies the legal impossibility test to a hypothetical scenario.

### Renunciation

**Renunciation** (or abandonment) may also be a defense to attempt. At common law, renunciation was not a defense to attempt. Thus, the crime of attempt was complete once the defendant formed the intent to commit a crime and engaged in acts in furtherance of that crime; however, the Model Penal Code and other jurisdictions do allow for the defense of renunciation when the defendant voluntarily abandoned the intent to commit the crime. The Model Penal Code requires proof of a "complete and voluntary renunciation of the criminal purpose." Involuntary renunciation, however, is not a defense to attempt.

**Involuntary renunciation** occurs when the defendant abandons the intent to commit a crime due to intervening causes. Thus, a defendant in such a case does not voluntarily renounce because of a change of heart, but rather is prevented from carrying out the

> **Pause for Thought 9–2**
>
> Consider the following: Johnny is a shy kid who desperately wants to be perceived as cool by his older brother's friends. The older guys are always in trouble and tend to live on the wild side. In an effort to impress them, Johnny decides to tell them he has stolen several video games from a local electronics store. Johnny tells his brother Joe that he has stolen 10 games but will need assistance selling them on the street. Joe agrees to get his friends to help in exchange for a share of the money. Joe and his friends arrive at the place where Johnny says the games are located—a nice hiding spot in the woods behind the electronics store. They begin to retrieve the games, but are stopped by an officer. In fact, the games are not stolen, but were given to Johnny by his friend, Frank. Are Joe and his friends guilty of attempted receipt of stolen property?
>
> **Scenario Solution**
>
> Joe and his friends may be charged with attempted receipt of stolen property. Despite the fact that the property was not actually stolen, Joe and his friends believed it was. This case illustrates factual impossibility. Although the defendants were mistaken about the stolen character of the games, they intended to obtain stolen goods and committed acts beyond mere preparation.

> **Pause for Thought 9–3**
>
> Consider the following: Jerry decides to go hunting on Saturday, September 15. Hunting season officially begins on October 1. After several hours in the woods, Jerry notices a huge deer. He then shoots the deer. After seeing the deer move after being shot, Jerry approaches the animal to assess things. Once there, Jerry is furious. His friends have placed a fake deer in his line of sight as a joke. Jerry has not killed a real animal. Is Jerry guilty of attempting to shoot a deer out of season?
>
> **Scenario Solution**
>
> Even though Jerry may be guilty of some other hunting violation, he is not guilty of hunting out of season. Even if Jerry completed all the acts necessary to violate the hunting law, it is legally impossible to commit this crime because the deer was not real; as such, it could never constitute the elements of the crime.

intent. Examples of intervening causes include the following:

- Arrival of law enforcement or other witnesses which thwart the defendant's actions
- Defendant is unable to complete the act due to technical problems
- Defendant elects to commit the crime at a later date

## Solicitation

The offense of **solicitation** occurs when an individual commands, encourages, or requests another to commit a crime. Solicitation was a crime at common law and may occur when an individual hires, counsels, or otherwise encourages another to commit a crime. Solicitation, like attempt, requires proof of specific intent that the crime be carried out.

Whether solicited acts actually occur is immaterial to a charge of solicitation. Additionally, there is no requirement that the person being solicited accept or engage in an act in furtherance of the solicitation. The request itself is sufficient to establish solicitation. In most jurisdictions, the crime that is solicited must be a felony for the solicitation to be punishable; however, if the crime actually occurs, the solicitor may be charged as a principal or as an accessory before the fact. Additionally, after another agrees to commit the crime and completes an overt act, the parties may be charged with conspiracy. Solicitation, also known as incitement, is distinguishable from attempt in that solicitation involves acts often thought of as merely preparatory. At common law, as well as most modern jurisdictions, the solicitor may be charged as a principal or accessory before the fact if the intended crime actually occurs. The legal elements required for the crime of solicitation appear in Figure 9–4.

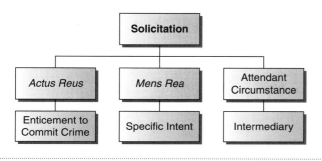

**Figure 9–4** Solicitation.

## Intermediaries

In many solicitation cases, the defendant engages the services of an **intermediary**, meaning the defendant does not directly solicit, instead using a third party to solicit another to commit a crime. Does the presence of an intermediary prevent the original actor from being successfully prosecuted for solicitation? It depends. The Model Penal Code does not require direct communication between the original solicitor and the individual encouraged or requested to commit the crime. Rather, it is acceptable for the prosecution to merely establish that the conduct of the solicitor was "designed to effect such communication" (MPC 5.02[2]). In some jurisdictions that do not use the Model Penal Code, however, direct communication between the solicitor and the individual who will commit the crime often required.

## Defenses to Solicitation

Renunciation, as described earlier, was not a defense to solicitation at common law. The Model Penal Code (5.02[3]) does recognize renunciation, however, as a defense to solicitation. In order to raise the renunciation defense under the Model Penal Code, the defendant must establish a complete and voluntary abandonment of criminal intent. The solicitor must further persuade the solicited party not to commit the offense or otherwise prevent him from committing the crime. Pause for Thought 9–4 illustrates how the defense of renunciation can be legally applied.

## Conspiracy

The crime of conspiracy originally developed because of the belief that the likelihood of success and risk of harm increases when two individuals agree to commit a crime. Referred to as "concert in criminal purpose," **conspiracy** usually occurs in secret. Individuals meet and organize in secret, which enables the parties to plan their crime.

At common law, conspiracy was a misdemeanor. Common law conspiracy prohibited agreements to commit acts that were injurious to public health, welfare, or morals. In contrast, modern conspiracy statutes require an agreement to commit a crime. Conspiracy can be classified as a felony or misdemeanor, depending on the nature of the target crime. The required elements for the crime of conspiracy can be examined in Figure 9–5.

### Agreement Between Parties

The offense of conspiracy seeks to address the concerted efforts of individuals to commit crimes. Common law conspiracy required proof of a true

---

### Pause for Thought 9–4

Consider the following: Darla Doolittle decides that it is finally time to "off" her husband. Darla frequently complains to friends that her hubby refuses to share money and will not take her out to eat regularly. Fed up with the state of the marriage, Darla decides to enlist her brother, Danny Doolittle, to assist her with the crime. Danny is responsible for luring Mack, Darla's husband, to an abandoned house, where Danny will then shoot Mack. After making arrangements with Danny, Darla changes her mind. She contacts Danny but to her surprise, he is unwilling to change the plan. He indicates that he hates Mack and wants to kill him. On the agreed-on date, Danny and Mack arrive at the abandoned house; however, undercover officers appear from the woods just as Danny reaches for his gun. Unbeknownst to Danny, Darla contacted law enforcement, explained what had occurred, and asked for their help to protect Mack. Is Darla guilty of solicitation to commit murder?

### Scenario Solution

Darla may raise the defense of renunciation if the jurisdiction uses the Model Penal Code. Darla clearly renunciated her criminal intent and took actions to prevent the killing of her husband by contacting law enforcement.

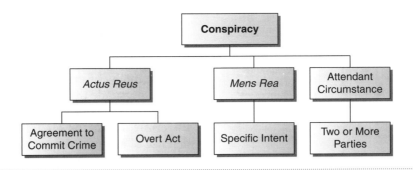

**Figure 9-5** Conspiracy.

agreement between the parties. The linchpin of conspiracy is the agreement between the parties. In the majority of jurisdictions, there must be at least two parties to the agreement. This is referred to as the **bilateral theory** of conspiracy. Although there may be more than two parties, there must be at least two. These parties are known as **conspirators**.

The Model Penal Code endorses a unique view of conspiracy. Sections 503-504 follow the **unilateral theory** of conspiracy, which allows conviction for conspiracy even if only one party believes an agreement exists and specifically intends to commit a crime; therefore, in cases in which an individual reaches an agreement with an undercover law enforcement officer or informant, the unilateral theory allows the individual to be charged with conspiracy.

There are significant differences among states regarding the proof necessary to establish an agreement under modern conspiracy statutes. In most states, there is no need to establish a formal agreement between the parties. Rather, proof of the agreement can be inferred from the circumstances. This brings us to the difference between express and implied agreements. Express proof of an agreement may include a formal written or verbal agreement between the parties. Very few parties to a criminal conspiracy place their intentions in a written contract or other such document. Thus, the law allows proof of an agreement between the parties to be implied from the circumstances. This allows prosecutors to establish an agreement by circumstantial evidence. Whether the intended crime actually occurs is irrelevant in conspiracy cases. Conspiracy is a separate and distinct crime. Thus, a conviction does not depend on the successful commission of the target crime.

## Specific Intent

As with attempt and solicitation, the crime of conspiracy requires proof that the accused specifically intended the target crime to occur. Unlike general intent, specific intent cannot be inferred from the circumstances. For example, in a conspiracy case in which the target crime is first-degree murder, the prosecution must prove that each of the co-conspirators possessed the specific intent to kill.

What if one of the parties is insane or mentally defective? Can that person knowingly enter into an agreement to commit a crime? It depends. If the individual is so mentally defective that he or she is unable to form the specific intent to commit the crime, he or she could not be convicted of conspiracy; however, if the agreement occurred in a state that follows the unilateral theory of conspiracy, then the other party could be convicted of conspiracy despite the mental illness of the co-conspirator. Moreover, what about those individuals who know a crime will be committed but do not participate? Is simple knowledge of a crime sufficient to establish conspiracy? Generally, mere knowledge, without proof that the party has specific intent that the crime will occur and will participate in some manner is insufficient; however, there is case law to suggest that mere knowledge may be sufficient in certain circumstances. For example, mere knowledge may be sufficient where the actor is a seller of dangerous goods or knows that the goods will be used in a serious crime and encourages the commission of the crime (or continues his or her involvement beyond the transaction). Pause for Thought 9-5 applies the specific intent principle to a hypothetical example.

## Crimes or Lawful Objectives by Unlawful Means

In the majority of jurisdictions, conspiracy requires proof that the parties specifically intended to commit a crime; however, others expand the definition of conspiracy to include individuals who agree to accomplish a lawful objective by unlawful means. This was also allowed at common law. With the latter, there is specific intent to engage in illegal activity even though the ultimate goal is not a crime, such as if two individuals agree to fraudulently obtain names of individuals from whom they may solicit charitable donations. It is not a crime to solicit charitable dona-

> **Pause for Thought 9–5**
>
> Consider the following: An individual visits a pawn shop to purchase a firearm. During the transaction, the individual tells the pawn shop owner that he intends to shoot and kill his wife when he gets home. The owner continues with the sale, does nothing, and later reads about the murder in the newspaper. Is the owner guilty of conspiracy to commit murder?
>
> **Scenario Solution**
>
> The pawnshop owner may be charged with conspiracy to commit murder. He is a seller of dangerous goods and knows that the goods may be used in a serious crime. Although the owner did not encourage the crime, he did nothing to prevent it either and therefore may be charged with conspiracy in some jurisdictions.

tions (assuming the charity exists and the individuals are actually authorized to solicit for the charity), however, if these individuals engaged in fraud to obtain a list of potential donors, they may be guilty of conspiracy to commit fraud.

## Overt Act

In order to be found guilty of conspiracy, most jurisdictions (including the federal government) require proof that parties engaged in an *overt act* in furtherance of the conspiracy. As with attempt, an **overt act** is something beyond mere preparation. In jurisdictions that do not require an overt act as a conspiracy element, the agreement is the *actus reus* of the offense.

## Special Considerations

Generally, the commission of the target offense and conspiracy are separate and distinct offenses. For example, a group of individuals may be charged with both (1) conspiracy to commit murder and (2) murder. The conspiracy charge addresses the agreement between the parties coupled with the specific intent and an overt act, whereas the murder charge addresses the actual killing. Thus, conspiracy does not merge with the target offense on completion. This is referred to as the merger doctrine.

An exception to the merger doctrine is **Wharton's Rule**, which applies in cases in which the target crime of the conspirators requires at least two or more individuals. Wharton's Rule has traditionally been applied to offenses such as abortion, adultery, bribery, incest, and dueling. Let us consider adultery. Adultery remains a criminal offense in many jurisdictions, but by its very definition requires the participation of two individuals. In order to prosecute individuals for adultery, a third party must be involved in the agreement to commit the crime. Stated differently, Wharton's Rule holds that "an agreement to commit a crime cannot be prosecuted as conspiracy when the crime itself is of such a nature as to necessarily require the participation of two persons for its commission" (*United States vs. Figueredo*, 1972). Thus, where the crime itself requires the concerted action of two individuals, those parties cannot be charged with conspiracy because conspiracy is an inherent element of the target offense. Wharton's Rule does not apply in cases where extra individuals are involved (e.g., five individuals agree to engage in adulterous acts).

The Rule of Consistency is another legal doctrine that applies in conspiracy cases. The underlying logic of this rule is related to the bilateral theory of conspiracy. The **Rule of Consistency** provides where all alleged conspirators but one are acquitted of conspiracy, the remaining alleged conspirator may not be convicted. If co-conspirators are tried separately, however, most such jurisdictions would not apply the rule of consistency. If there are separate trials for the alleged co-conspirators, however, other factors unrelated to the existence of a conspiracy may have resulted in the acquittal. For example, factors such as quality of the evidence or presentation of the prosecutors, demeanor of witnesses, jury composition, and other factors may have contributed to a not guilty verdict. Because the Model Penal Code follows the unilateral theory of conspiracy, the rule of consistency would not apply; therefore, one party could be convicted of conspiracy despite the acquittal of other co-conspirators.

A criminal conspiracy is not complete until all objectives are accomplished. Thus, criminal liability for all co-conspirators continues; however, what are the limits to co-conspirator liability? In general, co-conspirators are criminally liable for all acts that are (1) in furtherance of the conspiracy and (2) reasonably foreseeable as a consequence of the conspiracy. This two-pronged approach is the prevailing view in the majority of American jurisdictions. Pause for

Thought 9–6 illustrates the proper legal interpretation for the culpability of those who conspire to commit crimes.

In contrast, the minority view suggests that when parties agree there will be no violence but where violence nonetheless occurs, the co-conspirator should not be guilty of the violent acts because of the absence of intent.

## Defenses to Conspiracy

In the majority of American jurisdictions, abandonment or renunciation is not a defense to conspiracy. Thus, after the agreement occurs and an overt act is committed, the crime of conspiracy is complete. In a very few jurisdictions and under the Model Penal Code, one may use withdrawal from a conspiracy as a defense in limited cases. In these jurisdictions, there must be a complete and voluntary withdrawal from the conspiracy.

## Evidentiary Considerations

**Hearsay** is testimony from one person about what another person said or did. Essentially, then, the witness providing information really has no first-hand knowledge to assist the court's search for truth. As such, even cross-examination would prove fruitless given that any in-depth examination of the person's knowledge would be minimal. The introduction of hearsay evidence has been met with great resistance in American courts because there simply is very little opportunity to ascertain the truth of those statements. For example, suppose Suzy testifies at trial that Johnny told Bob the traffic light was red. The opposing attorney strongly objects to Suzy's testimony on the grounds that Johnny's statements to Bob are hearsay and therefore inadmissible.

Unless a hearsay exception exists to allow admission of such statements, the trial judge would sustain an objection, thereby refusing to admit the testimony. There are many instances, however, where statements that otherwise would be considered hearsay are permissible in court proceedings. Hearsay exceptions generally reflect a willingness among courts to allow such testimonial evidence for public policy reasons. In other words, the importance of the evidence to a legal proceeding outweighs its potential risk or prejudice that may result from its admission.

Other statements also are "exempt" from the hearsay rule despite the fact that they walk and talk like hearsay statements. Statements made by co-conspirators, for example, qualify for exempt status (or designation as non-hearsay). For evidence to be exempt, three criteria must be satisfied. First, the statement(s) must be made by a co-conspirator of a party opponent. Second, the statement(s) must be made during the course of the conspiracy. Third, the statement(s) must have been made in furtherance of the conspiracy. Allowing such statements as evidence in criminal trials considerably enhances the ability of the prosecution to establish the existence of a criminal conspiracy and the identity of the participants. Additionally, with this exemption, the prosecution is able to introduce evidence of secret communications among co-conspirators.

## Conspiracy and Solicitation and Federal Law

Federal law addresses both conspiracy and solicitation in Title 18, Chapter 19 of the United States Criminal Code (§§371-373). The chapter specifically highlights conspiracies to commit offenses or to

---

### Pause for Thought 9–6

Consider the following: Suppose you are in a jurisdiction that follows the majority rule. A group of friends agrees to use a gun to hold up a convenience store; however, it is agreed that no shots will be fired. Upon entering the convenience store, one of the group members demands money from the clerk. The clerk pulls a gun and fires a shot. Out of fear, one group member shoots the clerk. Are the remaining co-conspirators guilty of felony murder?

### Scenario Solution

Yes. Because the jurisdiction follows the majority rule, the co-conspirators are criminally liable for all acts in furtherance of the conspiracy that were reasonably foreseeable as a consequence of the conspiracy. Here, the prosecutor would argue the agreement to engage in armed robbery is a dangerous offense. The use of a weapon increases the likelihood of self-defense by the victim. Moreover, the potential that someone may be hurt or killed during the armed robbery should be reasonably foreseeable.

defraud the United States (§371) and to impede or injure officers (§372). Although titled Conspiracy, the statute concurrently regulates solicitation as a component of conspiracy. As such, solicitation at federal law (§373) stipulates

> Whoever, with intent that another person engage in conduct constituting a felony that has as an element the use, attempted use, or threatened use of physical force against property or against the person of another in violation of the laws of the United States, and under circumstances strongly corroborative of that intent, solicits, commands, induces, or otherwise endeavors to persuade such other person to engage in such conduct, shall be imprisoned not more than one-half the maximum term of imprisonment . . . prescribed for the punishment of the crime solicited . . . ; or if the crime solicited is punishable by life imprisonment or death, shall be imprisoned for not more than twenty years.

## Party Liability

As with solicitation and conspiracy, the law recognizes that multiple parties collaborating in criminal purpose enhances the likelihood of success. Based on this philosophy, the consequences for those who assist with the commission of a criminal offense (referred to as an accomplice or accessory) also are culpable under the legal doctrine of party liability. The nature of their participation will determine the degree of liability, but some basic principles guide legal examinations of such criminal assistance.

An **accomplice** is an individual who assists, solicits, aids, or abets the perpetrator before the commission of the crime or those who fail to exercise a legal duty to prevent the crime. In order to be criminally liable, the accomplice must know their actions would assist or otherwise aid an individual with the commission of a crime. Examples of activities that may result in accomplice liability are facilitating drug transactions, purchasing instruments to be used in a crime, luring a victim to a crime, restraining a victim during an assault or killing, or driving a getaway vehicle; however, it is essential for the prosecution to establish that the accomplice intended that the crime be committed.

In the majority of jurisdictions (including states which follow Model Penal Code 2.06), accomplices are charged as principals. For example, if an accomplice is the driver of a robbery getaway car, he may be charged with robbery even though he did not take property of another through fear or force. At common law, an accomplice present at the scene of the crime, and who assisted in this way, would be labeled as a **principal at the fact**. Moreover, federal law defines principals in Title 18, Chapter 1 of the United States Criminal Code (§2) as one who

> commits an offense against the United States or aids, abets, counsels, commands, induces or procures its commission, is punishable as a principal [or] willfully causes an act to be done which if directly performed by him or another would be an offense against the United States.

Generally speaking, an **accessory after the fact** represents those individuals who assist a perpetrator after commission of a crime. These individuals typically assist with concealing the crime or aiding with the escape of the perpetrator. In order to be charged as an accessory after the fact, the prosecution must establish that the individual acted with the intent to assist the principal in avoiding arrest, prosecution, conviction, or punishment. In the majority of jurisdictions, accessories after the fact are not charged as principals, but rather with lesser offenses accompanied by lesser punishment. Examples of acts that may result in prosecution as an accessory after the fact are assisting with disposal of a body, cleanup of a crime scene, or the escape of the principal.

A person at common law could not be convicted as an accessory after the fact unless the principal was convicted. Thus, if the principal was acquitted of the crime, there could be no conviction for an accessory after the fact. The common law standard has been abrogated in most American jurisdictions (including federal prosecutions) and now is deemed entirely separate and distinct. Thus, despite the acquittal of the principal, an individual may be charged with accessory after the fact. Thus, federal law defines an "accessory after the fact" in Title 18, Chapter 1 of the United States Criminal Code (§3) as

> whoever, knowing that an offense against the United States has been committed, receives, relieves, comforts or assists the offender in order to hinder or prevent his apprehension, trial or punishment, is an accessory after the fact . . . shall be imprisoned not more than one-half the maximum term of imprisonment . . . if the principal is punishable by life imprisonment or death, the accessory shall be imprisoned not more than 15 years.

At common law, an additional category of **accessory before the fact** existed to describe individuals not present at or near the scene of the crime (as with a getaway driver), but who did assist, aid, or abet before the commission of a crime. In those jurisdictions that retain this category, an accessory before the fact may be charged as a principal; however, the modern trend is to abandon this distinction and treat all accessories before the fact as accomplices and therefore chargeable as principals, also the approach found in the Model Penal Code (2.06). A prerequisite for charging individual accomplices to a crime is proof of specific intent. Thus, the prosecutor must establish that the accomplice specifically intended that the crime occur or that an accessory specifically intended to assist with escape or concealment of the crime.

## Defenses

Accomplices possess the same defenses that a principal may raise. For example, if the principal could raise a defense of insanity, mitigation, self-defense, or impossibility, so too, may the accomplice. The defense of abandonment is available to both accomplices and accessories. The defense of abandonment was available at common law and is also available under the Model Penal Code. In order to abandon the crime successfully, the accomplice or accessory must voluntarily renounce participation, communicate this to his or her co-conspirator, and wholly negate his or her participation or make efforts to prevent the crime.

## Summary

Solicitation, conspiracy, and attempt are examples of offenses that exist to prohibit conduct that precipitates or entices the commission of a crime. By criminalizing such acts, society seeks to prevent the eventual completion of crimes; however, crimes that are inchoate in nature are also extremely complex. As you can see, absent the completion of an intended crime, it can be quite difficult to determine whether the participants went far enough in the planning or preparation of the crime. When multiple actors engage in a crime, determining the extent of liability for each individual can also be difficult. Party liability is the legal doctrine that exists to determine whether individuals are guilty as a principal or as an accessory. This is a distinction that is extremely important as the penalties can be very different.

## Practice Test

1. A(n) _____ offense is defined as incipient and generally leads other crimes.
   a. gateway
   b. novice
   c. inchoate
   d. exploitation
   e. chain reaction

2. The _____ test focuses on that which is remaining for a defendant to accomplish to commit the intended crime.
   a. dangerous proximity
   b. physical proximity
   c. indispensable element
   d. probable desistance
   e. substantial step

3. The _____ test focuses on the ability of a defendant to carry out criminal acts.
   a. dangerous proximity
   b. physical proximity
   c. indispensable element
   d. probable desistance
   e. substantial step

4. The _____ test evaluates whether it is likely, in light of facts and circumstances, that a defendant will cease from commission of the criminal act.
   a. dangerous proximity
   b. physical proximity
   c. indispensable element
   d. probable desistance
   e. substantial step

5. The crime of attempt requires a defendant have _____ intent to commit a crime.
   a. general
   b. presumptive
   c. constructive
   d. transferred
   e. specific

6. _____ refers to an inability to commit some intended crime due to the existence of a material piece of knowledge absent to the defendant or beyond the control of the defendant.
   a. Mistake of fact
   b. Factual impossibility
   c. Intervening cause
   d. Legal impossibility
   e. Proximity

7. _____ exists when a defendant is mistaken about the law or charged with the wrong crime.
   a. Mistake of fact
   b. Factual impossibility
   c. Intervening cause
   d. Legal impossibility
   e. Proximity

8. _____ is insufficient to prove the elements required for the crime of attempt.
   a. Ability
   b. Physical proximity
   c. Mere preparation
   d. Dangerous proximity
   e. Specific intent

9. The _____ test is no longer used because of its strict requirement.
   a. last act
   b. indispensable element
   c. probable desistance
   d. equivocality
   e. substantial step

10. _____ solicitation occurs when individuals do not directly solicit, but instead use a third party to solicit another to commit a crime.
    a. Secondary
    b. Third party
    c. Second-degree
    d. Indirect
    e. Intermediary

11. _____ addresses efforts of individuals to get others to commit crimes.
    a. Renunciation
    b. Solicitation
    c. Inducement
    d. Attempt
    e. Conspiracy

12. _____ allows conviction for conspiracy even when only one party believes an agreement was made to commit a crime.
    a. Unilateral theory
    b. Single approach
    c. Agreement intention theory
    d. Soloist view
    e. Bilateral theory

13. _____ are persons who assist a perpetrator after the commission of a crime.
    a. Solicitors
    b. Assistants
    c. Partakers
    d. Accessories
    e. Conspirators

14. To charge with the crime of conspiracy, most jurisdictions (including the federal government) require proof that the parties engaged in a(n) _____ act in furtherance of the conspiracy.
    a. prudent
    b. physical
    c. reasonable
    d. criminal
    e. overt

15. _____ rule holds that an agreement to commit a crime cannot be prosecuted as conspiracy when the crime itself is of such a nature as to necessarily require the participation of two or more persons.
    a. Thompson's
    b. Gillian's
    c. Williams'
    d. Wharton's
    e. Johnson's

16. The _____ provides when all alleged conspirators, but one, are acquitted, the remaining alleged conspirator may not be convicted.
    a. Standard of Conspiracy
    b. Conspirator Rule
    c. Rule of Consistency
    d. Conspiracy Protection Act
    e. Acquittal Guideline Act

17. _____ requires proof that the defendant has voluntarily abandoned the intent to commit the crime.
    a. Unilateral theory
    b. Wharton's Rule
    c. Rule of Consistency
    d. Bilateral theory
    e. Renunciation

18. By its very nature, _____ evidence is considered unreliable.
    a. assumptive
    b. third-degree
    c. hearsay
    d. corroborative
    e. presumptive

19. At common law, an accomplice present at the scene of a crime and who assisted with the crime was labeled a _____.
    a. principal at the fact
    b. first-degree accomplice
    c. second-degree accomplice
    d. principal after the fact
    e. primary accomplice

20. A(n) _____ is an individual who assists, solicits, aids, or abets a perpetrator before a criminal commission or who fails to exercise their legal duty to prevent crime.
    a. partner-in-crime
    b. accomplice
    c. solicitor
    d. assistant
    e. associate

## References

Garner, B. A. (Ed.). (2009). *Black's law dictionary* (9th ed.). Eagan, MN: West Group.

# Defenses to Criminal Responsibility

**CHAPTER 10**

## Key Terms

- Affirmative defense
- Alibi
- Automatism
- Battered wife syndrome
- Blockburger test
- Castle doctrine
- Cognition
- Consent of the victim
- Continuing offense
- Criminal Lunatics Act of 1800
- Diminished capacity
- Duress
- Durham Rule
- Entrapment
- Excuse defense
- Guilty but mentally ill
- Inducement
- Infancy
- Insanity defense
- Insanity Defense Reform Act of 1984
- Intoxication
- Involuntary intoxication
- Irresistible impulse test
- Justification defense
- M'Naghten Rule
- Mental defect
- Mental disease
- Mistake of age
- Mistake of fact
- Mistake of law
- Necessity
- Perfect self defense
- Policeman at the elbow test
- Predisposition
- Premenstrual syndrome
- Retardation
- Stand your ground laws
- Statute of limitations
- Substantial capacity test
- Tolling
- Twinkie defense
- Volition
- Voluntary intoxication
- Waiver
- XYY chromosome abnormality

## Introduction

Individuals accused of criminal conduct in the U.S. criminal justice system are afforded several options before trial. In most cases, defendants negotiate a plea agreement with the prosecution to allow for more lenient sentencing. This is accomplished either through a guilty plea to a lesser offense or a sentencing recommendation from the prosecution to the trial judge. In other cases, the accused may simply choose to plead not guilty. In these cases, the burden of proof rests with the prosecution, and the guilt of the accused must be proved beyond a reasonable doubt. Additionally, defendants may choose to raise a defense based on their rights, duties, or state of mind during the commission of the offense. This chapter explores the various defenses that can be raised during a criminal trial and provides insight regarding their rationale.

## Legal and Moral Rationale for the Allowance of Defenses

The concept of criminal responsibility was developed under English common law through the concept of *mens rea* (guilty mind). Because many crimes at common law required intent as an element of the offense, *mens rea* became pivotal to establishing a defense. If a defendant could not distinguish right from wrong (or good from evil) at the time of the offense, the necessity of punishment was negated. Courts began to recognize several defenses that accounted for the mental capacity of the defendant as well as defenses that relied on mitigating circumstances surrounding the events of the offense. Such circumstances include rights to protect one's self, others, and property, as well as duties required by those acting in a law enforcement capacity. These defenses evolved from case law and statute.

## Affirmative Defenses: Justification and Excuse Defenses Distinguished

**Affirmative defenses** acknowledge the commission of a crime while concurrently offering some justification or excuse to negate culpability. **Justification defenses** are premised on the belief that an accused had a right or duty to commit what normally is a crime. Meanwhile, **excuse defenses** rest on the notion that *mens rea* associated with criminal wrongdoing should be negated (or mitigated) due to mental incapacity. With affirmative defenses, the burden of proof is shifted to the defendant. In these instances, the burden is based on the preponderance of the evidence rather than reasonable doubt. In some cases, the standard of proof may require clear and convincing evidence, which falls between the standards of reasonable doubt and preponderance of the evidence.

## Justification Defenses

In justification defenses, the accused admits committing the act but offers evidence that they had a right or duty to do so. In these cases, the defendant's state of mind is not at issue; rather, the primary issue is the circumstances under which the offense occurred. As mentioned previously, justification defenses are based on individual rights or duties. For example, a police officer has a duty to protect the public and can raise an affirmative defense to homicide if he or she kills an armed suspect. Similarly, individuals have the same right to defend themselves against an armed attacker. Justification defenses permitted under United States law are discussed later in further detail.

### Rights: Self Defense, Defense of Others, and Defense of Property

The right to defend one's self is a long-standing tradition under English common law. Generally, self defense justifies physical injury to an aggressor. Although states differ on the specific elements, similar themes emerge with regard to when a citizen can legally use force: (1) the citizen must *reasonably believe* that he or she is under some unlawful threat of physical injury or death, (2) force must be necessary to avoid the harm, and (3) the force used must be reasonable and proportionate to the anticipated harm. When these criteria are met, a citizen has the right to use force to repel the attacker. In general, state laws regulating self-defense are equally applicable to the defense of others; however, laws regarding the defense of property are slightly different and normally do not permit use of deadly force to protect possessions.

A reasonable belief of danger implies that a normal person under similar circumstances would perceive the same threat. As a consequence, individuals with mental health problems (such as paranoia or dementia) may not qualify for this defense because their mental capacity often precludes the assumption of reasonable perception. Unlawful threat of physical injury or death applies only in instances in which such threat is imminent (unavoidable) and not used in the context of law enforcement. For example, self-defense is not applicable in situations where a law enforcement officer is killed after commanding a suspect to drop his or her weapon. Moreover, the notion of an aggressor ordinarily implies that the defendant was not the party who initiated the conflict or threat. Usually, any initial aggression from a defendant precludes self-defense as an affirmative defense to assault or homicide. For example, if the defendant verbally abused and subsequently was attacked by a bar patron, he or she cannot ordinarily assert self-defense as a justification for ensuing injuries. The concept of reasonable force in self-defense requires that the force used in response to aggression be reasonably proportionate to the imminent threat and be used concurrently with the force used by an attacker. For example, if a citizen punches an attacker with enough force to render unconsciousness, then no further force is justified.

Although regulation of general force is somewhat consistent across the states, they differ significantly when it comes to the use of deadly force because of the high probability of losing human life. Most states allow the use of deadly force only when the threat of death, serious bodily injury, rape, or kidnapping is imminent. Additionally, state laws regarding the use of deadly force often include provisions concerning a duty to retreat. States that require such a duty mandate that deadly force be used only when retreat is not possible. Exhibits 10–1 and 10–2 illustrate and compare the deadly force statutes for Mississippi and Delaware.

The **Castle doctrine** ordinarily removes the duty to retreat pursuant to home invasions. States adhering to the castle doctrine do not require homeowners to retreat (to safety) if they believe their lives (or lives of others) were in immediate danger. Some states also extend this provision to automobiles. Similarly, **stand your ground laws** incorporate the castle doctrine, but are rooted in the concept that persons should be able to defend themselves regardless of immediate danger. There has been considerable debate regarding these laws, primarily because many states have

## Exhibit 10-1 Mississippi Laws Regarding the Use of Deadly Force

**Miss. Code Ann. (1972) §97-3-15**

(1) The killing of a human being by the act, procurement or omission of another shall be justifiable in the following cases:

    (e) When committed by any person in resisting any attempt unlawfully to kill such person or to commit any felony upon him, or upon or in any dwelling, in any occupied vehicle, in any place of business, in any place of employment or in the immediate premises thereof in which such person shall be;

    (f) When committed in the lawful defense of one's own person or any other human being, where there shall be reasonable ground to apprehend a design to commit a felony or to do some great personal injury, and there shall be imminent danger of such design being accomplished;

(4) A person who is not the initial aggressor and is not engaged in unlawful activity shall have no duty to retreat before using deadly force under subsection (1) (e) or (f) of this section if the person is in a place where the person has a right to be, and no finder of fact shall be permitted to consider the person's failure to retreat as evidence that the person's use of force was unnecessary, excessive or unreasonable.

## Exhibit 10-2 Delaware Laws Regarding the Use of Deadly Force

**11 Delaware Code §464**

(a) The use of force upon or toward another person is justifiable when the defendant believes that such force is immediately necessary for the purpose of protecting the defendant against the use of unlawful force by the other person on the present occasion.

(b) Except as otherwise provided in subsections (d) and (e) of this section, a person employing protective force may estimate the necessity thereof under the circumstances as the person believes them to be when the force is used, without retreating, surrendering possession, doing any other act which the person has no legal duty to do or abstaining from any lawful action.

(c) The use of deadly force is justifiable under this section if the defendant believes that such force is necessary to protect the defendant against death, serious physical injury, kidnapping or sexual intercourse compelled by force or threat.

(e) The use of deadly force is not justifiable under this section if:

    (1) The defendant, with the purpose of causing death or serious physical injury, provoked the use of force against the defendant in the same encounter; or

    (2) The defendant knows that the necessity of using deadly force can be avoided with complete safety by retreating, by surrendering possession of a thing to a person asserting a claim of right thereto or by complying with a demand that the defendant abstain from performing an act which the defendant is not legally obligated to perform except that:

        a. The defendant is not obliged to retreat in or from the defendant's dwelling; and

        b. The defendant is not obliged to retreat in or from the defendant's place of work, unless the defendant was the initial aggressor . . .

extended them to cover perceived threats both inside and outside of the home. The extent to which stand your ground laws remove stipulations regarding the use of deadly force varies from state to state. For example, in 2005, Florida incorporated a stand your ground law with several controversial components (Florida Statutes Title XLVI 776.014, 2008):

- The presumption of immediate danger is implied on unlawful entry of a home or vehicle.
- There is no duty to retreat *anywhere* if a person believes that he or she or others are in immediate danger.
- Persons using lawful deadly force are immune from prosecution and suit in civil courts.

Many in the criminal justice community claim that these laws promote vigilantism and value property over life. For example, Florida law permits the killing of an unarmed person attempting to enter another's home regardless of the immediacy or level of perceived danger; however, Florida's stand your ground laws apply only to persons not engaged in illegal activity. As such, it would not apply to a drug dealer who uses deadly force against someone breaking into his or her house to steal drugs. Pause for Thought 10-1 illustrates how self defense law is applied.

### Duties: Police Officers and Correctional Officers

Although rare, the use of deadly force often is required when police make an arrest. Although states differ regarding the use of deadly force by citizens, all states provide law enforcement officers with such authority. Traditionally, law enforcement officers were permitted to use deadly force to prevent suspected felons from fleeing arrest. Most state statutes reflected this tradition until 1985 when the U.S. Supreme Court issued their landmark decision. In *Tennessee vs. Garner* (1985), the court held that deadly force was only permissible when

> the officer has probable cause to believe that the suspect poses a threat of serious physical harm, either to the officer or to others, [and] it is not constitutionally unreasonable to prevent escape by using deadly force.

After *Tennessee vs. Garner*, states modified their statutes to conform to those provisions. Most (if not all) state laws regulating the use of deadly force include provisions specific to law enforcement officers (and citizens who may assist them).

Correctional officers, too, must sometimes use deadly force within their employment. Although self-defense and defense of others are available to correctional officers who cause serious injury or death to an inmate, another situation that requires an officer to use deadly force is that of capital punishment. Correctional officers are often assigned to act as executioners, and although state laws do not specifically address the use of deadly force in this context, general provisions regarding the use of deadly force are applicable to executions. For example, Mississippi law allows the use of deadly force when "committed by public officers or those acting by their aid and assistance, in obedience to any judgment of a competent court" (Miss. Code Ann. 1972 §97-3-15[1a]). Because of the gravity of executions, federal law permits both federal and state correctional officers to

---

### Pause for Thought 10-1

Consider the following: Jim and Richard agree to meet at the batting cages to practice for an upcoming softball game. Due to considerable heat and humidity, Jim's bat slips from his hands and strikes Richard in the leg. Richard becomes upset and punches Jim in the face several times. Jim pleads with Richard to stop hitting him and warns that he will hit back. Richard then picks up the bat and begins to swing the bat in the direction of Jim's head. Jim picks up a ball and strikes Richard in the head. Richard falls to the ground unconscious. Jim then grabs the bat from Richard's hands and proceeds to strike him repeatedly. Police and medical personnel arrive on the scene and determine that Richard is dead. Is Jim protected under Delaware self-defense laws?

### Scenario Solution

No, although Jim was initially protected under Delaware law when he threw the ball, Richard was unconscious when Jim initiated use of deadly force. Because Jim was not in immediate danger of death, serious physical injury, kidnapping, or sexual intercourse compelled by force or threat, his actions did not constitute self defense.

abstain from participating in executions if they object to the use of capital punishment on religious or moral grounds (18 U.S.C. §3597b). Pause for Thought 10–2 presents a hypothetical situation regarding the use of deadly force within the scope of law enforcement duties.

## Excuse Defenses

Similar to justification defenses, excuse defenses account for mitigating circumstances regarding an offense, but also account for circumstances of the offender; however, unlike justification defenses, excuse defenses are not based on rights or duties but rather the *mens rea* of the offender. If the accused cannot control his or her behavior, comprehend the consequences of his or her actions, or honestly believed that his or her actions were within the law, then an affirmative excuse defense is appropriate. As mentioned previously, these types of defenses shift the burden of proof to the defendant, thus the prosecution has no obligation to prove that the accused committed the crime. The prosecution must only offer evidence to contradict the defendant's affirmative defense.

### Imperfect Self-Defense

The use of necessary and reasonable force in defense of self or another or to prevent the commission of a violent felony is called **perfect self-defense** and is not criminal because of the legal justification with which force is used. There are times, however, when people possess a subjective (or personal) belief that circumstances warrant the use of force but in actuality are mistaken as to the objective circumstances. These situations are referred to as imperfect self-defense and present the legal system with a unique and regrettable duty to prosecute some conducts that possess no underlying desire to do wrong. Abused women often resort to deadly force and kill domestic abusers with the mistaken belief that the law permits such action. Should their mistaken killings be justifiable? To answer in the affirmative, one must conclude that the defendant possessed some right or duty to kill or that the killing was accidental. Neither of these circumstances exists, however, as the defendant possessed no duty, was mistaken as to their perceived right, and did kill intentionally, thus classification as an "accident" is a legal impossibility.

### Insanity

Insanity is a legal, not medical, term. An **insanity defense** requires proof that the accused possessed a state of mind that would render the person not responsible for wrongful actions. In contrast to portrayals on television and in movies, insanity defenses are seldom successful because jurors are quite cynical about its existence. Think for one moment about what your reaction would be upon hearing that an accused claims to be insane. Did you roll your eyes? Most people do, and thus the insanity defense is rarely employed when other reasonable defenses are available. Defendants who successfully argue insanity are not usually released to society, but rather are confined to a mental health facility. Subsequent release from the facility is contingent on a recommendation from a physician or mental health specialist—meaning that their confinement could be indefinite. Keep in mind, too, that faking insanity also has its perils, in that double jeopardy

---

### Pause for Thought 10–2

Consider the following: Joe is a police officer in Alabama. During routine patrol, he spots a suspected burglar exiting the window of a home. Joe exits his vehicle and approaches the suspect. After identifying himself as a police officer, Joe commands the suspect to get on his knees. As the suspect kneels, Joe hears a woman screaming, "He stabbed me! Help!" The screaming came from inside the house where the suspect exited. The suspect takes advantage of the distraction and begins to flee. Joe commands the suspect to stop, but he keeps running. If Joe decided to shoot the fleeing suspect, would his actions be legal under *Tennessee vs. Garner*?

### Scenario Solution

Yes, under the circumstances, Joe would be justified in shooting the fleeing suspect. The woman screaming for help inside the house certainly qualifies as probable cause regarding the suspect. Additionally, Joe identified himself as a police officer and warned the suspect to stop fleeing. *Tennessee vs. Garner* allows the use of deadly force to prevent escape when an officer has probable cause to believe the suspect poses a serious threat.

does not preclude a new trial in such cases. So, as you can see, the risks are great and the rewards potentially short lived. Not exactly the kind of odds sought when gambling for your freedom.

The insanity defense evolved in 18th century England. In 1724, an English judge re-sentenced a mentally incompetent person to life, explaining that the original death sentence should not have been issued because the accused could no more distinguish right from wrong than a "wild beast" (*Rex vs. Arnold*, 16 How. St. Tr. 695). Years later, the **Criminal Lunatics Act of 1800** created a separate verdict of *not guilty on account of insanity*; however, this verdict was used primarily to hold defendants accused of treason for indefinite periods of time in lieu of excusing the mentally ill from their actions (Memon, 2006).

*M'Naghten.* The insanity defense evolved in 1843 when Daniel M'Naghten attempted to murder Sir Robert Peel, a British official still considered by many the founding father of modern policing. M'Naghten failed to kill Peel, instead murdering his assistant. M'Naghten's lawyers argued he was delusional and could not distinguish right from wrong. The jury agreed and found M'Naghten not guilty on the grounds of insanity. The controversy over the verdict prompted the English Parliament to create a standard for jury instructions regarding the insanity defense. The standard became known as the **M'Naghten Rule** and stipulates that

> it must be clearly proved, that, at the time of the committing of the act, the party accused was laboring under such a defect of reason, from disease of the mind, as not to know the nature and quality of the act he was doing, or, if he did know it, that he did not know he was doing what was wrong. (*M'Naghten's Case*, 8 Eng. Rep. 718, 1843)

Subsequent to this ruling, the M'Naghten Rule (or right–wrong test) also was established as the standard insanity test in U.S. courts.

Opponents of the M'Naghten Rule argued that the wording of the statute was vague and did not provide a clear definition of insanity. Additionally, they argued that M'Naghten was based on mental cognition and did not account for other factors such as the defendant's emotional state of mind. Moreover, the M'Naghten Rule did not specify what constituted a disease of the mind and left the courts to decide what mental deficiencies were applicable under the statute. Although the M'Naghten Rule is still used by 25 states, inconsistencies among state rulings have perpetuated the development of additional tests for insanity. Interestingly, all eight states that border the Gulf of Mexico and/or Mexico currently use the more conservative M'Naghten Rule, and only two southern states (Tennessee and Arkansas) in their entirety have deviated from M'Naghten (instead choosing the Model Penal Code's Substantial Capacity Test). Likewise, not one New England state (comprised of Maine, New Hampshire, Vermont, Massachusetts, Rhode Island, and Connecticut) uses the M'Naghten Rule. Do you think there might be some philosophical differences among geographic regions of our country (The insanity defense among the states, 2009)?

*Irresistible Impulse.* In *Parsons vs. State* (81 Ala. 577, 2 So. 854, 1887), the concept of irresistible impulse was instituted to augment the M'Naghten Rule. Although M'Naghten provided a definition of insanity through evaluating the cognitive abilities of the accused, it neglected to address the issue of choice, or volition. The **irresistible impulse test** corrected that omission and stipulated that individuals were insane when they could not control their actions because of some mental incapacity which made them incapable of distinguishing right from wrong. Specifically, the court defined the test as:

> (1) Where there is no capacity to distinguish between right and wrong as applied to the particular act, there is no legal responsibility; (2) Where there is such capacity, a defendant nevertheless is not legally responsible if, by reason of the duress of mental disease, he has so far lost the power to choose between right and wrong as not to avoid doing the act in question, so that his free agency was at the time destroyed, and, at the same time, the alleged crime was so connected with such mental disease in the relation of cause and effect, as to have been the product or offspring of it solely.

Simply put, the irresistible impulse test uses the guidelines set forth in M'Naghten but adds a component to test for the inability to make rational choices stemming from mental disease. The concept of irresistible impulse further evolved as it was adopted by other states after *Parsons*. As it did, the **policeman at the elbow test** became a benchmark for assessing irresistible impulse. This version tested whether the accused would have committed the same offense in the presence of a police officer (*People vs. Hubert*, 1897; *U.S. vs. Kunak*, 1954). The purpose of this test was to ascertain the defendant's awareness of the consequences of his or her actions: would they have acted in the same manner knowing they would be caught?

*Durham.* In 1954, the U.S. Court of Appeals for the District of Columbia temporarily changed the definition of insanity. In *Durham vs. U.S.*, the court held that the right–wrong (M'Naghten) and irresistible impulse tests disregarded current psychological and psychiatric opinions regarding behavior and mental illness. Specifically, the court stated that

> the right–wrong test is inadequate in that (a) it does not take sufficient amount of physic realities and scientific knowledge, and (b) it is based upon one symptom and so cannot validly be applied in all circumstances . . . also inadequate in that it gives no recognition to mental illness characterized by brooding and reflection and so relegates acts caused by such an illness to the application of the inadequate right–wrong test.

The court's new insanity standard stated that "an accused is not criminally responsible if his unlawful act was the product of mental disease or mental defect" (*Durham vs. U.S.*, 1954). The court's decision became known as the **Durham Rule** and drew clear distinctions between **mental defect** (a permanent, unchanging condition) and **mental disease** (a condition that could improve or worsen over time), as well as increased the scope of decreased mental capacity.

Although the Durham Rule was applauded by mental health professionals, critics claimed it was too broad and placed too much confidence in psychologists and psychiatrists as expert witnesses. Moreover, the Durham Rule failed to include a specific watermark for the capacity to distinguish right from wrong (whereas M'Naghten stipulated that the accused must have no capacity to make that distinction). As a result, most states refused to abandon the M'Naghten Rule and/or irresistible impulse test. In 1972, the Durham Rule was eliminated by the same district court that promoted its use (see *U.S. vs. Brawner*, 1972); however, the Durham Rule is alive and well in the State of New Hampshire (The insanity defense among the states, 2009).

*Substantial Capacity.* Comprised of legal scholars and professionals, the American Law Institute (ALI) proposed its own insanity standard in the Model Penal Code. The substantial capacity test (or ALI Standard) sought to address the void between the M'Naghten Rule and Durham Rule. The **substantial capacity test** accounted for both cognition and uncontrollable behavior (similar to when M'Naghten was augmented by irresistible impulse) as well as decreased the standard for mental capacity required to appreciate or control one's actions. As stated in the Model Penal Code

> (1) A person is not responsible for criminal conduct if at the time of such conduct as a result of mental disease or defect he lacks substantial capacity either to appreciate the criminality (wrongfulness) of his conduct or to conform his conduct to the requirements of the law. (2) As used in this article, the terms "mental disease or defect" do not include an abnormality manifested only by repeated criminal or otherwise antisocial conduct (§4.01).

The substantial capacity test borrowed from both the M'Naghten and irresistible impulse tests in recognizing the role that mental illness plays in perceptions and actions. The ALI essentially created two independent prongs to assess legal insanity: (1) **Cognition**—Did the accused possess substantial mental capacity to distinguish his or her actions as right or wrong? (2) **Volition**—Did the accused possess substantial mental capacity to conduct himself or herself in accordance with the law? Moreover, it applied substantial capacity as the standard for mental deficiency, which mediated the standards of the M'Naghten Rule (no capacity to distinguish right from wrong) and the Durham Rule (no standards of capacity specified).

Substantial capacity was generally accepted as the appropriate insanity test after release of the Model Penal Code in 1962. Many states incorporated the new standard as their definitive test for insanity—either through case law or statute. Federal courts, too, adopted the ALI standard as their insanity benchmark. In *Freeman vs. U.S.* (1966), the U.S. Court of Appeals for the 2nd Circuit held that the M'Naghten Rule was not an appropriate insanity test, even when augmented by a component of irresistible impulse. Moreover, the court stated the importance of selecting the correct test for insanity:

> The criminal law . . . is an expression of the moral sense of the community. The fact that the law has for centuries regarded certain wrong-doers as improperly punished is a testament to the extent to which that moral sense has developed. Thus, society has recognized over the years that none of the asserted purposes of the criminal law—rehabilitation, deterrence, and retribution—is satisfied when the truly irresponsible, those who lack substantial capacity to control their actions, are punished.

Because of the Supreme Court's abstinence from endorsing any test of insanity, the court decided to adopt substantial capacity as the insanity standard for the Second Circuit. Today, the substantial capacity test is used nearly as often as M'Naghten, with 21 states relying on its legal guidance (The insanity defense among the states, 2009). Exhibit 10-3 details how the substantial capacity test was applied in the John Hinckley, Jr. case.

*Post-Hinckley Insanity Defenses.* Hinckley's acquittal shocked the nation. Citizens, legal professionals,

> **Exhibit 10-3  John Hinckley, Jr.**
>
> John W. Hinckley, Jr. attempted to assassinate President Ronald Reagan on March 30, 1981, in Washington, DC. Hinckley, a failed songwriter and heir to a modest family oil business, became obsessed with Jodie Foster after repeated viewings of the movie *Taxi Driver*. He eventually followed Foster to Connecticut (she was attending Yale University) where he tried unsuccessfully to win her affection. Hinckley's attempt on Reagan's life was intended to attract Foster's attention and to impress her. Thankfully, his aim was poor and he only injured Reagan (albeit because of a ricochet). He also wounded a Secret Service agent, a police officer, and James Brady—a press secretary who would later earn national attention for supporting gun control and inspiring legislation known as the Brady Bill. Hinckley's defense was based on the ALI's substantial capacity test—specifically, his lawyers argued that he lacked substantial mental capacity to conform his conduct to the requirement of the law. In 1982, Hinckley was found not guilty on all counts by reason of insanity and remanded into the custody of St. Elizabeth's Hospital in Washington, DC. Hinckley remains there today, with limited visitation privileges that allow him to leave hospital grounds and travel to his parent's home.

and politicians were outraged and called for a review and reform of the substantial capacity test. As a result, the **Insanity Defense Reform Act of 1984** provided strict guidelines for using insanity as a defense in federal courts. Codified under 18 U.S.C. §17, the new provisions stated that a person is legally insane when as a result of mental disease or defect

> was unable to appreciate the nature and quality or the wrongfulness of his acts. Mental disease or defect does not otherwise constitute a defense. . . . The defendant has the burden of proving the defense of insanity by clear and convincing evidence.

The statute made sweeping changes to the test and burden of proof for insanity. First, the act eliminated the volition (irresistible impulse) prong of the ALI test, which notably was the component used in Hinckley's defense. Second, the magnitude of mental incompetence required to qualify as insanity was changed from "lacking substantial capacity" to "severe mental disease or defect." Essentially, this shift reflected the more rigorous standards of the M'Naghten Rule. Although the burden of proof for insanity had always been placed on the defense, the act mandated a standard of clear and convincing evidence rather than the traditional standard of preponderance of the evidence.

*Guilty but Mentally Ill.* In 1975, Michigan passed legislation creating the verdict **guilty but mentally ill** (GBMI). Although not technically an affirmative defense, GBMI warrants discussion. When the accused pleads GBMI, he or she offers evidence of mental illness to negate punishment associated with an offense and seeks treatment for the mental illness. Defendants who plead this defense assert that they were mentally ill at the time of the offense, yet not to the extent of insanity. The defendant is still subject to the same sanctions as a person found guilty; however, a GBMI verdict allows the judge (and sometimes jury) to consider the defendant's mental illness when calculating his or her sentence. Moreover, this verdict usually attaches some form of mandatory mental health treatment to the sentence.

Subsequent to Hinckley's acquittal, four states (Idaho, Kansas, Montana, and Utah) completely abolished insanity as an affirmative defense, opting instead to offer criminal defendants the opportunity to proclaim that they are guilty but mentally ill (The insanity defense among the states, 2009). Nine other states also developed such statutes and allow them to be considered by juries along with the insanity option. Adoption of GBMI standards signaled a growing apprehension regarding insanity acquittals. GBMI has been contested in various jurisdictions on differing constitutional grounds (such as due process and cruel and unusual punishment) but has been upheld on nearly every challenge; however, "court rulings have affirmed that GBMI is essentially *no different than a conventional guilty plea or verdict*. It does not guarantee a right to treatment for a mentally ill defendant, and it does not imply any diminished responsibility for the crime" (Coleman, 1999).

## Diminished Capacity

**Diminished capacity** refers to a defense that offers evidence that the accused did not possess the necessary *mens rea* to meet a specific element of intent. A defendant who employs a diminished capacity defense argues that their mental capacity was not sufficient to form the requisite intent. Because diminished capacity attacks the element of intent, a successful defense will result in either acquittal or conviction for a lesser offense. For example, if an individual is charged with murder (which requires intent), he or she could argue diminished capacity to limit the conviction to manslaughter (which does not require intent).

Diminished capacity is an accepted defense under federal law and some states. In *U.S. vs. Fishman* (1990), a U.S. District Court held that the diminished capacity defense was not barred by the Insanity Defense Reform Act (18 U.S.C. §17). Specifically, the court stated that "diminished capacity is simply a label that identifies evidence introduced by a defendant to support a claim that he did not commit the crime charged because he did not possess the requisite *mens rea*." In this ruling, the court clearly indicated that diminished capacity and insanity defenses are independent. Diminished capacity has also been used to create some very interesting defenses (see forthcoming Controversial Diminished Capacity Defenses).

Diminished capacity often is argued when a specific mental deficiency impacts a person's behavior. For example, veterans of recent wars in Afghanistan and Iraq who experienced traumatic combat events may suffer from posttraumatic stress disorder (PTSD), which often alters the mood and/or behavior of a person. Some veterans have claimed diminished capacity resulting from PTSD as a defense in both military and civilian courts. This defense is based on the notion that PTSD reduces a person's ability to distinguish right from wrong, therefore impacting one's reactions to various trigger situations; however, in most (if not all) instances, the result of a successful PTSD-based diminished capacity defense is a reduction in charge or sentence rather than acquittal (Robson, 2008).

*Infancy.* The concept of **infancy** was established under English common law to absolve children of criminal responsibility based on the notion of presumption. Under common law, a child under the age of 7 years could not be charged with a crime because he or she was conclusively presumed to lack the mental capacity to form criminal intent. Children between the ages of 7 and 14 years were rebuttably presumed to lack the mental capacity to form intent; therefore, children of these ages could be prosecuted as adults if the state could demonstrate that a child had the capacity to distinguish right from wrong. Children over the age of 14 years were presumed to possess the mental capacity to distinguish right from wrong and therefore were capable of forming criminal intent; however, evidence could be presented by the defense to demonstrate that the accused lacked such capacity—similar to an insanity defense.

The concept of infancy is somewhat moot under United States law. Development of the juvenile justice system eliminated the need for an infancy defense by providing children with additional protection from the criminal justice system. Moreover, the common law infancy defense was based on the argument that age is an indicator of mental capacity—a concept currently believed to be fallacy. Under U.S. law, children under the age of 18 years are considered juveniles and are not normally prosecuted in criminal courts. Exceptions to this rule may occur when a juvenile is a repeat offender or has committed a serious offense. In these cases, a **waiver** is issued transferring a juvenile to criminal court. Other exceptions may occur when state statutes delineate certain crimes that are exclusively under a criminal court's jurisdiction, such as capital murder.

*Intoxication.* **Intoxication** refers to diminished mental capacity caused by alcohol or drug use. Although it is highly unlikely that an intoxication defense will result in acquittal, some defendants choose to use intoxication as a mediating factor regarding intent. For crimes such as first-degree murder, where intent is specific rather than general, the *mens rea* of the accused is a vital element. Defendants often try to negate the formation of specific intent through diminished capacity resulting from intoxication. Federal case law and most state laws distinguish voluntary and involuntary intoxication as separate and independent defenses.

**Voluntary intoxication** refers to the purposeful ingestion of alcohol or drugs. Legal tradition holds that individuals exercise free will when ingesting such substances, and thus, that same free will also governs subsequent actions. Under common law, criminal behavior resulting from drunkenness was not excused and, in fact, often was punishable by death. United States case law contains few instances in which voluntary intoxication was successfully proffered as a form of diminished capacity, and many states have outright refused to accept voluntary intoxication as a form of diminished capacity. Some states have even outright abolished voluntary intoxication as a defense through their statutes. Although these statutes have been challenged on the basis that they violate the due process clause of the U.S. Constitution, they have passed the scrutiny of the U.S. Supreme Court (*Montana vs. Egelhoff*, 1996). The few states that consider voluntary intoxication as a form of diminished capacity accept the defense only as a mitigating factor, meaning that complete acquittal is not possible.

**Involuntary intoxication** can occur through deception or fraud or by accident. Common examples of involuntary intoxication include:

- Tainted food or drink—such as when rohypnol, GHB (gamma hydroxybutyric acid), or other date-rape drug is slipped to an unsuspecting victim
- "Spiking the punch"—placing an intoxicating substance in the central distribution point of communally available nonalcoholic beverages
- Force or coercion—ingesting an intoxicating substance under duress or threat of harm
- Unforeseeable interactions—neglecting to read the prescription label concerning alcohol and/or other substances that should be avoided in combination with a drug
- Unforeseeable reactions—allergic or otherwise abnormal reaction to a substance

Involuntary intoxication is a viable defense only when an accused was unable to differentiate right from wrong at the time of the offense. For example, if an unsuspecting person drinks spiked punch and subsequently commits murder, involuntary intoxication would not automatically be a valid defense. It is highly improbable that any amount of consumed alcohol would diminish a person's mental capacity to the point where he or she would not comprehend the severity of murder. Evidence presented by the defense must indicate (1) diminished capacity which negated the defendant's ability to distinguish right from wrong (2) at the time of the offense and (3) resulted from involuntary ingestion of an intoxicating substance. Exhibit 10–4 illustrates Tennessee's intoxication law.

*Mental Retardation.* Retardation is now an uncommon term, replaced by socially sensitive phrases like mentally handicapped, challenged, or disabled. Regardless of its political incorrectness, however, **retardation** is still used in most statutes to refer to delayed mental development, cognitive abilities, communication skills, or a limited comprehension of health and safety. States differ on what constitutes mental retardation, but still must follow federal standards (18 U.S.C. §4241, §4246) and case law (*Drope vs. Missouri*, 1975). These guidelines stipulate that (1) the defense may request a competency hearing at any time prior to sentencing, (2) the

---

### Exhibit 10–4 Tennessee Laws Defining Intoxication

(a) Except as provided in subsection (c), intoxication itself is not a defense to prosecution for an offense. However, intoxication, whether voluntary or involuntary, is admissible in evidence, if it is relevant to negate a culpable mental state.

(b) If recklessness establishes an element of an offense and the person is unaware of a risk because of voluntary intoxication, the person's unawareness is immaterial in a prosecution for that offense.

(c) Intoxication itself does not constitute a mental disease or defect within the meaning of § 39-11-501 [statute defining insanity as an affirmative defense]. However, involuntary intoxication is a defense to prosecution, if, as a result of the involuntary intoxication, the person lacked substantial capacity either to appreciate the wrongfulness of the person's conduct or to conform that conduct to the requirements of the law allegedly violated.

(d) The following definitions apply in this part, unless the context clearly requires otherwise:

(1) "Intoxication" means disturbance of mental or physical capacity resulting from the introduction of any substance into the body;

(2) "Involuntary intoxication" means intoxication that is not voluntary; and

(3) "Voluntary intoxication" means intoxication caused by a substance that the person knowingly introduced into the person's body, the tendency of which to cause intoxication was known or ought to have been known. (Tennessee Code 39-11-503).

burden of proving incompetence rests with the defense, and (3) the preponderance of evidence will guide the court's decision. Furthermore, (4) if the court determines the defendant is mentally incompetent to the point where they cannot (a) understand the charges against them or the consequences of such charges or (b) assist in their own defense, (5) then the defendant must be remanded to a mental health facility until judged competent to stand trial or until (a) a 4-month period has expired and a reasonable additional time period for improvement is estimated or (b) the charges are dismissed or (c) the court determines that no improvement in competence will likely occur, and a civil commitment hearing is necessary to permanently house the defendant in a mental health facility. Put simply, mentally incompetent defendants cannot be criminally convicted because they do not understand the consequences of their actions. Persons unable to fathom such actions also are incapable of distinguishing right from wrong and thus cannot form criminal intent. Most states define mental retardation as a developmental disorder rather than a disease or defect of the mind. Mental illness, mental retardation, and insanity are quite different—although they can each negate the element of specific intent in a diminished capacity defense.

*Other Controversial Defenses.* Because of the general nature of the diminished capacity defense and its acceptance under federal law, several related defenses have surfaced in recent years. Although these defenses were mostly unsuccessful, they illustrate the scope of diminished capacity as an affirmative defense and demonstrate the willingness of the courts to entertain new scientific theories. Additionally, they are a testament to the creativity and dedication of defense attorneys.

**Automatism** refers to an involuntary action. Courts have differed as to how automatism should be presented as a defense to the commission of a crime—is it an insanity defense or a separate defense altogether? Most courts regard automatism as independent from insanity and require evidence to indicate that the defendant was unaware of the actions which constituted the offense. Common examples of automatism include sleepwalking (somnambulism), blackouts (from a condition other than intoxication), and sleep deprivation; however, a defense of automatism must include evidence that the defendant acted involuntarily—not merely memory loss (*McClain vs. Indiana*, 678 N.E.2d 104, 1997).

**Battered woman syndrome** describes an extreme emotional state caused by the cyclical pattern of domestic violence. Currently, the term *battered person syndrome* has been substituted to include children and victims of sexual assault. Battered person syndrome in the context of self-defense assumes that the accused used (or attempted to use) deadly force as a means of protection. Unlike traditional self-defense arguments, the element of imminent serious physical injury or death is not present. Battered woman syndrome was introduced as an element of self defense in Dr. L.E. Walker's *The Battered Woman* (see *Ibn-Tamas vs. U.S.*, 1979). Subsequent to the book's release in 1979, Dr. Walker began offering testimony at trial proposing that battered women constantly perceive themselves in imminent danger, which prevents them from leaving abusive relationships. Regarding homicide, battered woman syndrome was slow to gain acceptance within the scientific community (see *Ibn-Tamas vs. U.S.*, 1983); however, it did gain some support from the U.S. Supreme Court:

> Although traditional self-defense theory may seem to fit the situation only imperfectly . . . the battered woman's syndrome as a self-defense theory has gained increasing support over recent years. (*Moran vs. Ohio*, 1984)

Individuals exposed to abuse or neglect could also argue diminished capacity resulting from exposure to severe trauma (e.g., PTSD).

The **XYY chromosome abnormality** defense was first connected to criminal behavior in the 1960s through research on male inmates. These "supermales" possessed an extra *Y* chromosome and exhibited characteristics such as above-average height and moderate to severe acne outbreaks. Research linking aggressive behavior and the *XYY* genetic mutation was initially accepted but then summarily dismissed in the scientific community, primarily because of the existence of *XYY* individuals in the general population who failed to commit crimes or even exhibit abnormal behavior. In 1966, Richard Speck killed eight nurses in Chicago and attempted to use the *XYY* abnormality as a form of diminished capacity. The defense not only was unsuccessful, but inappropriate as well—it was later found that Speck did not even have the genetic mutation. In the United States, the *XYY* defense has never resulted in acquittal.

**Premenstrual syndrome** (PMS) has been used as an element of diminished capacity to an assortment of criminal offenses but with limited success. Although symptoms of severe PMS include depression and thoughts of suicide, such cases are rare and do not qualify as a mental illness, disorder, or defect. Because PMS is a function of hormonal imbalance rather than mental defect or disease, it is difficult to use as a component of diminished capacity. The use of PMS as an excuse has been more successful in England, France, and Canada than the United States (Davidson, 2000).

The **Twinkie defense** refers to diminished capacity resulting from the mass consumption of junk food. Proven to be a myth, the defense stems from the trial of Dan White for the murder of San Francisco Mayor George Moscone and Supervisor Harvey Milk in 1978. The defense argued diminished capacity to negate the intent element attached to murder. Psychiatrists offered testimony on White's behalf pointing to dramatic changes in his lifestyle and behavior as indicators of mental illness. One of the changes was increased consumption of junk food—before the murder, White had an affinity for health food. The jury recognized the testimony as evidence of diminished capacity and returned a verdict of voluntary manslaughter rather than murder. Citizens of San Francisco were outraged, and their anger was fueled by media misrepresentation of the evidence presented to the jury. The media misreported the ingestion of junk food as the cause of diminished capacity rather than a symptom of mental illness, and the term *Twinkie defense* was born. Interestingly, California subsequently eliminated diminished capacity as a method for negating intent (California Penal Code §§25-29), even though it was arguably the first state to recognize diminished capacity as a defense.

## Mistake of Fact and Law

Affirmative defenses are not always based on refuting intent through mental defect or disease. Although the phrase "ignorance of the law is no excuse" generally holds true, ignorance or mistake can be used as an affirmative defense. Generally, mistake of fact and mistake of law are used to negate specific elements of a criminal offense. Most often, they are used to refute specific intent. These defenses are not applicable to strict liability offenses, where the *actus reus* (criminal act) itself defines the offense.

**Mistake of fact** can be used as a defense when the offense resulted from honest mistake. For example, a cab driver charged with aiding and abetting an escaped inmate could argue mistake of fact to negate criminal culpability. Courts have generally upheld the mistake of fact defense when the accused had an honest belief that such actions were not criminal.

Furthermore, some states authorize mistake of fact as a defense by statute. For example, the Texas Penal Code (§8.02a) states,

> It is a defense to prosecution that the actor through mistake formed a reasonable belief about a matter of fact if his mistaken belief negated the kind of culpability required for commission of the offense.

There are two instances where the mistake of fact defense generally is not permitted. First, it is not applicable when the act was intended to cause harm to a specific person but instead harmed another. Suppose that a car thief steals an SUV from a parking lot and later realizes a child was in the back seat. He cannot argue mistake of fact as a defense to kidnapping because his initial intent was to commit a crime, and thus, he or she is responsible for consequences subsequent to that action. Second, mistake of fact does not apply to criminal negligence because such acts do not account for intent. For example, if Bob left a loaded firearm in the presence of several children and one of those children accidentally shot and killed another, he could not argue mistake of fact as a defense to criminally negligent homicide.

Mistake of fact is often discussed within statutory rape crimes. Most states have classified statutory rape as a strict liability offense and therefore do not recognize mistake with respect to a victim's age. Some states, however, do recognize such a defense. Usually, **mistake of age** defenses are defined by statute. One of the earliest examples of a successful mistake of age defense occurred when the California Supreme court (*People vs. Hernandez*, 393 P.2d 673, 1964) held that mistake of age was a viable defense to charges of statutory rape unless otherwise specified by statute. Examples of state statutes regarding mistake of age defenses are presented in Exhibit 10–5.

**Mistake of law** is a defense that relies on the genuine and honest belief of the accused that he or she acted in accordance with the law. Whereas mistake of fact involves a defendant's perceptions of everyday occurrences, mistake of law is based on the defendant's belief that his or her actions were consistent with the law. In *U.S. vs. Barker* (1976), the court held that mistake of law

> generally will not excuse the commission of an offense. A defendant's error as to his *authority* to engage in a particular activity, if based upon a mistaken view of legal requirements (or ignorance thereof), is a mistake of *law*.

> ### Exhibit 10–5  Mistake of Age Statutes
>
> **Delaware**
>
> (a) Mistake as to age.—Whenever in the definition of a sexual offense, the criminality of conduct or the degree of the offense depends on whether the person has reached that person's sixteenth birthday, it is no defense that the actor did not know the person's age, or that the actor reasonably believed that the person had reached that person's sixteenth birthday.
>
> (e) Teenage defendant. —As to sexual offenses in which the victim's age is an element of the offense because the victim has not yet reached that victim's sixteenth birthday, where the person committing the sexual act is no more than 4 years older than the victim, it is an affirmative defense that the victim consented to the act "knowingly" as defined in § 231 of this title. Sexual conduct pursuant to this section will not be a crime. This affirmative defense will not apply if the victim had not yet reached that victim's twelfth birthday at the time of the act. (11 Delaware Code §762).
>
> **New York**
>
> Notwithstanding the use of the term *knowingly* in any provision of this chapter defining an offense in which the age of a child is an element thereof, knowledge by the defendant of the age of such child is not an element of any such offense and it is not, unless expressly so provided, a defense to a prosecution therefore that the defendant did not know the age of the child or believed such age to be the same as or greater than that specified in the statute. (New York Penal Law §15.20(3)).

In this instance, the court hypothesized that mistake of law would be applicable

> if a private person is summoned by a police officer to assist in effecting an unlawful arrest, his reliance on the officer's authority to make the arrest may be considered reasonable as a matter of law. (*U.S. vs. Barker*, 1976)

In general, the courts have held that the mistake of law defense "is extremely limited and the mistake must be objectively reasonable" (*U.S. vs. Moore*, 1980). Mistake of law is rarely used as a defense because of its narrow guidelines. It has been used successfully, however, against bigamy charges. Although state laws concerning marriage are hardly consistent, none allows bigamy. Pause for Thought 10–3 applies mistake of law to what otherwise would be a criminal offense.

Mistake of law has also been used to challenge a city ordinance requiring convicted felons to register (*Lambert vs. California*, 1957). In this case, the court held that notice is a fundamental principle of due process and that the accused could not be charged

> ### Pause for Thought 10–3
>
> Consider the following: A woman reports her husband missing, but the police are unsuccessful in determining his whereabouts. Ten years later she meets another man and petitions the court to have her husband declared dead. The court grants her request, and she then marries her fiancé. Two weeks later, her husband appears and claims that he had been lost in the jungle. Under state law, she is now married to two men. Does the woman have a viable defense if the state charges her with bigamy?
>
> ### Scenario Solution
>
> Yes, the woman's second marriage was an honest mistake. She relied on the legal authority of the court to marry her second husband, and as such, she is entitled to use mistake of law as a defense to any forthcoming bigamy charge.

with failure to register if notice to do so had not been given. Here, the court expressed its reluctance to follow the "ignorance is no excuse" tenet if it violated due process of the law.

## Consent of the Victim

**Consent of the victim** is a defense that negates culpability when the victim, in advance, voluntarily consented to nonserious bodily harm. Consent of the victim is a viable defense when it is an element of a crime against persons (such as rape or theft); consent has no bearing on crimes against the public order (such as gambling and intoxication). When the accused proffers consent as a defense, two general elements must be established. First, the victim must have the capability and authority to give consent. This element is especially important when considering charges such as statutory rape or sexual assault by persons of authority. For example, a 12-year-old child cannot consent to sex with an 18-year-old adult, but two 18-year-old adults can consent to sex with one another. Conversely, an 18-year-old high school student cannot consent to sex with a 22-year-old teacher.

The second consent element pertains to the manner of its obtainment. Simply put, the consent must have been voluntarily granted. Force, fraud, and other means of unreasonable or unlawful coercion are unacceptable means of obtaining consent; however, the coercion must be, for lack of a better term, reasonably unreasonable. For example, if Jill consents to sexual intercourse with Bill under the threat of death, the consent was clearly obtained unreasonably. Conversely, if Barry tells Jill that he will never call her again if she does not consent to sex, no true threat ever really evolved.

Consent also is applicable (but rarely used) in defense of injuries caused through sanctioned sporting events, such as football and boxing; however, this consent only applies to activities inherent in the event and does not cover malicious behavior. For example, a football player who breaks another player's leg while tackling him is covered under consent. Conversely, a player who break's another player's leg during a fight on the field is not covered.

## Duress

An affirmative defense of **duress** (or coercion) implies an accused committed a criminal act in response to another's volition (or will). Put simply, the defendant argues that someone made them commit the crime. Generally, federal courts have intimated that duress can only be argued as a defense when coercion was accomplished through threat of imminent physical harm or death (*Shannon vs. U.S.*, 1935; *U.S. vs. Housand*, 1977). Additionally, duress is only applicable if no other alternatives are available. Perhaps the best definition of duress was provided in *Shannon vs. U.S.* (1935):

> Coercion which will excuse the commission of a criminal act must be immediate and of such nature as to induce a well-grounded apprehension of death or serious bodily injury if the act is not done. One who has full opportunity to avoid the act without danger of that kind cannot invoke the doctrine of coercion.

Duress is generally not applicable in instances where the defendant placed himself or herself in a situation in which duress is likely or probable. For instance, a gang member would not be able to use duress as a defense to larceny because he or she knew the risks inherent to gang involvement—principle among them being the threat of reprisal for nonparticipation.

Although case law generally has been consistent regarding the permissiveness of duress as an affirmative defense, states differ as to specific elements regarding its use. Some states allow a duress defense only when implied physical harm was directed toward the defendant; other states include close relatives as permissible targets, and a few states include any such third party. Additionally, states differ as to which crimes are excused. Many states have codified duress as an affirmative defense under statute or penal code (Exhibit 10–6).

Under English common law, duress (or coercion) was a viable defense for crimes other than murder—the moral decision under duress was always to sacrifice one's own life rather than to take an innocent one. Even today, the defense of duress is not available in any state against a murder charge; however, some statutes allow evidence of duress as a mitigating factor of *mens rea*, which, in a jury trial, could result in a lesser charge or more lenient sentence. Additionally, common law afforded women the privilege of assumed duress if she committed a felony (other than murder or high treason) while her husband was present. Little attention was given to this tradition under United States law, although some states have seen fit to exclude explicitly the presumption by statute (Utah Code §76-2-302).

## Necessity

Whereas duress assumes unlawful coercion from another person, the **necessity** defense assumes that existing conditions (in nature or otherwise) caused an accused to commit a criminal act. A defense of necessity requires that "physical forces beyond the actor's control rendered illegal conduct the lesser of two evils" (*U.S. vs. Bailey*, 1980). This defense is only successful when the accused can demonstrate there were no reasonable alternatives to the criminal conduct.

### Exhibit 10-6 State Statutes Codifying Duress

**Tennessee**

(a) Duress is a defense to prosecution where the person or a third person is threatened with harm that is present, imminent, impending and of such a nature to induce a well-grounded apprehension of death or serious bodily injury if the act is not done. The threatened harm must be continuous throughout the time the act is being committed, and must be one from which the person cannot withdraw in safety. Further, the desirability and urgency of avoiding the harm must clearly outweigh the harm sought to be prevented by the law proscribing the conduct, according to ordinary standards of reasonableness.

(b) This defense is unavailable to a person who intentionally, knowingly, or recklessly becomes involved in a situation in which it was probable that the person would be subjected to compulsion. (Tennessee Code §39-11-504).

**Utah**

(1) A person is not guilty of an offense when he engaged in the proscribed conduct because he was coerced to do so by the use or threatened imminent use of unlawful physical force upon him or a third person, which force or threatened force a person of reasonable firmness in his situation would not have resisted.

(2) The defense of compulsion provided by this section shall be unavailable to a person who intentionally, knowingly, or recklessly places himself in a situation in which it is probable that he will be subjected to duress.

(3) A married woman is not entitled, by reason of the presence of her husband, to any presumption of compulsion or to any defense of compulsion except as in Subsection (1) provided. (Utah Code §76-2-302).

---

Pause for Thought 10–4 illustrates an application of the defense of necessity.

Although both federal and state courts have historically considered duress and necessity as independent defenses, some case law illustrates otherwise (see *U.S. vs. Bailey*, 1980).

One interesting necessity relationship is that with escape (from incarceration). Considering that prison conditions are often less than desirable, it is not surprising inmates attempt to escape. Courts have ruled that necessity (or duress) is a viable affirmative defense for escape, but many conditions must exist to successfully use the defense. A California appellate court (in 1974) ruled that necessity is an appropriate defense to escape provided that (1) the prisoner is faced with specific threat of death, forcible sexual

### Pause for Thought 10–4

Consider the following: A mother and child are traveling on a deserted road in the middle of winter when their vehicle breaks down. There are no other vehicles traveling on the road, and the only sign of civilization is a gas station (with no telephone) closed until the following morning. The child is thirsty and cold, and the mother also begins to feel the effects of the weather. The mother decides to break a window at the gas station to retrieve water, food and clothing to survive the night. Does the mother have a viable excuse to a charge of burglary?

### Scenario Solution

Yes, considering that no other options were available, the mother could successfully argue necessity as a defense to a charge of burglary (or other theft offense).

attack, or substantial bodily injury in the immediate future; (2) there is no time for a complaint to the authorities or there exists a history of futile complaints which make any result from such complaints illusory; (3) there is no time or opportunity to resort to the courts; (4) there is no evidence of force or violence used toward prison personnel or other "innocent" persons in the escape; and (5) the prisoner immediately reports to the proper authorities when he or she has attained a position of safety from the immediate threat (*People vs. Lovercamp*, 43 Cal. App. 3d 823, 1974). As a general rule, federal (and most state) courts follow these guidelines when considering necessity as an escape defense (*Johnson vs. State*, 379 A.2d 1129, 1977). Moreover, it is vital that the defendant surrender once the imminent threat passes. Given that escape is a continuing offense (discussed later in this chapter), an inmate who absconds from custody longer than necessary to avoid harm has discarded the use of this defense. All other existing conditions, no matter how deplorable, are irrelevant if the inmate does not promptly surrender to authorities after escape from immediate danger (*U.S. vs. Bailey*, 1980).

### Entrapment

Criminal activity often is not conducted in the presence of law enforcement. In response, officers may "go undercover" and pose as criminals to obtain evidence for arrest and conviction. Other times, however, they may use a confidential informant to gather evidence. When law enforcement participates in criminal activity, entrapment may be a defense under certain conditions. Essentially, the **entrapment** defense argues that police were responsible for encouraging or enticing one to commit a crime. An entrapment defense asserts two elements: inducement and predisposition. **Inducement** refers to police actions that present a person with an opportunity to engage in crime, whereas **predisposition** refers to past or present behavior. For entrapment to apply, evidence must be presented that (1) the crime in question was induced by a government official (e.g., a police officer) or agent of the government (e.g., a confidential informant) and (2) the defendant was not predisposed to commit such criminal activity and likely would have refrained from committing the crime under normal conditions. In most cases, inducement alone is not enough to sustain entrapment, and evidence of the defendant's predisposition must be presented. In cases in which inducement stems from egregious police behavior, however, a defense of "outrageous government conduct" may be presented.

Although entrapment to some degree violates due process, it was not recognized at common law nor considered a constitutional defense. Some states, however, have codified entrapment as an affirmative defense. For example, Delaware defines entrapment and its appropriate use as occurring when (a) the accused engaged in the proscribed conduct because the accused was induced by a law-enforcement official or the law-enforcement official's agent who is acting in the knowing cooperation with such an official to engage in the proscribed conduct constituting such conduct which is a crime when such person is not otherwise disposed to do so. The defense of entrapment as defined by this Criminal Code concedes the commission of the act charged but claims that it should not be punished because of the wrongdoing of the officer. . . . (b) The defense . . . is unavailable when causing or threatening physical injury is an element of the offense charged and the prosecution is based on conduct causing or threatening such injury to a person other than the person perpetrating the entrapment (11 Del Code §432).

Other states rely on case law to guide entrapment defenses. Although entrapment had previously been argued in state and federal courts, *Sorrells vs. U.S.* (1932) established the necessary elements for entrapment, thus affirming it as a valid defense. The U.S. Supreme Court stated that

> it is unconscionable, contrary to public policy, and to the established law of the land to punish a man for the commission of an offense of the like of which he had never been guilty, either in thought or in deed, and evidently never would have been guilty of if the officers of the law had not inspired, incited, persuaded, and lured him to attempt to commit it.

A more recent case involving child pornography illustrates the court's reluctance to exclude entrapment as a defense, even though the defendant's previous behavior indicated a predisposition to inappropriate activity (*Jacobson vs. U.S.*, 1992). Pause for Thought 10–5 details the outcome of a landmark entrapment decision of the U.S. Supreme Court.

## Other Defenses

Other defenses are similar to affirmative defenses, in that they do not necessarily deny the commission of a criminal act; however, these defenses are not based on duties, rights, or excuses. One such defense, alibi, is based on impossibility. Others are based on statutory or constitutional law that precludes prosecution under certain conditions.

## Alibi

The defense of alibi is common in U.S. courts primarily because it is an instrument used to reduce rea-

> **Pause for Thought 10–5**
>
> Consider the following case: Keith Jacobson (in 1987) was arrested subsequent to a sting operation aimed at curbing consumption of child pornography. Three years earlier, Jacobson ordered magazines containing photographs of naked boys and as a result was placed on the company's mailing list. In that same year, the Child Protection Act of 1984 (18 U.S.C. §2252A) was passed, increasing restrictions on the production, distribution, and consumption of child pornography. Jacobson's initial purchase, although deplorable, was not illegal. Subsequent to passage of the act, law enforcement officers obtained Jacobson's name from the publisher's mailing list and (as part of the sting operation) began to solicit him for purchases. Jacobson ignored the solicitations for some 26 months but ultimately relented. He was arrested, indicted, and convicted for violation of the new child pornography laws. His entrapment appeal reached the Supreme Court in 1991.
>
> **Case Decision**
>
> In 1992, the court reversed his conviction on grounds that (1) evidence of predisposition only indicated a pattern of behavior which was not illegal at the time of his initial purchase and (2) the government's case did not provide substantial evidence to indicate that his illegal purchase was independent of the government's repeated solicitations.

sonable doubt. An **alibi** defense offers evidence (usually witness testimony) that the accused could not have committed the offense in question because he or she was in another physical location. Alibi defenses are advantageous to the defendant, in that the burden of proof regarding the offense is beyond a reasonable doubt and rests with the prosecution. The defendant only needs to present evidence of an alibi to establish reasonable doubt with the jury. Although this defense has advantages, it is not perfect. If the alibi evidence is provided by a witness, the jury must weigh the credibility of that witness against the prosecution's evidence. As such, the character of the witness becomes an important factor to establishing an alibi defense. Additionally, if the witness is close to the defendant, such as a relative, spouse, or friend, his or her testimony may be regarded as doubtful by the jury. Moreover, the alibi witness risks charges of perjury if the prosecution successfully impeaches his or her testimony.

Generally, alibi defenses are most successful when evidence indicates the accused was in another location prior to and during the commission of the offense. Although a defendant's behavior subsequent to the crime may be introduced as evidence (e.g., upset or angry), the physical location after the offense is not pertinent unless it was far away from where the crime was committed. Establishing the time of an offense is critical to an alibi defense, and defense attorneys work hard to dispute eyewitness and expert testimony, as well as physical evidence, to create reasonable doubt regarding such times; however, current trends in forensic science as well as increased use of video surveillance (by both the police and public) have increasingly diminished the ways in which to dispute the time of an offense.

## Constitutional and Statutory

As mentioned earlier, several defenses rely on statutory or constitutional law to preclude prosecution for an offense—even when the evidence indicates guilt beyond a reasonable doubt (or the accused admits guilt). These defenses include (but are not limited to) double jeopardy, speedy trial, and statute of limitations. Although other defenses also are applicable under U.S. law, these three are the most commonly established defenses in federal and state courts.

### Double Jeopardy

The 5th Amendment to the U.S. Constitution states that "[no] *person* [shall] *be subject for the same offense to be twice put in jeopardy of life or limb.*" The legal history and procedural considerations regarding double jeopardy are lengthy and intricate, and thus beyond the scope of this text. Although these elements are important and pertinent to criminal law in general, present discussion regarding double jeopardy will be limited to basic fundamental principles. Generally, we examine double jeopardy as it precludes prosecution for multiple charges stemming from one criminal episode.

Double jeopardy can be used as a defense against multiple prosecutions, but not against prosecution for separate offenses. Historically, courts have cycled through different tests regarding the application of double jeopardy. In its first application, the **Blockburger test** stipulated that one criminal act could constitute two or more separate offenses only if "each provision requires proof of an additional fact which the other does not" (*Blockburger vs. U.S.*, 1932). Under the Blockburger test, only elements of the offenses are considered. For example, a defendant could not be charged with both vehicular manslaughter and reckless driving because reckless driving is a lesser included element of vehicular manslaughter.

The U.S. Supreme Court shifted its double jeopardy and successive prosecutions position in *Illinois vs. Vitale* (1980). Rather than deviating from the *Blockburger* elements, the test added a component. The court stated that two offenses, even if considered separate under Blockburger standards, would not be separated if the evidence for both was identical. In this case, the defendant was convicted of failing to slow down (to avoid an accident) and then subsequently convicted of manslaughter because of a death caused in the accident. The court ruled that the evidence for the first offense (failing to slow down) was identical to the evidence used to prove the manslaughter charge and therefore constituted double jeopardy. The court solidified this test (known as the same-evidence test) in *Grady vs. Corbin* (1990). After the *Corbin* and *Vitale* decisions, states began to differ regarding their preferred method for testing separate offenses. As a result, the U.S. Supreme Court again shifted its position and reverted to the same standards applied under *Blockburger* (*U.S. vs. Dixon*, 1993).

Double jeopardy generally does not bar prosecutions in multiple jurisdictions (*Bartkus vs. Illinois*, 1959). A defendant can be charged with the same offense in two different state courts or a state court and federal court. Suppose an individual kidnaps a child in Louisiana and transports him or her across the state line into Mississippi. Because kidnapping is both a state and federal crime, the offense could be prosecuted in Louisiana, Mississippi, and Federal court; however, some states have policies (both official and unofficial) that bar prosecution for an offense already successfully prosecuted in another jurisdiction. Whether this is a result of overworked courts or an attempt to avoid an acquittal based on double jeopardy is debatable. This provision does not apply to multiple courts within one state, meaning double jeopardy is applicable when the offense occurred in two jurisdictions within one state.

In criminal trials, jeopardy attaches, or is applied, after the jury is sworn (*Downum vs. U.S.*, 1963). This means that a defendant is covered under double jeopardy even if no testimony is presented to a jury, such as cases which result in early mistrial; however, defendants may only argue double jeopardy as a defense to subsequent prosecution for the same offense if (1) the prosecution requests a mistrial and (2) the defense raises an objection. Hence, double jeopardy does not always attach with a mistrial. If the defendant requests and is granted a mistrial, then the prosecution is not precluded from pursuing the case again unless dismissed with prejudice. Additionally, double jeopardy is not applicable in cases which result in a hung jury. One interesting facet of double jeopardy is the element of jeopardy itself. If a verdict of acquittal was a result of serious judicial misconduct (such as a bribe), then the defendant was never truly in jeopardy, and therefore, a defense of double jeopardy would not be permitted (*U.S. ex rel. Aleman*, 1997; *Aleman v. Illinois*, 1998).

## Speedy Trial

The right to *enjoy a speedy and public trial* is guaranteed under the 6th Amendment to the U.S. Constitution. Although it is unlikely that the framers of the Constitution could have predicted the volume of court cases today, they nonetheless saw the wisdom in protecting citizens from malicious prosecutors. Because there is no specific time period defined in the amendment, the individual States and federal government have had to construct their own definitions. Federal law states that trial must commence no later than seventy days after indictment in federal court (18 U.S.C. §3161c[1]). Some states have codified time limits for the commencement of trials within their rules of criminal procedure, whereas others rely on more subjective case law which outlines certain factors to be considered. The right to a speedy trial can be waived by the defendant, but doing so precludes the use of speedy trial as a defense.

*Barker vs. Wingo* (1972) is considered the primary case law regarding speedy trials. The court stated that any inquiry into a speedy trial claim necessitates a functional analysis of the right in the particular context of the case:

> The right of a speedy trial is necessarily relative. It is consistent with delays and depends upon circumstances. It secures rights to a defendant. It does not preclude the rights of public justice. (*Beavers vs. Haubert*, 1905)

In *Barker*, the court developed a four-pronged test to determine subjectively when speedy trial rights have been violated. The test considered (1) defendant's assertion of the right, (2) prejudice to the petitioner,

(3) length of delay, and (4) the reason(s) for delay. The test elements are then weighed to ascertain whether the defendant's rights were violated. In *Barker*, the court held that a lengthy delay did not outweigh the fact that (1) the defendant did not want a speedy trial and (2) there was minimal prejudice to the defendant as a result of the delay (Exhibit 10–7).

## Statute of Limitations

Whereas the right to a speedy trial focuses on the time between indictment (or arrest) and trial, a **statute of limitations** is concerned with the time frame between the commission (or discovery) of a crime and arrest or indictment. English common law did not recognize these provisions, nor are such rights bestowed in the U.S. Constitution—they are purely legislative provisions. The federal government and most states have enacted laws placing time restrictions of varying lengths on the prosecution of certain crimes. Generally, murder has no statute of limitations.

Federal law regarding statutes of limitations is lengthy and covers a wide array of offenses. For most noncapital crimes, the statute of limitations is 5 years. Also, crimes against children are afforded a lengthier time—10 years after the offense or during the life of the child (18 U.S.C. §3283). This means that an offense such as kidnapping or sexual abuse of a child has no statute of limitations so long as the child is alive. If the child is dead, then the statute reverts to the 10-year time limit. As stated before, there is no federal statute of limitations regarding capital offenses (offenses that carry the death penalty) (18 U.S.C. §3281).

Although statutes of limitations define the period between the commission of the crime and arrest or indictment, there arises a problem regarding offenses that (by design) are concealed from the eyes of the criminal justice system. Concealing assets, ongoing fraud, and other similar crimes are considered **continuing offenses**, which do not end until the activity is discovered by authorities. As such, continuing offenses are not applicable under statutes of limitations until discovered. For example, if a banker intentionally conceals assets before filing bankruptcy, the offense of fraud is continuous until he or she confesses or those activities are discovered by law enforcement personnel. As mentioned previously, escape is also considered a continuing offense and does not end until the absconded individual is apprehended.

The **tolling** of a statute, or pause in the time limit imposed by a statute of limitations, is triggered by certain events or circumstances. These events may include but are not limited to (1) the issuance of an arrest warrant or the filing of an information, (2) an initiated prosecution, (3) absconding from justice (fleeing fugitive), and (4) concealing an offense. States differ as to what tolls a statute of limitation, but nearly all contain some element of the above conditions. Pause for Thought 10–6 illustrates the proper interpretation of statutes of limitation.

Exhibit 10–8 details the statutes of limitations for two states.

## Summary

This chapter provided a comprehensive overview of the defenses available in the criminal justice system; however, the list of defenses is hardly exhaustive, as the topic of defense can easily fill a separate text! The evolution of defense is a continuous process, and students are encouraged to remain aware of such evolving changes in law. Additionally, this chapter provided a realistic portrait of defense under U.S. law. Television shows and movies often portray defenses in unrealistic terms, where bad guys "get off on technicalities" and good guys avoid imprisonment for killing bad guys. In reality, most defendants

### Exhibit 10–7 Byron De La Beckwith

Byron De La Beckwith was an admitted racist and member of the Ku Klux Klan who was unsuccessfully prosecuted for the assassination of civil rights leader Medgar Evers in Jackson, Mississippi. The two initial trials in 1964 resulted in mistrials due to less-than-unanimous verdicts from exclusively white juries. Thirty years later, Beckwith was successfully prosecuted for the death of Medgar Evers. Beckwith appealed his conviction on the grounds that his right to a speedy trial was violated, due to the 30-year lapse between prosecutions. The Mississippi Supreme Court disagreed and upheld the conviction (*Beckwith vs. State*, 707 So. 2d 547, 1997).

### Pause for Thought 10-6

Consider the following: A state imposes a 5-year statute of limitations on fraud. In 1990, Tim obtains a false credit card in someone else's name and makes several purchases. He stops using the card in 1993, at which time he obtains a new fraudulent card. In 2001, the local police discover Tim's 1990 fraud and charge him accordingly. Does the statute of limitations on fraud prevent Tim from being charged?

### Scenario Solution

No, the statute of limitations for the fraud in question would begin in 2001 (with the discovery of the crime) and end in 2006 because Tim concealed his fraud and the offense was continuous. If Tim were to flee the state after his arrest, the "clock would stop" on the statute and not begin again until he reentered the state.

---

### Exhibit 10-8 State Laws Regarding Statutes of Limitations

**Mississippi**

The passage of time shall never bar prosecution against any person for the offenses of murder, manslaughter, aggravated assault, kidnapping, arson, burglary, forgery, counterfeiting, robbery, larceny, rape, embezzlement, obtaining money or property under false pretenses or by fraud, felonious abuse or battery of a child as described in Section 97-5-39, touching or handling a child for lustful purposes as described in Section 97-5-23, sexual battery of a child as described in Section 97-3-95(1)(c), (d) or (2), or exploitation of children as described in Section 97-5-33. A person shall not be prosecuted for conspiracy, as described in Section 97-1-1, or for felonious assistance program fraud, as described in Section 97-19-71, unless the prosecution for such offense be commenced within five (5) years next after the commission thereof. A person shall not be prosecuted for any other offense not listed in this section unless the prosecution for such offense be commenced within two (2) years next after the commission thereof. Nothing contained in this section shall bar any prosecution against any person who shall abscond or flee from justice, or shall absent himself from this state or out of the jurisdiction of the court, or so conduct himself that he cannot be found by the officers of the law, or that process cannot be served upon him (Miss. Code Ann. § 99-1-5).

**Vermont**

(a) Prosecutions for aggravated sexual assault, murder, arson causing death, and kidnapping may be commenced at any time after the commission of the offense.

(b) Prosecutions for manslaughter, sexual assault, lewd and lascivious conduct, sexual exploitation of children, grand larceny, robbery, burglary, embezzlement, forgery, bribery offenses, false claims, fraud under subsection 141(d) of Title 33, and felony tax offenses shall be commenced within six years after the commission of the offense, and not after.

(c) Prosecutions for sexual assault, lewd and lascivious conduct and lewd or lascivious conduct with a child, alleged to have been committed against a child 16 years of age or under, shall be commenced within the earlier of the date the victim attains the age of 24 or six years from the date the offense is reported, and not after. For purposes of this subsection, an offense is reported when a report of the conduct constituting the offense is made to a law enforcement officer by the victim.

(d) Prosecutions for arson shall be commenced within 11 years after the commission of the offense, and not after.

(e) Prosecutions for other felonies and for misdemeanors shall be commenced within three years after the commission of the offense, and not after (Vermont Statutes §4501).

never see the inside of a courtroom, and those that do rarely use creative means to defend themselves. Common defenses (such as alibi, statutes of limitations, or self-defense) most often are employed during the course of investigation—before a person is even charged with a crime.

## Practice Test

1. The _____ test was the first effort by the U.S. Supreme Court to determine separate offenses under the double jeopardy clause.
   a. Frye
   b. Blockburger
   c. Substantial capacity
   d. Irresistible impulse
   e. M'Naghten

2. _____ is a defense that asserts that unlawful coercion from another induced the commission of a criminal act.
   a. Duress
   b. Necessity
   c. Diminished capacity
   d. Justification
   e. Consent

3. Statutes of limitations do not apply to _____ until discovered.
   a. strict liability offenses
   b. excuse defenses
   c. mistakes of law
   d. continuing offenses
   e. mistakes of fact

4. Constitutional protection from multiple prosecutions for the same offense is known as _____.
   a. statutes of limitations
   b. duress
   c. double jeopardy
   d. *demurrer*
   e. entrapment

5. The _____ was used at common law to assess legal insanity.
   a. Blockburger test
   b. substantial capacity test
   c. diminished capacity test
   d. ALI Standard
   e. M'Naghten Rule

6. A _____ defense assumes that an accused was honestly in error about certain extra-legal circumstances preceding the offense.
   a. mistake of law
   b. mistake of age
   c. infancy
   d. mistake of fact
   e. entrapment

7. Mental _____ is a condition that will not improve or deteriorate over time.
   a. defect
   b. disease
   c. incompetence
   d. illness
   e. deficiency

8. _____ refers to intoxication by fraud or coercion and is a viable affirmative defense if the accused had no criminal intent prior to the intoxication.
   a. Voluntary intoxication
   b. Entrapment
   c. Involuntary intoxication
   d. Insanity
   e. Inducement

9. _____ refers to an involuntary action(s) by a defendant that results in a criminal offense.
   a. Tolling
   b. Insanity
   c. Intoxication
   d. Automatism
   e. Retardation

10. A(n) _____ defense asserts some justification or excuse for the commission of a criminal act.
    a. affirmative
    b. statutory
    c. insanity
    d. alibi
    e. negative

11. _____ refers to decreased mental competency that mitigates responsibility for a criminal act through the element of *mens rea*.
    a. Substantial capacity
    b. Diminished capacity
    c. Insanity
    d. Mistake of law
    e. Duress

12. The _____ doctrine removes the traditional duty of a homeowner to retreat (to safety) from their home when their lives (or lives of others) are in immediate danger.
    a. Castle
    b. Blockburger
    c. Substantial capacity
    d. M'Naghten
    e. Frye

13. _____ is a defense that asserts that government officials or agents induced the accused to commit an offense that under normal circumstances would not have been committed.
    a. Duress
    b. Coercion
    c. Justification
    d. Mistake of law
    e. Entrapment

**14.** Some states require a(n) _____ before using deadly force in the defense of self and others.
   a. alibi
   b. consent
   c. inducement
   d. duty to retreat
   e. waiver

**15.** _____ of a statute of limitations refers to "stopping the clock" on prosecutorial time limits and may be triggered by flight from prosecution.
   a. Termination
   b. Inducement
   c. Tolling
   d. Ticking
   e. Waiver

**16.** _____ is a defense that asserts that an accused committed a criminal act because of extenuating yet natural circumstances beyond their control.
   a. Necessity
   b. Duress
   c. Automatism
   d. Coercion
   e. Alibi

**17.** _____ places a limit on the time allowed between commission of a crime and arrest or indictment of the defendant.
   a. Waiver
   b. Mistake of law
   c. Right to speedy trial
   d. Mistake of fact
   e. Statute of limitations

**18.** _____ placed limits on the police use of deadly force against fleeing felons.
   a. *Barker vs. Wingo*
   b. *Aleman vs. Illinois*
   c. *Downum vs. U.S.*
   d. *Tennessee vs. Garner*
   e. *People vs. Lovercamp*

**19.** _____ refers to past or present behavior of an accused that indicates whether such person was inclined to commit certain acts.
   a. Predisposition
   b. Duress
   c. Inducement
   d. Cognition
   e. Volition

**20.** _____ refers to the standard of mental competence used in the ALI Standard.
   a. Insanity
   b. Diminished capacity
   c. Substantial capacity
   d. Infancy
   e. Compulsion

## References

Coleman, S. (1999). *Mentally ill criminals and the insanity defense: a report to the Minnesota legislature* (p. 17). St. Paul: Center for Applied Research and Policy Analysis.

Davidson, M. J. (2000). Feminine hormonal defenses: premenstrual syndrome and postpartum psychosis. *The Army Lawyer, July 2000.*

The insanity defense among the states. Retrieved on September 13, 2009, from http://criminal.findlaw.com/crimes/more-criminal-topics/insanity-defense/the-insanity-defense-among-the-states.html.

Memon, R. (2006). Legal theory and case law defining the insanity defense in English and Welsh law. *Journal of Forensic Psychiatry & Psychology, 17*(2), 230–252.

Robson, S. (2008, August 21). Using PTSD as a defense. *Stars and Stripes.* Retrieved on June 22, 2009, from http://www.stripes.com.

# Organized Crime and Terrorism

**CHAPTER 11**

## Key Terms

Antiterrorism and Effective Death Penalty Act of 1996
Aviation and Transportation Security Act of 2001
Bank Secrecy Act of 1970
Criminal enterprise
Domestic terrorism
Espionage
Foreign Intelligence and Surveillance Act of 1978
Foreign Intelligence and Surveillance Court
Homeland Security Act of 2002
International terrorism
Jihad
La Cosa Nostra
Misprision of treason
Money laundering
Money Laundering Control Act of 1986
Narco terrorism
Omnibus Crime Control and Safe Streets Act of 1968
Organized crime
Organized Crime Control Act of 1970
Political-dissident terrorism
Protect America Act of 2007
Quasi-political terrorism
Racketeering
Religious-extremist terrorism
RICO Act of 1961
Sabotage
Sedition
Smith Act of 1940
State-sponsored terrorism
Structured transactions
Terrorism
Treason
USA PATRIOT Act of 2001
Usury
Vigorish

## Introduction

This chapter provides a comprehensive overview of criminal law applications to organized crime and terrorism. Although not often associated with each other, these entities do possess similarities such as complicated and problematic definitions, ethnicity and nationality-based participation, multiple actors, and unique methods of prosecution. Additionally, both organized crime and terrorism have influenced controversial legislation which has altered traditional law enforcement techniques. This chapter also explores the histories and structures of various organized crime groups and terrorist organizations.

## Organized Crime

The following section will provide readers with a broad organized crime foundation, beginning with common misconceptions about and a brief history of organized crime in the United States. Legal issues regarding organized crime, as well as common offenses committed within the scope of organized criminal activity, also are explored. The section concludes with a discussion about emerging issues in organized crime.

## Misconceptions of Organized Crime in America

Defining organized crime can seem like an exercise in futility. Stop for a moment and form a mental image of the individuals in the business of organized crime. What do you see? Pinky rings and silk suits? Most Americans have a preconceived notion of organized crime, most of which are incorrect. To understand how criminal law functions with respect to organized criminal enterprises, it is important to understand the history of organized crime in America.

## History of Organized Crime in America

Traditional organized crime in America likely began in the Five Points district of New York in the mid 1800s. Immigration was at an all-time high, with most neighborhoods occupied by individuals with cohesive ethnic backgrounds. Poverty and disease were rampant, and living conditions were deplorable. Small gangs formed as a means of survival, protection, and economic advancement, eventually evolving into larger ethnic-based organizations. Most notably, the Irish, Italian, and Jewish communities were the visible leaders of the organized crime movement.

### Tammany Hall

Although Italian immigrants are primarily credited with initiating American criminal syndicates, the Irish mob is the eldest of such organizations. By 1860, Irish immigrants established a foothold in American politics through membership in Tammany Hall, an Irish political organization founded in the early 1800s in the bars and taverns of New York. Its primary purpose was to further the interests of the mostly poor Irish community. In 1868, the head of Tammany Hall, William March (Boss) Tweed, was elected to the New York state senate. The years after his election produced corruption at the highest levels, including the pilfering of New York City's treasury. The Tammany Hall political machine dominated New York politics until the 1930s. Thus, Tweed's wedding of organized crime with political corruption places him as a prime candidate for marking the real beginning of the American organized crime movement.

### The Mafia

The evolution of Italian-American organized crime is steeped in mystery, and no two versions of the story are the same. Details aside, the current incarnation of Italian-American organized crime began with two groups: the Mafia, a loosely associated group of Sicilian land owners, and the Camorra, a group of criminal syndicates sprouting from Italian prisons. Both factions were represented among Italian immigrants, and Americanized versions of each group began to appear in the early 1900s. By 1930, two major crime families emerged in New York, under the patriarchal guidance of Joe (The Boss) Masseria and Salvatore Maranzano.

Prohibition provided ample opportunity for these Italian groups to profit from the importation and sale of illegal alcohol, but internal competition resulted in the Castellammarese War. The ensuing and bloody conflict ended through the efforts of Charles (Lucky) Luciano, who engineered the murder of both bosses and corralled Italian crime syndicates into one group led by bosses of each family—the Commission. Luciano revolutionized Italian organized crime by establishing partnerships with Jewish criminals such as Meyer Lansky and Benjamin (Bugsy) Siegel, bolstering his influence both during and after the war. Essentially, this marked the beginning of the modern Mafia, also known as **La Cosa Nostra** (*Our Thing*).

### Jewish Organized Crime

The origin of Jewish organized crime is murky at best. The earliest and most notable organized Jewish criminal activity began with Arnold "The Brain" Rothstein, known for his involvement in prostitution, gambling, narcotics, and a host of other crimes in the early 1900s. Rothstein also was responsible for fixing the 1919 World Series—in which eight Chicago Black Sox players were paid to throw games. Rothstein also had ties with Meyer Lansky and Benny "Bugsy" Siegel, two notable Jewish criminals who aligned themselves with the Italian Mafia. Although the existence of a "Jewish Mafia" has never been proven, there is little doubt that groups of Jewish criminals engaged in organized crime activities. In the tradition of the Lansky/Siegel-Luciano partnership, these groups were not above working with other ethnic-based gangs. For example, Dutch Schultz, whose gang was mostly comprised of Jewish criminals, ran one of the largest numbers rackets in New York during the 1930s, a feat that would not have been possible if the Harlem (black) gangs had not participated.

## Nature of Organized Crime

Before exploring a legal definition of organized crime, we first must establish a working definition. **Organized crime** can be categorized as ongoing criminal activity perpetrated by individuals belonging to semi-exclusive groups. Organized crime often occurs in conjunction with legitimate business. Although this definition may seem overly broad, it includes four essential elements of organized activity:

- Ongoing criminal activity refers to crimes that take place over extended periods of time. Simply put, criminal activity is considered "business as usual."
- The term *individuals* indicates that more than one person is involved in criminal activity.
- Organized crime groups are semiexclusive, which means that a large portion of criminal syndicates base their membership eligibility on ethnicity, nationality, or physicality (location). Other groups may base eligibility on previous criminal behavior or require prospective mem-

bers to engage in a serious crime (such as murder) as a rite of passage.
- Legitimate businesses often act (knowingly and unknowingly) as fronts for organized crime groups. Mostly, these are cash-based businesses, such as casinos, bars, restaurants, or groceries. Illegal income is then "washed" or "laundered" through these businesses to mask its origin. Organized crime groups have also infiltrated labor unions to manipulate business contracts and control pension funds.

## Legal Issues in Organized Crime

To more fully understand organized crime, we will explore the history of legal responses to its activities. While some laws (such as Prohibition) inadvertently increased organized crime activity, other laws have specifically combated organized crime by providing law enforcement with advanced investigation and prosecution tools. This section provides a brief overview of the legal issues regarding organized crime.

### Prohibition (1920–1933)

The era of Prohibition was marked by the passage of the Volstead Act and the ratification of the 18th Amendment to the U.S. Constitution (1919–1920). As touched on in Chapter 7, the purpose of the Volstead Act of 1919 was to regulate the manufacture, importation, exportation, and possession of alcohol in the United States. Primarily, however, the Volstead Act defined what constituted intoxicating liquor. Shortly after the Volstead Act was passed, the 18th Amendment was ratified. Section 1 of the 18th Amendment states:

> After one year from the ratification of this article the manufacture, sale, or transportation of intoxicating liquors within, the importation thereof into, or the exportation thereof from the U.S. and all territory subject to the jurisdiction thereof for beverage purposes is hereby prohibited.

The 18th Amendment also gave concurrent jurisdiction to the federal government and the States to enforce alcohol violations.

Organized crime syndicates, realizing that demand for alcohol would dramatically increase, quickly seized the opportunity and established bootlegging operations throughout the nation. The competition for business quickly turned deadly, and organized crime was thrust into the spotlight as a major problem in the United States, particularly in larger cities like New York and Chicago. The resulting chaos spawned the birth of the Federal Bureau of Investigation.

Prohibition ended in 1933, when the 21st Amendment to the U.S. Constitution was ratified. The 21st Amendment repealed the 18th Amendment and gave exclusive jurisdiction over alcohol to state and local governments. The 18th Amendment is unique in that it is the only constitutional amendment to be repealed. Although prohibition had ended (at least on a federal level), many states chose to prohibit the manufacture, sale, and transportation of alcohol, allowing organized crime syndicates to continue bootlegging operations. Organized crime did not decrease when prohibition ended, however, and groups specializing in the illegal transportation and sale of alcohol simply increased their involvement in other illicit activities, such as narcotics, prostitution, and illegal gambling.

### Omnibus Crime Control and Safe Streets Act of 1968

In 1967, the President's Commission on Law Enforcement and Administration of Justice issued a report warning of the proliferation of organized crime in America. A logical response to this growing problem, the **Omnibus Crime Control and Safe Streets Act of 1968** was the first piece of American legislation to specifically define organized crime: unlawful activities of the members of a highly organized, disciplined association engaged in supplying illegal goods and services, including but not limited to gambling, prostitution, loan sharking, narcotics, labor racketeering, and other unlawful activities. The act not only provided a legal definition of organized crime but further created and funded subagencies in the Department of Justice to combat organized criminal activity. Moreover, this act regulated the Federal Bureau of Investigation's wiretapping activity.

### Organized Crime Control Act of 1970

The **Organized Crime Control Act of 1970** provided prosecutors and law enforcement agencies with two valuable tools to combat organized crime. First, this act gave prosecutors the authority to (1) hold uncooperative witnesses in jail and (2) protect witnesses and their families during and after trial. Second, the act contained a provision known as the Racketeer Influenced and Corrupt Organizations Act (RICO).

### RICO

The **RICO Act of 1961** provides law enforcement agencies and prosecutors with creative ways to investigate, indict, and convict individuals involved in organized crime. Specifically, RICO stipulates that individuals associated with an enterprise (formal or informal) and who commit certain offenses twice during a 10-year period are guilty of **racketeering**. The act also provided that other members of such a group can also be prosecuted for racketeering—provided their association with the offenders can be proven in court. The RICO Act (18 U.S.C. §1961) defines a **criminal enterprise** as "any individual, partnership, corporation, association, or other legal entity, and any union or group of individuals associated in fact although not a legal entity."

The list of RICO offenses is quite lengthy and includes state crimes such as murder and bribery, as well as federal crimes such as wire fraud and witness tampering. The RICO Act also includes provisions for freezing assets during investigations and asset seizure under civil proceedings. Most states have enacted similar legislation to apply the provisions of RICO under state law and enable state-level prosecutions for racketeering. RICO also permits prosecutors to charge conspiracy under the umbrella of racketeering, thus easing the burden of proof for the prosecution in that no overt act (other than the two initial offenses) must be proven. Simply put, individuals with knowledge of the offenses and with proven ties to the offenders can be charged with racketeering and subject to criminal penalties of up to 20 years in prison and a $25,000 fine for each count. Pause for Thought 11–1 illustrates how RICO is applied at federal law.

## Offenses Associated with Organized Crime

Although no offense is specific to organized crime, some crimes repeatedly occur in association with organized crime, such as counterfeiting, money laundering, and usury (loan sharking). Other crimes commonly associated with organized crime (prostitution, illegal gambling, and narcotics sales) are addressed elsewhere in this book.

## Counterfeiting

Counterfeiting is defined by federal law (18 U.S.C. §470) as (1) making, dealing, or possessing any counterfeit obligation or other security of the United States or (2) making, dealing, or possessing any plate, stone, analog, digital, or electronic image, or other thing, or any part thereof, used to counterfeit such obligation or security. Counterfeiting currency has long been a lucrative business, primarily because of the reliability, strength, and worldwide acceptance of the American dollar. As previously mentioned, many organized crime groups use legitimate, cash-based businesses as fronts for their organizations. These businesses often circulate counterfeit money, with the volume of transactions making the origin of counterfeit money difficult to trace. Illegal gambling establishments controlled by organized crime groups have also used counterfeit money to pay out winnings. Although some organized crime groups engage in counterfeiting as a means to increase profit, counterfeit operations often stand alone as a sole source of income for criminal organizations.

Counterfeiting operations directed at U.S. currency are not limited to American soil, and the number of international counterfeiting rings has steadily increased over the last 20 years. Colombian drug cartels have used counterfeit U.S. currency to finance operations, and some evidence indicates that North Korea has used such currency to finance arms purchases. The reality is that international counterfeiting operations are especially difficult to detect and prosecute, primarily because of jurisdictional issues and extradition procedures. The U.S. Secret Service, and thereby the U.S. Department of Homeland Secu-

### Pause for Thought 11–1

Consider the following: John and Bill are members of The Dukes, a social club whose members gather weekly. They are under investigation for attempting to bribe a health inspector who examined the kitchen at a restaurant owned by Bill. After the inspector refused the bribe, John and Bill called the inspector's house and left threatening messages on his answering machine. Is it possible for federal prosecutors to indict John and Bill under the RICO Act?

### Scenario Solution

Yes, John and Bill belonged to an enterprise and committed two offenses (attempted bribery and witness intimidation) covered under the RICO Act within a 10-year period. Additionally, even if John and Bill did not belong to a social club, they would still be considered an enterprise under the definition provided in the RICO Act (Exhibits 11–1 and 11–2).

> **Exhibit 11-1** *NOW v. Schedler et al.*
>
> **510 U.S. 249 (1994)**
>
> In 1994, the U.S. Supreme Court held that motive for economic advancement (profit) was not an element of racketeering. In essence, racketeering charges could be filed against anyone committing predicate offenses under RICO, even if they were not concerned with making a profit. The court stipulated that an enterprise can affect commerce even if it is not economically motivated. Put simply, the Court held that disruption of business qualifies as interfering with interstate and foreign commerce. Although this ruling was in response to a civil suit brought by the National Organization for Women against various antiabortion activists, the ruling still applied to federal prosecutions under the RICO statutes.

> **Exhibit 11-2** *Waucaush v. U.S.*
>
> **380 F.3d 251 (6th Cir. 2004)**
>
> In 2004, the U.S. Court of Appeals held that predicate acts are not applicable as RICO offenses unless they affect interstate commerce. In other words, murdering a rival gang member to control his turf is only a predicate offense to RICO if the prosecution can prove that the offense affected interstate commerce.

rity, has legal jurisdiction over counterfeiting operations involving U.S. currency (18 U.S.C. §3056).

## Money Laundering

**Money laundering** refers to the process by which illegal income is "washed" or disguised as legitimate. Federal law (18 U.S.C. §1956) defines this offense as:

> Whoever, knowing that the property involved in a financial transaction represents the proceeds of some form of unlawful activity, conducts or attempts to conduct such a financial transaction which in fact involves the proceeds of specified unlawful activity . . . knowing that the transaction is designed in whole or in part to conceal or disguise the nature, the location, the source, the ownership, or the control of the proceeds of specified unlawful activity; or to avoid a transaction reporting requirement under State or Federal law.

Money laundering is common among organized crime groups because of the extralegal sources of their income. As mentioned at the beginning of this chapter, organized crime groups often use legitimate businesses to launder proceeds from illegal activities. Money laundering is a relatively complicated process involving three stages: placement, layering, and integration. First, placement involves depositing illegal income into a bank account through **structured transactions**, designed to break up large amounts of money into multiple smaller deposits—primarily because all deposits in the U.S. greater than $10,000 must be reported to the federal government. Second, layering involves multiple transactions to move illegal income among various accounts and holdings. Lastly, integration occurs when money is placed in legitimate areas of the economy, such as stocks or legitimate business ventures.

Two pieces of legislation primarily address money laundering in the United States. First, the **Bank Secrecy Act of 1970** mandated financial institutions to report transactions exceeding $10,000 and implemented a reporting system for suspicious activity. Second, the **Money Laundering Control Act of 1986** augmented the Bank Secrecy Act of 1970 and criminalized the use of structured transactions for the purpose of avoiding detection. Criminal penalties for money laundering include up to 20 years in prison and fines up to $500,000. The federal government can also pursue money laundering cases in civil court to seize associated assets.

## Usury

**Usury** (also known as loan sharking) refers to excessive interest rates applied to loans. Organized crime syndicates often favor usury as an income means due to (1) the nature of their customers and (2) their reputations for violence. Organizations that operate illegal gambling establishments often provide "loans" to customers with interest rates exceeding 25%. Moreover, those who receive the loans understand that failure to remit payment may result in bodily harm. Although definitions for usurious interest differ from state to state, portions of the RICO Act (18 U.S.C. §1961) address usury in the following manner:

> Unlawful debt means a debt incurred or contracted in gambling activity which was in violation of the law of the U.S., a State or political subdivision thereof, or which is unenforceable under State or Federal law in whole or in part as to principal or interest because of the laws relating to usury, and which was incurred in connection with the business of gambling in violation of the law of the U.S., a state or political subdivision thereof, or the business of lending money or a thing of value at a rate usurious under State or Federal law, where the usurious rate is at least twice the enforceable rate.

Interest on a usurious loan often is called **vigorish** (or vig). Most loan-sharks apply larger vigs for smaller loans and smaller vigs for larger loans. The charge of usury is mostly used in conjunction with other RICO offenses.

## Emerging Issues in Organized Crime

In the past 20 years, the face of organized crime has changed dramatically. Russian organizations, primarily in New York, have gained a reputation for ruthlessness and violence that rivals that of any other organized criminal enterprise. Additionally, there has been an increase in the number of web-based organized crime groups, most notably e-mail scams based out of Nigeria. As technology advances, so do the operational methods and capabilities of criminal syndicates. Most offenses associated with techno-crimes (both U.S. based and international) are covered under laws regarding wire fraud (18 U.S.C. §1343 and §1961).

## Organized Crime and Gangs

Throughout this chapter, we have examined the definition of organized crime and common attributes of organized crime syndicates. Several questions come to mind when thinking about organized crime: What about street gangs? How do we classify them? What is the difference between a street gang and an organized criminal enterprise? The answers to these questions are intertwined into one concept: organizational activity.

Most street gangs are social in nature, and criminal activity within the context of the gang is often a function of individual preference rather than membership. Gangs engage in violence, prostitution, narcotics sales, and other offenses associated with organized crime, but the extent of involvement of the actual "gang" is rarely organized; however, as gangs become increasingly powerful, they may evolve into an organized criminal enterprise (Exhibit 11–3).

## Terrorism

Under federal law, terrorism is delineated into two basic categories: domestic and international (18 U.S.C. §2331). **Domestic terrorism** refers to acts occurring within the United States or its territories, whereas **international terrorism** comprises acts occurring outside the territory of the United States. In general, the legal definition of terrorism rests on the intent behind a dangerous act. The U.S. Code lists three motives underlying terrorist acts: (1) intimidation or coercion of a civilian population, (2) influencing government policy by intimidation or coercion, and (3) affecting the conduct of a government by mass destruction,

### Exhibit 11–3 The Conservative Vice Lords

In the early 1970s, the Vice Lords, one of the largest Chicago street gangs, changed their name to the Conservative Vice Lords to bolster their legitimacy. They began initiating community improvements and organizing semipolitical events to raise community awareness. Oddly enough, they applied for and received a grant from the Rockefeller Foundation (who had no prior knowledge of the group's history). The Conservative Vice Lords were also involved in other activities such as recruiting other gangs and narcotics sales. Subsequently, a federal investigation regarding the misuse of grant funds resulted in the indictment and conviction of several of its leaders. Although the Vice Lords began as a common street gang, their activities evolved into the appearance of an organized criminal syndicate.

assassination, or kidnapping (18 U.S.C. §2331). In this sense, the only thing that separates a common murderer from a terrorist is motive. The U.S. Code (22 U.S.C. §2656f[d]) also provides some generalized terrorism definitions for reporting purposes:

1) "international terrorism" means terrorism involving citizens or the territory of more than 1 country;
2) **terrorism** means premeditated, politically motivated violence against noncombatant targets by subnational groups or clandestine agents.

Although these terrorism definitions seem relatively sound, they only apply to U.S. law. The United Nations has struggled for many years to establish a universal definition of terrorism, but have made little headway. The difficulty of establishing a universal terrorism definition lies in perspective. One famous saying, whose initial authorship is disputed, posits that "one man's terrorist is another man's freedom fighter." Although this expression is clever, it is misleading. No one has ever claimed to be a terrorist, whereas many have claimed to be freedom fighters (or revolutionaries). Terrorism is not an occupation or goal, but rather a method of manipulation.

## Types of Terrorism

There are many ways to classify terrorist acts. No two organizations or texts seem to agree on any taxonomy system for terrorism or terrorists; however, differences in classification are really about semantics (naming the type) rather than tangible differences among groups. Primarily, terrorism can be classified into four categories: political-dissident, state-sponsored, religious-extremist, and quasi-political. Not all terrorist acts are exclusively classified and may fall into two or more categories. Additionally, it is necessary to understand that a terrorist group's motivation may change over time.

## Political-Dissident

**Political-dissident terrorism** (also known as insurgency) occurs when citizens of a nation or state attack the government or society as a result of conflicting ideals. Political-dissident terrorism is common in third-world countries where governments are unstable or corrupt and regime changes frequently occur. Recent examples include the insurgency movement in Iraq following the capture of Saddam Hussein, and the Taliban-led movement in Afghanistan following the Soviet-Afghan War. Primarily, these groups use terrorism to influence government policy and intimidate the civilian population. They resort to terrorist methods because of their small numbers and limited access to money, weapons, and equipment.

Even though political-dissident terrorism is most common among third-world countries, it is not geographically specific. The United States has experienced its share of political-dissident terrorism. Timothy McVeigh and Terry Nichols were convicted of the 1995 bombing of the Murrah Federal Building in Oklahoma City. Their motives were reputedly tied to disapproval of government actions regarding the destruction of the Branch Davidian compound in Waco, Texas, and the Ruby Ridge incident. Theodore "Ted" Kaczynski, also known as the Unabomber, waged a 17-year bombing campaign by mail and other methods because of his disdain for industrial progress in the United States. Again, these individuals are "terrorists" because of their motives, not their actions.

### State-Sponsored

**State-sponsored terrorism** refers to terrorist acts perpetrated by a government. These acts can be directed at its citizens or citizens of another state or country. State-sponsored terrorism is extremely difficult to combat, primarily because governments are responsible for, approve of, or encourage terrorist acts. Most state-sponsored terrorism is directed at citizens within its borders, and the majority of acts are primarily used for controlling or manipulating the population; however, the United States designates state-sponsors of terrorism by observing their actions in other countries and willingness to extradite terrorist suspects. The U.S. State Department currently lists four countries as state-sponsors of terrorism: Cuba, Iran, Sudan, and Syria. It should be noted, however, that the U.S. State Department listed North Korea as a state-sponsor of terrorism until 2008 (U.S. State Department, 2009).

### Religious-Extremist

**Religious-extremist terrorism** refers to acts committed by individuals who espouse a hardline view of their religion. These individuals/groups believe they have an moral justification for committing terrorist acts. Religious-extremists are not necessarily concerned with political affairs, and their primary goal is to eliminate actions in contrast to their religious beliefs. Common examples of this type of terrorism are individuals who bomb abortion clinics, or those who wage holy war, or **jihad**, under jihadi-Salifi Islamic principles. These individuals do not seek political change per se, but rather purification of the world. For example, the goal of individuals who bomb abortion clinics is to discourage the operation of the clinics rather than influencing government policy regarding abortion (i.e., overturning *Roe vs. Wade*). In some cases, such as Islamic extremists, political and religious motivations are synonymous because of the intermingling of religion and government in many countries.

### Quasi-Political

**Quasi-political terrorism** refers to terrorist acts committed by individuals with no intrinsic desire for political or religious change, but rather a desire to instill fear in another group. These individuals (or groups) use terrorist tactics as methods of subversion, intimidation, and manipulation. For example, the Klu Klux Klan bombed black churches and lynched members of the black community during the Civil Rights Movement in the 1960s to discourage minorities from voting. Drug cartels in Central and South America often engage in **narco terrorism** to eliminate competitors in the narcotics industry or intimidate government officials. These groups may have some degree of underlying religious or political justification for their actions, but they do not seek political change at the government level.

## Legal Issues in Terrorism

The legal response to terrorism in the United States has been swift and broad. Laws aimed at terrorism provide investigators with increased surveillance and enhanced investigatory powers, as well as proscribe harsh punishments for individuals convicted of engaging in terrorist activities. This section provides a brief overview of the laws designed to combat terrorism.

### The Foreign Intelligence and Surveillance Act (1978, 2000)

The **Foreign Intelligence and Surveillance Act of 1978** (FISA) granted the federal government powers to conduct electronic surveillance and physical searches in the United States without a warrant, provided those persons of interest were not U.S. citizens (50 U.S.C. Ch. 36). There are provisions in FISA that regulate the surveillance of U.S. citizens, and warrants for these individuals are issued by a judicial body created by FISA known as the **Foreign Intelligence and Surveillance Court** (FISC). Court proceedings are sealed, and the public is not privy to any information regarding their activities.

The **Protect America Act of 2007** amended original FISA provisions and provided the federal government with authority to monitor electronic communications of American citizens without a warrant, provided no one citizen was the focus of investigation. This act addressed several FISA provisions aimed at monitoring foreign powers outside the United States. The Protect America Act was scheduled to expire under sunset provisions in early 2008 but was renewed under the FISA Amendments Act of 2008.

### Antiterrorism and Effective Death Penalty Act (1996)

The **Antiterrorism and Effective Death Penalty Act of 1996** grants the federal government with authority to prosecute certain crimes normally pursued in state courts, provided the motivation for the crimes was to "coerce, intimidate, or retaliate against a government or a civilian population," 18 U.S.C. §2332(d). The provisions of the act address homicide, attempt or conspiracy to commit homicide, and other violent acts. Specifically, the provisions provide federal officials with the authority to pursue terrorists outside of the United States who kill, harm, or attempt to harm American citizens. Moreover, it allows federal prosecutors to seek the death penalty for terrorists who kill American citizens outside of the jurisdiction of the United States.

Under the act, murder can be punished with the death penalty. Meanwhile, conspiracy to commit murder is punishable up to life imprisonment, and an attempt to commit murder has been deemed worthy of up to 20 years of imprisonment. Turning our attention to manslaughter, voluntary manslaughter can receive up to 10 years in prison, whereas involuntary manslaughter is punishable up to 3 years in prison. Finally, other violent offenses, regardless of whether harm occurs, can be punished with 10 years in prison.

### USA PATRIOT Act (2001)

The Uniting and Strengthening America by Providing Appropriate Tools Required to Intercept and Obstruct Terrorism Act of 2001 (**USA PATRIOT Act of 2001**) was signed into law following the 9/11 attacks. The act provided the federal government with broad surveillance authority and includes provisions allowing for the liberal use of sneak-and-peek search warrants and roving wire taps, as well as liberal access to records and documents of U.S. citizens (such as business and library records). The act also modified U.S. law regarding terrorism, attacks against transit systems, attacks involving weapons of mass destruction, and money laundering.

Some provisions of the act have been successfully challenged in federal courts, primarily because their application as a criminal investigative tool, rather than an intelligence gathering tool, violated the Fourth Amendment to the U.S. Constitution (*Mayfield vs. U.S.*, 2007). The USA PATRIOT Act is controversial and viewed by many as an erosion of American civil liberties. The original act was scheduled to expire under sunset provisions at the end of 2005. The majority of the original act's provisions were reauthorized under the USA PATRIOT Act Additional Reauthorizing Amendments Act of 2006 (Exhibits 11–4 and 11–5).

## Aviation & Transportation Security Act (2001)

The **Aviation and Transportation Security Act of 2001** created the Transportation Security Administration (TSA) and vested certain TSA employees with the powers of a federal law enforcement officer. This act also included provisions that strengthened airport security, increased cockpit integrity, and increased the presence of Federal Air Marshals on high risk flights. Additionally, the act mandated the presence of TSA officers at most airports and granted them increased authority to screen passengers and baggage. TSA was housed under the U.S. Department of Transportation until the passage of the Homeland Security Act of 2002.

## Homeland Security Act (2002)

The **Homeland Security Act of 2002** also was a response to the 9/11 attacks. It created the Department of Homeland Security and included provisions integrating over twenty federal agencies under the Homeland Security umbrella: the Transportation Security Administration, U.S. Coast Guard, U.S. Secret Service, and the Bureau of Alcohol, Tobacco, and Firearms are but a few of the agencies transferred to the Department of Homeland Security. The Federal Emergency Management Agency also was incorporated under Homeland Security.

The Homeland Security Act eliminated the Immigration and Naturalization Service and created new agencies responsible for immigration, customs, and border protection. The Bureau of Customs and Border Protection was created to oversee border security and perform many administrative functions previously handled by the Immigration and Naturalization Service. The Bureau of Immigration and Customs Enforcement was established as the primary enforcement agency responsible for immigration and customs laws. Both agencies act in a federal law enforcement capacity. The purpose of the Homeland Security Act was to increase cooperation between federal agencies by routing all intelligence and investigatory information through one centralized location. By transferring control of these agencies to one department, the federal government enabled an elevated level of information sharing which increased availability of intelligence.

---

### Exhibit 11–4 Sneak-and-Peek Warrants

A sneak-and-peek search warrant under provisions of the USA PATRIOT Act allows federal law enforcement officials to enter property without prior notification to the owner(s). There is no time limit for delayed notification specified by U.S. law, and the warrants are applicable to any federal crime. Also known as covert entry or surreptitious entry search warrants, sneak-and-peek warrants are only issued when (1) immediate notification may have an "adverse result" (i.e., flight from prosecution, destruction of evidence, etc.), (2) the seizure of "tangible property" is deemed reasonable by the court, and (3) notice of execution can be given "within a reasonable period" of time. [18 U.S.C. 3103a(b)]

---

### Exhibit 11–5 Roving Wiretaps

Roving wiretaps refer to surveillance warrants allowing investigators to monitor all lines of communication used by an individual. Before 1988, electronic surveillance warrants applied to specific lines of communication only (i.e., one residence or telephone line). Roving wiretap warrants allow for the broad surveillance of individual communication and therefore are very controversial because of their enhanced potential for violating the Fourth Amendment's prohibition against unreasonable searches and seizures.

## State Laws Regarding Terrorism

Although federal law addresses terrorism at length, many states have also incorporated terrorism laws into their statutes. Most state statutes use federal laws as a model, and their definitions of terrorism are similar in many ways (Exhibit 11-6).

## Other Terrorism-Related Crimes

The following section explores offenses that are similar to terrorism in that they are important to homeland security. Treason, sedition, espionage, and sabotage are serious crimes against the United States and the respective statutes in which they are defined provide serious penalties for committing such offenses.

### Treason

Treason is an unusual offense, in that it is the only crime defined in the U.S. Constitution. Under English common law, treason was liberally defined and commonly used against individuals who publicly criticized the monarchy. The framers of the Constitution were mindful of this trend, as they themselves were classified as traitors by the British, and therefore ensured that the definition of treason would be succinct and not open to liberal interpretation. Article III Section 3 of the U.S. Constitution states that

> Treason against the United States shall consist only in levying War against them, or in adhering to their Enemies, giving them Aid and Comfort. No person shall be convicted of Treason unless on the Testimony of two witnesses to the same overt Act, or on confession in open Court.

Article III Section 3 adds that "Congress shall have Power to declare the Punishment of Treason, but no Attainder of Treason shall work Corruption of Blood, or Forfeiture except during the Life of the Person attainted." Treason also is defined under federal law (18 U.S.C. 2381) as:

> Whoever, owing allegiance to the U.S., levies war against them or adheres to their enemies, giving them aid and comfort within the U.S. or elsewhere, is guilty of treason and shall suffer death, or shall be imprisoned not less than five years and fined under this title but not less than $10,000; and shall be incapable of holding any office under the U.S.

Several important factors should be considered when discussing treason. First, its Constitutional definition was only intended to limit the scope of its application and punishment. The codified definition of treason is slightly more specific, but does not conflict with the Constitutional definition. Additionally, the statute provides specific punishments for treason. Together, the U.S. Constitution and U.S. Code guide treason prosecutions. The elements of **treason** are (1) allegiance; (2) an overt act, defined as levying war,

---

### Exhibit 11-6

#### Alabama (13A Code of Ala. 10-152a)

A person is guilty of a crime of terrorism when, with intent to intimidate or coerce a civilian population, influence the policy of a unit of government by intimidation or coercion, or affect the conduct of a unit of government by murder, assassination, or kidnapping, he or she commits a specified offense.

#### Arizona (13 A.R.S. 2301c(12))

"Terrorism" means any felony, including any completed or preparatory offense, that involves the use of a deadly weapon or a weapon of mass destruction or the intentional or knowing infliction of serious physical injury with the intent to either: (a) Influence the policy or affect the conduct of this state or any of the political subdivisions, agencies or instrumentalities of this state [or] (b) Cause substantial damage to or substantial interruption of public communications, communication service providers, public transportation, common carriers, public utilities, public establishments or other public services.

#### Minnesota (Minn. Stat. 609.714)

As used in this section, a crime is committed to "further terrorism" if the crime is a felony and is a premeditated act involving violence to persons or property that is intended to: (1) terrorize, intimidate, or coerce a considerable number of members of the public in addition to the direct victims of the act; and (2) significantly disrupt or interfere with the lawful exercise, operation, or conduct of government, lawful commerce, or the right of lawful assembly.

adherence to enemies, and giving aid or comfort to enemies; and (3) confession of two witnesses to the overt act in open court. The element of allegiance is usually established if the individual charged with treason is a U.S. citizen (residing inside or outside of the territorial United States), resident alien, or other person residing in the United States owing temporary allegiance. Persons with dual or multiple citizenships owe allegiance to the United States when they reside in its territories. An overt act is established when an individual engages in activities intended to (1) levy war on the United States or (2) give relief to enemies of the United States. The third element, testimony of two witnesses in open court, is intended to further reinforce the absoluteness of the overt act.

Few treason cases have been pursued in the United States, and even fewer have resulted in conviction. Most cases were associated with the Civil War and World War II. For example, "Tokyo Rose" and "Axis Sally," notable for anti-American propaganda activities during World War II, were both successfully prosecuted for treason. Adam Gadahn, the first American indicted for treason in over 50 years, was charged in 2006 with aiding and giving comfort to al-Qaeda. He is currently number 2 on the FBI's Most Wanted Terrorists List (Most Wanted Terrorists, 2009). Most individuals who commit acts that could be interpreted as treasonous are charged with lesser offenses, primarily because the elements of those offenses are less restrictive. Common offenses of this type include misprision of treason, sedition, espionage, and sabotage.

## Misprision of Treason

**Misprision of treason** refers to concealing or suppressing knowledge of treasonous activities. Simply put, any individual that overtly conceals a treasonous act, or by omission of fact covertly conceals a treasonous act is guilty of misprision of treason. Specifically, the U.S. Code (18 U.S.C. §2382) provides

> Whoever, owing allegiance to the U.S. and having knowledge of the commission of any treason against them, conceals and does not, as soon as may be, disclose and make known the same to the President or to some judge of the U.S., or to the governor or to some judge or justice of a particular State, is guilty of misprision of treason and shall be fined under this title or imprisoned not more than seven years, or both.

## Sedition (Smith Act 1940)

**Sedition** refers to communication (speech or print) designed to advocate the overthrow of the federal government. Prior to 1940, there were several versions of sedition law; however, each of those provisions eventually expired. The **Smith Act of 1940** (also known as the Alien Registration Act) codified sedition into U.S. law (18 U.S.C. §2385). Although the word "sedition" is not specifically mentioned, the statute clearly addresses seditious conduct as

> Whoever knowingly or willfully advocates, abets, advises, or teaches the duty, necessity, desirability, or propriety of overthrowing or destroying the government of the U.S. or the government of any State, Territory, District or Possession thereof, or the government of any political subdivision therein, by force or violence, or by the assassination of any officer of any such government.

Seditious conspiracy also is a federal crime. The federal statute (18 U.S.C. §2384) contains language similar to treason and sedition statutes:

> If two or more persons in any State or Territory, or in any place subject to the jurisdiction of the U.S., conspire to overthrow, put down, or to destroy by force the Government of the U.S., or to levy war against them, or to oppose by force the authority thereof, or by force to prevent, hinder, or delay the execution of any law of the U.S., or by force to seize, take, or possess any property of the U.S. contrary to the authority thereof, they shall each be fined under this title or imprisoned not more than twenty years, or both.

The charge of sedition was common during the 1940s and 1950s, primarily as a tool to fight communism; however, the U.S. Supreme Court limited the application of sedition law in *Yates vs. U.S.* (1957). This was likely a result of the Court's reluctance to restrict speech which delineates, rather than advocates, alternate forms of government. Consequently, sedition prosecutions severely declined after the *Yates* ruling.

## Espionage

**Espionage** is another term for spying. Individuals who gather, transmit, or deliver intelligence information to unauthorized parties (or attempt to do so) are guilty of espionage. Moreover, if an individual loses sensitive information due to gross negligence or fails to report such losses, he or she can be charged with espionage. Generally, espionage is prosecuted under one of two statutes. Although the statutes are similar, there are two important differences. First, 18 U.S.C. §793 primarily addresses the act (or attempt) of *any* gathering, transmittal, or loss of sensitive information, whereas 18 U.S.C. §794 is concerned with the act (or attempt) of gathering, transmittal, or delivery of sensitive information to *any* foreign government. Second, the penalty for violating §793 is an unspecified fine and not more than 10 years in prison, whereas the penalty for violating §794 is an indeterminate prison sentence (up to life) or death.

Individuals engaged in treasonous activities are often charged with espionage, primarily because (1) both espionage statutes (§§793–794) have elements that are easier to prove than the elements of treason and (2) §794 carries the same penalties as treason (indeterminate prison sentence or death). There have been several high-profile espionage cases in the United States, such as the Rosenbergs, Aldrich Ames, and Robert Hanssen (Exhibits 11–7, 11–8, and 11–9).

### Sabotage

**Sabotage** refers to any act that purposely hinders or attempts to hinder the defense capabilities of the U.S. government. Specifically, U.S. law mandates the following activities as sabotage: trespasses upon, injury to, interference with, or destruction of fortifications, harbor defenses, or defensive sea areas; destruction of war material, war premises, or war utilities; intentional production of defective war material, war premises, or war utilities; destruction of national defense material, national defense premises, or national defense utilities; and intentional production of defective national defense material, national defense premises, or national defense utilities (18 U.S.C. §§2152–2156). Moreover, U.S. citizens engaging in (or aiding and abetting) sabotage also can be charged with treason; however, as mentioned earlier, treason's elements are somewhat difficult to prove, and other charges (such as sabotage) often are substituted. In some cases, saboteurs (and those aiding saboteurs) are charged with treason because of the audacity and heinousness of the crimes (*Ex Parte Quirin*, 1942; *U.S. vs. Haupt*, 1945).

## Summary

This chapter provided a foundation for understanding the evolution of organized crime and legal issues associated with detecting and prosecuting organized crime activity, as well as a basic understanding of terrorism and crimes associated with terrorist activities. Law enforcement agencies tasked with investigating organized crime and terrorisism are extremely disadvantaged regarding preventive efforts. Individuals that perpetrate these offenses, both organized crime syndicates and terrorists, are devoted to their organi-

---

### Exhibit 11–7 Julius and Ethel Rosenberg

Julius and Ethel Rosenberg were members of the American Communist Party who were convicted of espionage in 1951. Julius Rosenberg was employed by Emerson Radio as an engineer and was involved in a number of government projects related to the atomic bomb. His wife, Ethel, assisted him in preparing communications which were later passed to their Soviet handlers. The Rosenbergs were convicted under 50 U.S.C. §§32(a), 34 (earlier provisions of the same law found in 18 U.S.C. §794) and sentenced to death. Their motion for a stay of execution was denied by the U.S. Supreme Court (*Rosenberg vs. U.S.*, 346 U.S. 273, 1953), and they were executed in 1953. They were the first (and only) civilians executed for espionage activities during the Cold War.

---

### Exhibit 11–8 Aldrich Ames

Aldrich Ames was an employee of the Central Intelligence Agency from 1962 to 1994. For a time, he worked in counterintelligence and was in charge of analyzing Soviet intelligence capabilities and activities. He began spying for the Soviet Union in 1985, and his betrayal led to the deaths of several highly placed Soviet agents who were spying for the United States. Ames and his wife, Rosario, were arrested for espionage and conspiracy to commit espionage in 1994. Ames received a life sentence, and his wife received 5 years. Ames filed a motion to vacate his conviction in 2000 but was denied (*Ames vs. U.S.* 155 F. Supp. 2d 525, 2000).

> **Exhibit 11–9  Robert Hanssen**
>
> Robert Hanssen worked for the Federal Bureau of Investigation from 1976 to 2001. Hanssen's primary duties involved counterintelligence and technology, which gave him access to the most sensitive information available. Hanssen began spying for the Soviet Union in 1979, shortly after beginning his career. Hanssen provided Soviet intelligence services with detailed information regarding U.S. surveillance capabilities and Soviet agents who were spying for the United States. Much of the information corroborated intelligence information provided by Aldrich Ames, and as a result, many Soviets who were spying for the United States were imprisoned or executed. In 2001, Hanssen plead guilty to 1 count of conspiracy to commit espionage, 13 counts of espionage, and 1 count of attempted espionage. Although he was eligible to receive the death penalty, Hanssen's plea arrangement with the court secured him a life sentence.

zations and causes, and often operate within the scope of legitimate activities. Moreover, these individuals regard secrecy as a primary principle of operation, which further hinders detection and prevention. It is encouraging, however, that American law enforcement has evolved at a near-even pace with the criminals committing these offenses. Law enforcement agencies have increasingly used new and creative tools to fight organized crime and terrorism, such as RICO statutes and the USA PATRIOT Act. Although the constitutionality of the statutes has been debated, their value is undeniable when used within proper legal parameters. Additionally, increased support for these statutes within case law has further strengthened their effectiveness.

## Practice Test

1. _____ terrorism refers to terrorist acts perpetrated outside the United States.
   a. Domestic
   b. Intranational
   c. Quasi-political
   d. International
   e. State-sponsored

2. _____ is the legal term for spying.
   a. Espionage
   b. Sabotage
   c. Treason
   d. Misprision of treason
   e. Insurgency

3. _____ transactions are small deposits used in money laundering to disguise large sums of cash.
   a. Summative
   b. Structured
   c. Semiannual
   d. Sanitized
   e. Structured

4. Excessive interest on a usurious loan is known as the _____.
   a. vigorous
   b. vigorish
   c. vigandish
   d. value
   e. vulgaris

5. _____ terrorism refers to terrorist acts committed by individuals with no intrinsic desire for serious political or religious change.
   a. Insurgency
   b. Political-dissident
   c. Religious-extremist
   d. Quasi-political
   e. Domestic

6. The _____ Act enables prosecutors to indict individuals belonging to a criminal enterprise for racketeering even if there is no proof they committed an overt act.
   a. Volstead
   b. RICO
   c. USA PATRIOT
   d. FISA
   e. Homeland Security

7. The Department of _____ was created in 2002 to integrate the investigative and intelligence capabilities of over twenty agencies under one umbrella organization.
   a. Homeland Security
   b. Investigation
   c. Intelligence
   d. Interior
   e. Counterterrorism

8. _____ is an element of treason and applies to U.S. citizens, resident aliens, and other individuals.
   a. Allegiance
   b. Loyalty
   c. Duty
   d. Honor
   e. Tribute

9. _____ terrorism is also known as insurgency and refers to terrorist acts perpetrated by citizens against their government.
   a. Quasi-political
   b. Direct-action
   c. Political-dissident
   d. Indirect-action
   e. Quasi-dissident

10. _____ is the practice of disguising the nature and origin of illegal income.
    a. Bracing
    b. Illicit staging
    c. Leveraging
    d. Money laundering
    e. Illegal banking

11. _____ refers to communication designed to advocate the overthrow of the federal government.
    a. Treason
    b. Sabotage
    c. Espionage
    d. Sedition
    e. Misprision of treason

12. _____ apply to individuals rather than specific locations.
    a. Sneak and peek search warrants
    b. Roving wire taps
    c. Surreptitious entry search warrants
    d. Surreptitious wire taps
    e. Roving search warrants

13. Individuals who commit certain offenses defined under the RICO Act twice during a 10-year period can be charged with _____.
    a. corruption
    b. collusion
    c. racketeering
    d. counterfeiting
    e. reconnoitering

14. _____ refers to lending money at excessive interest rates and often involves the threat of bodily harm as incentive to remit payment.
    a. Money laundering
    b. Usury
    c. Vigorish
    d. Counterfeiting
    e. Kiting

15. _____ is the only crime defined in the U.S. Constitution.
    a. Sedition
    b. Espionage
    c. Murder
    d. Treason
    e. Sabotage

16. _____ search warrants do not require that prior notification be given to a suspect AND do not specify any time period of notification after entry.
    a. Sneak-and-peek
    b. No-knock
    c. Secret
    d. Omnibus
    e. Undisclosed

17. The _____ Act created secret courts to review sneak and peek search warrants and roving wire tap requests for intelligence and counterintelligence investigations.
    a. FISC
    b. FISA
    c. Volstead
    d. RISA
    e. USA PATRIOT

18. Individuals engaged in the drug trade often use _____ to eliminate their competitors or intimidate government officials.
    a. narco-subversion
    b. narco-insurgency
    c. narco-sedition
    d. narco-terrorism
    e. narco-diplomacy

19. The U.S. State Department is responsible for labeling countries engaging in acts of terror designed to control their citizens, also known as _____ terrorism.
    a. religious-extremist
    b. political-dissident
    c. indirect
    d. bottom-up
    e. state-sponsored

20. _____ is a broad definition of ongoing criminal activity perpetrated by individuals belonging to semiexclusive groups.
    a. Organized crime
    b. Usury
    c. Money laundering
    d. Racketeering
    e. Loan sharking

## References

FBI Most Wanted Terrorists. Retrieved November 16, 2009, from http://www.fbi.gov/wanted/terrorists/fugitives.htm.

U.S. State Department, Office of the Coordinator for Counterterrorism. Retrieved November 18, 2009, from http://www.state.gov/s/ct.

# CHAPTER 12

# White Collar Crime

## Key Terms

Adulteration
Check kiting
Churning
Clean Air Act of 1970
Clean Water Act of 1972
Environmental Protection Agency
False advertising
Federal Trade Commission
Float
Food and Drug Administration
Food, Drug, and Cosmetic Act
Identity theft
Identity Theft and Assumption Deterrence Act of 1998
Identity Theft Penalty Enhancement Act of 2004
Insider trading
Knowing endangerment
Mail fraud
Misbranding
National Stolen Property Act
Pinto Papers
Securities and Exchange Commission
Securities fraud
Short-sale orders
Stop-loss orders
Substantial product hazard
Till skimming
Toxic Substances Control Act of 1976
White collar crime
Wire fraud

## Introduction

Generally, **white collar crime** represents offenses perpetrated within the scope of legitimate business. Whereas offenders involved in typical crimes (i.e., robbery, burglary, narcotics offenses) tend to reside among the lower class (or blue collar), white collar criminals are typically middle-to-upper class with positions of authority. Most white collar crimes involve manipulation and/or deception of legitimate businesses or their customers. Historically, white collar crime was viewed as nonviolent; however, recent attention to injuries (and deaths) resulting from environmental crimes, as well as violations of laws within food and drug industries, have changed that perception.

Although white collar crime typically occurs at the corporate level, it is not necessarily limited to large businesses. Cashiers (such as in gas stations or fast food restaurants) who take money from registers and sales persons who file false reports of inventory loss are both considered white collar criminals. Most such offenders, however, are not prosecuted under typical white collar federal laws, but rather with larceny/theft state codes.

Several themes guide an understanding of white collar crime. First, there is often a relationship between an offender's socioeconomic status and classification as a white collar criminal. Corporate criminals are most often tried in federal court and are therefore eligible for federal prison, whereas most "blue collar" offenders who commit "white collar" crimes are charged with state-level offenses and therefore are likely to spend time in state correctional institutions. Many citizens feel that federal offenders go to "country club prisons" and will spend most of their sentence golfing or playing tennis. This perception likely arose during the 1980s, when the Prison Camp in Eglin, Florida, was dubbed "Club Fed" because of its perceived amenities. Although it is true that inmates of the Eglin camp often act as groundskeepers for the golf course at Eglin Air Force Base, they do not play golf (*Forbes*, 2004). Without question, federal institutions often have better facilities, but they are still prisons. Moreover, privileges frequently associated with federal incarceration (such as furloughs and freedom to roam the grounds) have progressively been restricted or eliminated. This offender disparity likely results from the fact that white collar crime within corporations is much farther reaching and is therefore investigated and prosecuted more often at the federal level.

Second, white collar crime is usually void of violence or physical injury to victims. Although most white collar crime is perpetrated through fraud for the purpose of financial gain, and many times does not even involve actual contact with a victim, some offenses do result in physical harm. As mentioned, environmental crimes, the production/distribution of harmful products, and violations of food and drug regulations often result in injury. Additionally, offenses that obscure product risk may cause injury. Some offenses, too, can indirectly cause injury, such as when a family is rendered homeless as a consequence of a fraudulent real estate transaction. It is important, then, to remain vigilant to the reality that white collar crime can (and often does) result in physical injury.

A third (and final) theme to consider is that technological advancement has increased the prevalence and scope of white collar crime. Identity theft and credit card fraud are relatively new offenses that are problematic to detect, investigate, and prosecute. Moreover, white collar crime has become an international endeavor, with offenders around the world often perpetrating massive schemes involving credit card theft and exchange of stolen identities.

## Corporate Crime and Liability in America

In 2004, CEO Kenneth Lay was indicted on charges stemming from the collapse of Enron, an energy corporation whose initial performance dominated the stock market. Enron's subsequent bankruptcy resulted in widespread accusations of fraudulent accounting practices and insider trading. Both Enron's financial collapse and criminal charges against their accounting firm, Arthur Andersen, marked the beginning of one of the largest prosecutions of white collar crime in America. In 2006, Kenneth Lay was convicted of 10 counts of securities and wire fraud and of making false statements. He died shortly thereafter and as such was never sentenced, but several other Enron employees cooperated with investigators in exchange for reduced sentences.

Corporate crime often is overlooked as serious, even though its consequences can affect hundreds or thousands of employees and affiliated persons. Government has worked incessantly to prevent corporate crime, with numerous agencies specifically designated to regulate domestic and foreign commerce. These agencies (e.g., the Securities and Exchange Commission, the Internal Revenue Service, and the Federal Trade Commission) are tasked with ensuring that U.S. corporations comply with federal laws regarding domestic and foreign commerce, financial reporting, antitrust regulations, and tax compliance. Meanwhile, other agencies (i.e., the Environmental Protection Agency and the Food and Drug Administration) seek to ensure that U.S. corporations comply with federal laws regulating the health and safety of Americans. White collar crimes can be prosecuted under the administrative statutes of such agencies or the criminal statutes within the U.S. Code. With white collar crime, the nature of the offense—not of the offender—dictates the venue and scope of investigation, prosecution, and punishment (i.e., state vs. federal and administrative vs. criminal).

## Tax Evasion

During the 1920s, Al Capone dominated the organized crime scene in Chicago. Federal officials were determined to halt Capone's activities, but their investigations were thwarted by a myriad of problems. Capone employed middle men to manage the day-to-day operations of his syndicate, of which most assets were not even in his name. Elliot Ness and the "Untouchables" pursued Capone diligently but could never connect him with a crime. In the end, Capone was charged with tax evasion and failure to file tax returns—charges that carry federal prison time. Although the media has portrayed Ness as the ultimate hero in this saga, his role was overdramatized. It is likely that federal accountants played a larger role in connecting Capone to his illicit income and in doing so sparked the birth of serious white collar crime investigation.

The Internal Revenue Service regulates tax collection and enforcement of taxation laws as defined in the Internal Revenue Code (Title 26, U.S. Code). Federal tax offenses are delineated under §7201 et seq. Of the 18 separate statutes defining tax offenses, only 5 are discussed. The remaining statutes pertain to administrative violations and thus are not commonly used within the context of criminal law.

Tax evasion is probably the most well-known offense among these crimes. The overt act proscribed in this law is the attempt to evade or defeat a tax and/or payment. Keep in mind, however, that the act need not be successful to complete this white collar offense. Tax Evasion (§7201) is defined as follows:

> Any person who willfully attempts in any manner to evade or defeat any tax imposed by this title or the payment thereof shall, in addition to other penalties provided by law, be guilty of a felony and, upon conviction thereof, shall be fined not more than $100,000 ($500,000 in the case of a corporation), or imprisoned not more than 5 years, or both, together with the costs of prosecution.

A second tax offense worthy of discussion (under Title 26) is the willful failure to file a tax return, provide tax information, or pay taxes (§7203). This statute warns that any person who purposely neglects to file a tax return or pay federal taxes is guilty of a misdemeanor and eligible for a fine of $25,000 and/or up to 1-year imprisonment. Corporations can also be charged under this statute and are eligible for a fine of $100,000 (but not prison). Offenses defined under §7206 and §7207 outline the consequences associated with aiding the preparation and submission of fraudulent tax returns, as well as concealment of taxable assets through rendering false information to the Internal Revenue Service. Penalties include fines ranging from $10,000 to $500,000 and 1 to 3 years in prison.

Perhaps most interesting, statute §7214 addresses offenses committed by federal employees in conjunction with any law under Title 26, including but not limited to extortion, bribery (including accepting gifts), collusion, conspiracy, and fraud. Authorized penalties include job termination, $10,000 fine, up to 5 years imprisonment, and restitution. The statute also makes it criminal for any internal revenue agent to be involved directly or indirectly in a business that manufactures tobacco or produces liquor. The penalty for either offense is job termination, along with a fine if the agent had interest in liquor production. Pause for Thought 12–1 illustrates how tax evasion is handled at federal law.

## False Advertising

Consumers often are lured into purchasing goods and services at excessive costs because of **false advertising**. Businesses use a variety of such methods, including knowingly and purposely advertising sale prices on goods not in stock, increasing regular prices to advertise an inflated "sale" price, and using surreptitious placement of "on sale" tags near goods which are regularly priced. Under U.S. law, the **Federal Trade Commission** has jurisdiction over false advertising offenses. Individuals convicted of false advertising are eligible for both criminal and civil penalties. The U.S. Code (15 U.S.C. §55a[1]) defines false advertisement as

> an advertisement, other than labeling, which is misleading in a material respect; and in determining whether any advertisement is misleading, there shall be taken into account . . . not only representations made or suggested by statement, word, design, device, sound, or any combination thereof, but also the extent to which the advertisement fails to reveal

---

### Pause for Thought 12–1

Consider the following: Jacob is informed by the Internal Revenue Service that he is under investigation for tax evasion. Agent Smith arrives 1 week later to discuss the investigation. During the conversation, Agent Smith mentions that he has four children in college and understands Jacob's financial difficulty. Although Jacob knows that his tax returns were prepared properly, he senses that Agent Smith will "back off" if properly compensated. As such, Jacob offers to donate $5,000 to Agent Smith's children for college tuition. Agent Smith accepts the money and concludes the investigation after a cursory examination of Jacob's tax records. Is Agent Smith guilty of a crime, even though Jacob's tax returns were correct?

### Scenario Solution

Yes, Agent Smith is guilty under 26 U.S.C. § 7214 because he accepted a gift from Jacob. In this case, the result of the investigation is not an element of the offense. Agent Smith would lose his job, face up to 5 years in prison, and pay a $10,000 fine.

facts material in the light of such representations or material with respect to consequences which may result from the use of the commodity to which the advertisement relates under the conditions prescribed in said advertisement, or under such conditions as are customary or usual.

Another statute makes it unlawful to disseminate false advertisement under federal law (15 U.S.C. §52a):

> It shall be unlawful for any false advertising person, partnership, or corporation to disseminate, or cause to be disseminated, any false advertisement—(1) By U.S. mails, or in or having an effect upon commerce, by any means, for the purpose of inducing, or which is likely to induce, directly or indirectly the purchase of food, drugs, devices, services, or cosmetics; or (2) By any means, for the purpose of inducing, or which is likely to induce, or indirectly, the purchase in or having an effect upon commerce, of food, drugs, devices, services, or cosmetics.

Simply put, the prosecution is only required to show proof of dissemination (act of presenting) to the public and does not require any proof of purchase based on the false advertisement. Similarly, individuals, partnerships, and corporations can be charged with false advertising—a misdemeanor punishable by a fine not to exceed $5,000 and/or imprisonment for up to 9 months. A second false advertising conviction warrants a fine not exceeding $10,000 and/or imprisonment for up to 1 year. Most (if not all) states have statutes regarding false advertising that follow federal standards. Liable individuals under these statutes include manufacturers, packers, distributors, and/or sellers of any product falsely advertised (15 U.S.C. 54b). Agents of dissemination are not liable under U.S. law for false advertising. Advertising agencies, newspapers, magazines, and radio and television stations also are not eligible for prosecution if they circulate false advertisement, unless they refuse to cooperate with the Federal Trade Commission in an investigation. Pause for Thought 12–2 illustrates how federal law addresses the crime of false advertising.

## Harmful Products

It is not uncommon to see product recall notices in newspapers, on television, or through other media outlets. This may seem odd considering that most products in the United States undergo rigorous testing for consumer safety. Moreover, it seems logical that manufacturers would do everything in their power to abide by consumer product laws to avoid expensive litigation. Although this is indeed true, the reality is that many corporations and manufacturing companies cut corners on product testing to increase profits. Individuals, partnerships, and corporations who knowingly manufacture, market, or distribute defective products that could be harmful to consumers are negligent and liable for prosecution under U.S. law.

Federal statutes regarding harmful products are lengthy and cover an array of issues. In fact, the U.S. Code contains an entire chapter devoted to consumer product safety (15 U.S.C. Chap. 47). Among other things, the statutes regulate the Consumer Product Safety Commission, banned substances (chemicals, poisons, etc.), and import/export guidelines. For the purposes of this chapter, we limit our examination to three specific sections of the consumer product safety statutes: §2064 (substantial

---

### Pause for Thought 12–2

Consider the following: The Computer Hut places an advertisement in a local newspaper for an "All Day" sale on laptop computers. The Computer Hut only has two of the $100 laptops in inventory, and Computer Hut managers purchase them before opening the store on the day of the sale. When consumers enter the store and ask about the laptops, store employees apologize and inform the consumers that they have already been sold. The employees then offer to sell consumers another "high-end" laptop at a discounted price of $800. Is this an instance of false advertising?

### Scenario Solution

Yes, most courts consider this false advertising. The advertisement certainly was misleading, in that the managers of Computer Hut purposely ensured that no "on-sale" laptops would be available in order to promote the sale of more expensive items. Thus, any manager or employee aware of the scheme could be charged with false advertising.

product hazards), §2068 (prohibited acts), and §2070 (criminal penalties). Under 15 U.S.C. §2064, a **substantial product hazard** is defined as follows:

> (1) a failure to comply with an applicable consumer product safety rule which creates a substantial risk of injury to the public, or (2) a product defect which (because of the pattern of defect, the number of defective products distributed in commerce, the severity of the risk, or otherwise) creates a substantial risk of injury to the public.

In examining this statute, we see two major elements that define a harmful product. First, any product that fails to meet consumer product safety rules and creates substantial risk to the public is considered harmful. Second, any product with a defect that creates substantial risk to the public is considered harmful. In short, products that (1) fail to meet consumer safety guidelines and/or (2) are defective and (3) pose a substantial risk to the public are considered substantial hazards.

Prohibited acts regarding substantial product hazards are defined in 15 U.S.C. §2068. There are a host of prohibited acts under this statute; however, two components delineate a generalized definition regarding substantial product hazards:

> (1) manufacture for sale, offer for sale, distribute in commerce, or import into the U.S. any consumer product which is not in conformity with an applicable consumer product safety standard under this chapter; (2) manufacture for sale, offer for sale, distribute in commerce, or import into the U.S. any consumer product which has been declared a banned hazardous product by a rule under this chapter.

In general, it is a crime to manufacture, sell (or offer for sale), distribute, or import any product that does not meet U.S. consumer safety standards and/or has been declared a hazardous product. This statute is applicable to individuals at multiple levels of the supply chain. A complete overview of §2068 is provided in Figure 12–1.

Individuals who commit acts prohibited by 15 U.S.C. §2068a are subject to a fine not exceeding $50,000 and/or up to 1 year in prison (§2070a). Additionally, directors, officers, or agents of a corporation who knowingly participate in prohibited acts are individually liable (§2070b). In short, any individual associated with the manufacture, distribution, sale, or importation of a substantially hazardous product that has knowledge of (1) the hazard and (2) notice of noncompliance from the Federal Trade Commission is

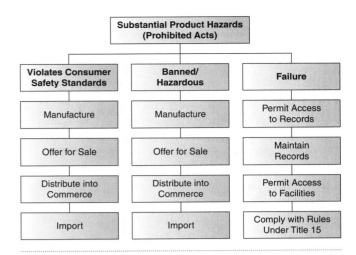

**Figure 12–1** Substantial Product Hazards

liable under U.S. law. Individuals who violate laws regarding consumer product safety are also liable for civil penalties (Exhibit 12–1).

## Food and Drug Administration

The **Food and Drug Administration** (FDA) is responsible for regulating all activities related to food and drug consumption in the United States. Specifically, their mission is to provide oversight regarding research, production, labeling, distribution, and sales of any food, drug, or cosmetic product in the United States (21 U.S.C. §393b). Provisions of the federal **Food, Drug, and Cosmetic Act** (21 U.S.C. Chap. 9) define prohibited acts (§331) and penalties (§333) for offenses related to food and drugs.

Most food and drug offenses involve adulteration or misbranding of a product. **Adulteration** refers to adding or removing a substance, compound, or other ingredient from a product. **Misbranding** is manipulating, destroying, or removing a label from a product. It is against federal law to introduce, deliver, or receive any misbranded or adulterated food, drug, or cosmetic. Like with other statutes, it is only applicable to interstate commerce. Most, if not all states have their own statutes that regulate intrastate commerce.

## Environmental Offenses

Today, most U.S. citizens make efforts to "go green" and are becoming increasingly mindful of the environment. Many individuals have altered their lifestyles to environment-friendly guidelines. Additionally, the federal government has instituted a variety of policies

> **Exhibit 12–1 The Pinto Papers**
>
> In the late 1960s, Ford Motor Company began production of the Pinto model. This small, affordable automobile had a design flaw that made the fuel tank extremely vulnerable in rear-end collisions. Multiple accidents in which the Pinto's fuel tank ruptured and caused numerous injuries (including deaths) led to several lawsuits against Ford. During the course of these litigations, a company memorandum regarding cost–benefit analysis was discovered. Later referred to as the **Pinto Papers**, the memo indicated that Ford was aware of the design flaw but determined that it was cheaper to settle possible lawsuits than to modify the design. Their cost–benefit analysis indicated that the expected cost of litigation—$50,000,000—was far less than fixing the design flaw—$121,000,000. Although Ford Motor Company was acquitted on all criminal charges, they suffered numerous defeats in civil court through compensatory and punitive damages. Substantial losses in litigation, negative media coverage, and pressure from federal agencies led to a massive recall. In the end, Ford suffered massive financial losses and a devastating blow to their reputation as an automobile manufacturer.

to curb pollution, such as vehicle emission standards; however, these efforts likely are rendered negligible when damage to the environment occurs at higher levels, such as improper sewage management or illegal disposal of toxic waste.

The **Environmental Protection Agency** has jurisdiction over the majority of cases involving environmental offenses and is responsible for enforcing laws regarding pollution, facility permits, and reporting. Primarily, offenses against the environment are considered threats to public health. Although there is some degree of criminal liability, most such cases are processed in civil court and result in fines and administrative sanctions; however, there are several statutes that criminalize certain environmental offenses. Most notable are the Clean Air Act of 1970, the Clean Water Act of 1972, and the Toxic Substances Control Act of 1976.

The **Clean Air Act of 1970** (42 U.S.C. §§7401-7642) provides the federal government with the power to regulate air pollution. Moreover, this act allows federal officials to grant enforcement powers to the state. As mentioned previously, these statutes provide a host of civil penalties for violating air pollution standards; however, certain provisions include criminal sanctions. The list of offenses is lengthy and includes (among other things) permit and construction violations, reporting violations, falsifying material statements, tampering with monitoring devices, and negligent release of pollutants into the ambient air. Every offense is punishable by fine (separate from those in civil proceedings), and most carry prison sentences ranging from 1 to 5 years.

The **Clean Water Act of 1972** (33 U.S.C. §§1251-1387) provides the federal government with the power to regulate discharge of material into navigable waters in the United States. In general, statutes under this act define and regulate permissible discharge (i.e., treated sewage) and provide criminal and civil penalties for any individual who violates the offenses defined therein. Specifically, offenses under this act include unlawful discharge of radiological, biological, or chemical warfare agents (§1311f); discharge of oil or other hazardous substances (§1321b[1]); unlawful discharge of sewage or sewage sludge (§1345e); and discharge of any other pollutant or hazardous substance (§1319c[1b], §1319c[2b]).

Under federal law, any violation of these statutes, either knowingly or by negligence, is punishable with criminal sanctions. Criminal penalties differ according to the severity and type of violation. Negligent offenses carry fines ranging from $2,500 to $25,000 per day in violation and a prison term of 1 year (§1319c[1]). Penalties are doubled for second-time offenders. Individuals who knowingly violate these statutes are eligible for fines ranging from $5,000 to $50,000 per day in violation and a prison term of 3 years (§1319c[2]). It also is against federal law to make false statements regarding permits, records, or compliance under the Clean Water Act (§1318c[4]). One interesting facet of this act is that it includes a separate offense regarding **knowing endangerment**. Under this statute, any person who commits one of the previously mentioned offenses and has knowledge that such places another person in danger of imminent death or serious bodily injury is subject to criminal penalties including a fine of $250,000 and up to 15 years in prison (§1319c[3a]). Additionally, an organization found guilty of knowing endangerment is subject to a fine (in criminal court) of $1,000,000.

The **Toxic Substances Control Act of 1976** (15 U.S.C. §§2601-2671) regulates the manufacture, distribution, and use of toxic chemicals in the United States. Specifically, these statutes pertain to chemical testing (§2603), production of new chemicals (§2604), labeling requirements (§2605), commercial use of chemicals (§2614[2]), maintenance of records (§2614[3a]), and issuance of reports (§2614[3b]). Individuals who commit an offense defined under these statutes are subject to fines up to $25,000 per day in violation and 1 year in prison.

## Securities Fraud and Insider Trading

Securities fraud and insider trading are relatively common white collar crimes. The **Securities and Exchange Commission** is responsible for overseeing securities transactions in the United States. **Securities fraud** may occur when persons (1) use a device, scheme or artifice to defraud, (2) make false or fraudulent statements, or (3) omit facts or information that affects activities regarding securities. In general, securities fraud and insider trading are addressed under 15 U.S.C. §78a et seq. Prosecution in criminal court is only possible if the illicit transactions involve interstate commerce and/or communication by mail or wire and national securities exchange facilities. Primarily, securities fraud refers to stock manipulation. The Securities and Exchange Commission's regulations are complicated and beyond the scope of this text; therefore, discussion only pertains to their function in general. Common illegal practices include (but are not limited to) the following:

- **Stop-loss orders**: used to determine thresholds for buying (at a low price) or selling (at a high price) securities; often used to illegally manipulate stock prices.
- **Short-sale orders**: the practice of selling securities for a high price without actually owning them; short sales are illegal when the security is not properly borrowed or the broker never returns a borrowed security.
- **Churning**: occurs when brokers affect multiple, frivolous trades of no consequence on behalf of investors; these excessive transactions generate commissions for the broker at the expense of the investor.

**Insider trading** occurs when securities are bought or sold based on knowledge that has not been made public. Generally speaking, insider trading is usually perpetrated by a corporate officer holding at least 10% of the available stock in their corporation; however, anyone who uses confidential or stolen information to affect the purchase or sale of a security can be charged with insider trading. Courts have held that insider trading is a deceptive device and therefore a component of fraud. Figure 12–2 illustrates prohibited acts delineated under statutes regulating securities.

Securities fraud and insider trading are also addressed under 18 U.S.C. §1348, which provides a general definition of securities fraud as well as the penalty for its violation. The offense is complete on attempt; thus, proof of a successful act is not required. The law reads as follows:

> Whoever knowingly executes, or attempts to execute, a scheme or artifice—(1) to defraud any person in connection with any security . . . or (2) to obtain, by means of false or fraudulent pretenses, representations, or promises, any money or property in connection with the purchase or sale of any security; shall be fined under this title, or imprisoned not more than 25 years, or both.

Pause for Thought 12–3 provides a scenario regarding the legal approach to insider trading.

## Mail Fraud

**Mail fraud** refers to any deceptive, fraudulent, or otherwise illegal activity occurring in conjunction with delivery or receipt of the U.S. mail. Generally speaking, mail fraud occurs within the context of

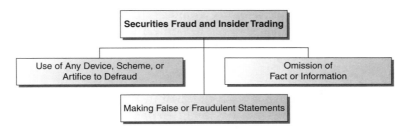

**Figure 12–2** Securities Fraud and Insider Trading

> ### Pause for Thought 12–3
>
> Consider the following: Bill works for a new (but large) computer software company whose stock prices have been relatively low since incorporation. Bill knows that the company will soon release a new software package to revolutionize the Internet. Because he already holds the maximum amount of stock allowed by the company, he is disappointed that his potential earnings will be limited. Bill loans his father the money to purchase a large portion of stock before the company's public announcement of the new software package. Several days later, the news regarding the software is made public, and the company's stock price soars. Bill's father then sells the purchased stock and splits the proceeds with his son. Did Bill and his father engage in insider trading?
>
> ### Scenario Solution
>
> Yes, Bill had knowledge of the software before public release that affected a purchase of stock. Thus, he can be charged with insider trading (even though his father purchased the stock). Bill's father also can be charged with insider trading because he acted on the same knowledge and profited from that transaction.

one or more additional offenses. The actual crime, then, is using the mail (via the U.S. Postal Service or other carrier) in the process of fraudulent activity. Additionally, this offense only requires proof of intent to defraud rather than a successful act. This statute is not applicable to offenses such as destruction of a mail receptacle or interference with the delivery of mail.

Mail fraud is codified under 18 U.S.C. §1341. Although the statute is rather verbose as to defining prohibited acts, the elements of mail fraud are rather simple: (1) a scheme or artifice to defraud (2) employed in conjunction with mail (3) to further the scheme. Mail fraud is punishable by fine and up to 20 years in prison. If the offense affects a financial institution, then punishment is increased to a fine of $1,000,000 and up to 30-years imprisonment. Additionally, any individual using a false name when sending or receiving mail while engaging in illicit activity can be punished by fine and up to 5-years imprisonment.

## Wire Fraud

The elements of wire fraud are almost identical to those of mail fraud; however, communication(s) must travel across state lines. **Wire fraud** can occur via telephone, radio, television, cable (i.e., the Internet), or other electronic means of communication. Wire fraud is codified as a criminal offense under 18 U.S.C. §1343 and is punishable by a fine and up to 20 years imprisonment. Just like with mail fraud, the punishment is increased to a fine of $1,000,000 and up to 30 years in prison if the offense affects a financial institution. Pause for Thought 12–4 provides a hypothetical scenario regarding mail fraud and wire fraud.

> ### Pause for Thought 12–4
>
> Consider the following: Mary calls Jim with the good news that he won a sweepstakes and is eligible to claim a cash prize of $100,000. She then tells Jim that he needs to send her a $100 money order to pay a special transaction fee, after which he can claim his winnings. Jim agrees and mails the money order. Mary cashes the money order and leaves the state, never to be heard from again. Obviously, Jim has been the victim of a telephone scam—but did Mary actually break any law?
>
> ### Scenario Solution
>
> Yes, Mary committed mail fraud because she used the mail in furtherance of a scheme to defraud Jim. If Mary and Jim resided in different states during their telephone communication, Mary also could be charged with wire fraud.

## Bad Checks

Few federal statutes specifically address offenses related to bad checks. The **National Stolen Property Act** does, however, allow federal prosecution for offenses in which a fraudulent or forged securities instrument having a value of $5,000 or more is transported, transmitted, bought, or sold within the confines of interstate commerce (18 U.S.C. §§2314-2315). Generally, however, most check-related offenses are prosecuted at the state level. Each state classifies check-related offenses differently: some as misdemeanors, others as felonies; some are fined, others punished with jail or prison time; and some require victim restitution, and others do not.

There are three basic types of check-related offenses: forgery (uttering), writing worthless checks, and check kiting. Some states do not differentiate uttering from worthless checks; however, most states have recognized the need to separately classify worthless checks. Because uttering was addressed in detail in Chapter 3, this section focuses on worthless check offenses and check kiting.

## Worthless Checks

Under common law, a forged check could be prosecuted as uttering or forgery; however, when an individual presents a check of his or her own that is known to be worthless, how is the offense to be classified? As mentioned previously, most states have incorporated statutes which specifically address worthless checks. Recognizing the fact that many individuals may bounce a check unintentionally, these statutes normally mandate a time period in which the individual can remit payment to the bank, business, or person to whom the check was issued. These periods differ among states and normally range anywhere from 7 to 30 days. Additionally, there are differences among state statutes as to when the offense is complete. Some laws stamp an offense as completed upon presentation of the check to the payee; others mandate that the offense is not complete until the check is presented to the bank at which the account originated. Regardless of these differences, one element remains constant: knowledge of insufficient funds to cover the amount of the check. Two examples of state statutes regarding worthless checks are presented in Exhibits 12–2 and 12–3.

## Check Kiting

**Check kiting** refers to the use of multiple businesses, banks, or other institutions in conjunction with issuing worthless checks. Simple check kiting schemes involve one individual and two or more banks. The individual submits a check from Bank A to Bank B for cash and then returns to Bank A to deposit the cash (usually to cover insufficient funds). The offender relies on the **float**, or time interval before the check is presented to the bank. The offender then returns to Bank A and deposits a check from Bank B, thereby inflating the account balance. The scheme may become cyclical if the offender continuously attempts to cover insufficient funds or uses the float to purposely inflate both account balances. Individuals engaging in these schemes are normally prosecuted under fraud or worthless check statutes.

Other more complicated check-kiting schemes involve multiple actors and multiple banks. These kiting rings often represent themselves as corporations or large businesses and are able to obtain large sums of

---

### Exhibit 12–2 Colorado (C.R.S. 18-5-205(2))

Any person, knowing he has insufficient funds with the drawee, who, with intent to defraud, issues a check for the payment of services, wages, salary, commissions, labor, rent, money, property, or other thing of value, commits fraud by check.

#### Penalties (C.R.S. 18-1.3-401(1III3A, 1VA))

<u>Class 2 misdemeanor</u> ($499 or less):

    3–12 months imprisonment and/or $250 to $1000 fine (Restitution permitted)

<u>Class 1 misdemeanor</u> ($500–$999):

    6–18 months imprisonment and/or $500 to $5000 fine (Restitution permitted)

<u>Class 6 felony</u> ($1000 or two prior convictions for check fraud):

    12–18 months imprisonment and/or $1000 to $100,000 fine (Restitution permitted)

> ### Exhibit 12–3 Maryland (Code Ann. Title 8 §8-103)
>
> *Prohibited—Issuing check with knowledge of insufficient funds:*
>
> A person may not obtain property or services by issuing a check if: (1) the person knows that there are insufficient funds with the drawee to cover the check and other outstanding checks; (2) the person intends or believes when issuing the check that payment will be refused by the drawee on presentment; and (3) payment of the check is refused by the drawee on presentment.
>
> **Penalties (Code Ann. Title 8 §8-106)**
>
> <u>Misdemeanor</u> ($99 or less)—Up to three months imprisonment and/or $500 fine
>
> <u>Misdemeanor</u> ($100–$499)—Up to eighteen months imprisonment and/or $100 fine
>
> <u>Felony</u> ($500 or more)—Up to fifteen years imprisonment and/or $1000 fine

money before the scheme is discovered. Kiting rings may change locations many times, which hinders detection, investigation, and prosecution because of varying investigative capabilities of law enforcement agencies and prosecutors. Although most states likely would charge individuals in check-kiting rings under fraud or racketeering statutes, it is possible to use the Federal Bank Fraud statute (18 U.S.C. §1344). On a final note, it is likely that instances of check kiting will decline in the future because of improvement in investigative technology and increased use of debit cards in lieu of checks as an immediate form of payment (Exhibit 12–4).

## Credit Card Theft/Fraud

Although crimes involving credit cards can be prosecuted under statutes relating to theft or fraud, many states have incorporated laws specifically targeting credit card fraud and/or theft. The use of a credit card involves three components:

- An individual who assumes responsibility for remitting payment to the creditor
- A creditor who assumes responsibility for remitting payment to a business or corporation
- A corporation or business that renders goods or services for credit

Under certain circumstances, credit card theft/fraud can be prosecuted under federal statutes relating to wire or mail fraud; however, individuals who use forged, stolen, or canceled credit cards to purchase goods and/or services are normally prosecuted under state law. Moreover, these offenses are still applicable to individuals in possession of only the credit card numbers (without the actual card). Two examples of state statutes regarding credit card theft and fraud are presented in Exhibits 12–5 and 12–6.

Common examples of credit card-related offenses include restaurant employees who alter the tip specified by a patron or retail employees who engage in **till skimming**, which ordinarily involve taking money from a cash register, overcharging customers, or underringing sales to compensate for cash losses. Offenders engaging in this type of till skimming are usually prosecuted for embezzlement. Alternative forms of till skimming compensate for the lost cash by overcharging several credit cards by small amounts to avoid immediate detection. For example, if a cashier adds $10 to a $200 charge, the customer may not notice the discrepancy. Other credit card

> ### Exhibit 12–4 The Check Kite Analysis System
>
> In 1989, the Federal Bureau of Investigation developed computer software designed to enhance their capabilities for detecting check kiting schemes. Because these schemes often involve multiple transactions among many banking institutions, detection and prosecution proved difficult. The Check Kite Analysis System (CKAS) employs database software to calculate actual balances across multiple bank accounts. The system has proven to be a valuable tool for Federal investigators and has successfully passed muster under judicial scrutiny (Turner & Albrecht, 1993).

### Exhibit 12–5  Florida (Title XLVI Florida Statutes 817.60)

A person who takes a credit card from the person, possession, custody, or control of another without the cardholder's consent or who, with knowledge that it has been so taken, receives the credit card with intent to use it, to sell it, or to transfer it to a person other than the issuer or the cardholder is guilty of credit card theft and is subject to the penalties set forth in s. 817.67(1). Taking a credit card without consent includes obtaining it by conduct defined or known as statutory larceny, common-law larceny by trespassory taking, common-law larceny by trick or embezzlement or obtaining property by false pretense, false promise or extortion.

**Penalty**

Misdemeanor (1st degree): Up to one year imprisonment and $1000 fine

---

### Exhibit 12–6  Code of Virginia § 18.2-195(1)

A person is guilty of credit card fraud when, with intent to defraud any person, he: (a) Uses for the purpose of obtaining money, goods, services or anything else of value a credit card or credit card number obtained or retained in violation of § 18.2-192 or a credit card or credit card number which he knows is expired or revoked; (b) Obtains money, goods, services or anything else of value by representing (i) without the consent of the cardholder that he is the holder of a specified card or credit card number or (ii) that he is the holder of a card or credit card number and such card or credit card number has not in fact been issued; (c) Obtains control over a credit card or credit card number as security for debt; or (d) Obtains money from an issuer by use of an unmanned device of the issuer or through a person other than the issuer when he knows that such advance will exceed his available credit with the issuer and any available balances held by the issuer.

**Penalty**

Class 1 misdemeanor ($200 or less in a six month period):

    Up to one year imprisonment and/or $2500 fine

Class 6 felony ($201 or more in a six month period):

    1 to 5 years imprisonment and/or $2500 fine

---

schemes involve Internet-based credit card theft rings. These individuals buy, sell, and trade stolen credit card numbers in chat rooms and discussion forums. Often, these stolen card numbers are used to purchase goods online, which are then shipped to several locations for immediate resale. These clearing houses act as distribution points for merchandise purchased with stolen credit card numbers, and their locations change frequently.

## Identity Theft

Identity theft has increased significantly over the past several years. Identity theft usually occurs in conjunction with schemes intended to defraud an individual, business, or financial institution to obtain money or property. Identity theft is also a major problem within the contexts of illegal immigration and terrorism. In general, **identity theft** refers to a series of offenses where a perpetrator represents himself or herself as another to obtain a line of credit, loan, property, money, or other good or service. Often, the victims of identity theft suffer financially because of damaged credit scores and stolen money.

The federal government has recognized the scope and severity of identity theft with regard to certain circumstances. Under the **Identity Theft and Assumption Deterrence Act of 1998**, any individual who produces, transfers, or possesses any false identification documents or other identifiers issued by federal, foreign, state, or local government is subject to penalties including a fine, forfeiture of personal property, and up to 1-year imprisonment (18 U.S.C. §1028a,b[5,6]). These identifiers include social security numbers, dates of birth, driver's licenses (and numbers), passports, immigration identification numbers, and other means of identification. The **Identity Theft Penalty Enhancement Act of 2004** amended federal law and created the offense of aggravated identity theft, which specified harsher penalties for identity theft occurring in conjunction with serious crimes, such as terrorism, illegal immigration, wire and mail fraud, and firearms purchases (§1028A). For example, the penalty for crimes committed in conjunction with terrorism increased to 30 years imprisonment (§1028b[4]).

As mentioned, many states recognize the propensity of identity theft and have enacted legislation to permit its prosecution at varying levels of government. As a result, offenders engaging in such theft face substantial prison time and fines. Additionally, state-level laws concerning identity theft are beneficial because federal law enforcement agencies often are burdened with investigating more violent crime and thus are often unable to pursue identity theft crimes effectively.

## Summary

This chapter examined the nature and extent of white collar crime in the United States. Although most media tend to focus on large-scale white collar crime (such as the Enron scandal), criminal justice scholars tend to agree that most white collar crime goes unnoticed among the public. Given that law enforcement agencies constantly deal with offenses which visibly impact communities (e.g., murder, burglary, and assault), it is not surprising that white collar crime receives relatively little attention. Additionally, many industrial white collar offenses are punished through administrative means in civil courts. As such, those perpetrators are often never charged with a crime. Thus, the true extent of white collar crime likely is far larger than estimates provided by various sources. On a final note, many white collar crime victims are corporations themselves, which further diminishes public attention and sympathy.

## Practice Test

1. The _____ is the length of time before a check is presented to its issuing bank.
   a. pass
   b. skim
   c. float
   d. make
   e. switch

2. _____ occurs in conjunction with delivery and receipt of written communication via any authorized depository, delivery, or receiving agency.
   a. Wire fraud
   b. Mail fraud
   c. Securities fraud
   d. Credit card fraud
   e. Identity theft

3. _____ refers to the alteration or destruction of a product label.
   a. Misbranding
   b. Adulteration
   c. Churning
   d. Kiting
   e. Shilling

4. Presentation of an advertisement to the public is referred to as _____.
   a. displayment
   b. disaggregation
   c. deception
   d. dissemination
   e. decimation

5. Information contained within the _____ indicated that Ford Motor Company ignored a design flaw of an automobile.
   a. Pinto Package
   b. Pinto Logs
   c. Policy Notes
   d. Policy Manual
   e. Pinto Papers

6. _____ has the same elements of mail fraud, but must involve communications that crossed state lines.
   a. Securities fraud
   b. Wire fraud
   c. Mail fraud
   d. Insider trading
   e. Identity theft

7. The _____ Agency is responsible for regulating pollution and other public health hazards caused by pollution.
   a. Homeland Security
   b. Federal Trade
   c. Environmental Protection
   d. Food and Drug
   e. Consumer Product

8. Adding, changing, or removing ingredients, components, or compounds from a food, drug, or cosmetic is known as _____.
   a. adulteration
   b. manipulation
   c. aggregation
   d. churning
   e. adulation

9. _____ was the moniker for the Federal Prison Camp in Eglin, Florida.
   a. Club Con
   b. Club Bed and Breakfast
   c. Prison for Preppies
   d. Disco Detention Center
   e. Club Fed

10. The _____ Act amended federal law to include severe penalties when identity theft was a component of more serious crimes such as terrorism and firearms purchases.
    a. Identity Theft and Assumption Deterrence
    b. Identity Theft Control
    c. Identity Theft Penalty Enhancement
    d. Identity Control Reform
    e. Terrorism Control

11. _____ refers to a broad range of offenses usually occurring within the scope of legitimate business and often perpetrated by employees.
    a. White collar crime
    b. Securities fraud
    c. Mail fraud
    d. Felonious larceny
    e. Fraud

12. An offense that occurs in conjunction with the purchase, sale, or exchange of securities in the United States can be classified as _____.
    a. fraud
    b. bank fraud
    c. securities fraud
    d. surreptitious fraud
    e. money laundering

13. _____ is the practice of affecting multiple frivolous trades of no consequence.
    a. Till skimming
    b. Short-selling
    c. Ordering a stop-loss
    d. Churning
    e. Circular trading

14. _____ requires proof of dissemination but no proof of purchase.
    a. Insider trading
    b. Identity theft
    c. False pretenses
    d. Fraud
    e. False advertising

15. Employees who remove money from a cash register and hide the loss using overcharges or underrings are engaging in _____.
    a. till skimming
    b. churning
    c. wire fraud
    d. insider trading
    e. short sales

16. The collapse of _____ sparked one of the largest white collar crime investigations in U.S. history.
    a. Ford Motor Company
    b. Enron
    c. U.S. Steel
    d. Shell Oil Company
    e. Exxon

17. Check _____ refers to submitting worthless checks to two or more banks to inflate account balances fraudulently.
    a. trading
    b. masking
    c. kiting
    d. gating
    e. dating

18. The Clean Water Act contains a(n) _____ provision enhancing penalties for offenders who had reason to believe their actions would result in death or serious injury.
    a. knowing endangerment
    b. negligence clause
    c. intent known
    d. knowing-negligence
    e. knowing intent

19. _____ refers to the unlawful taking or possession of any government-issued identifier, such as a social security number or driver's license number.
    a. Larceny
    b. False pretenses
    c. Collusion
    d. Fraud
    e. Identity theft

20. The _____ Administration is responsible for the oversight of food, drugs, and cosmetic products in the United States.
    a. Federal Trade
    b. Food and Drug Regulation
    c. Food and Drug
    d. Consumer Protection
    e. Drug Enforcement

# References

Forbes. (2004). *Best Places to Go to Prison*. Retrieved on November 18, 2009, from http://www.forbes.com/2002/09/12/bestprisonslide_2.html?thisSpeed=30000.

Turner, J. S., & Albrecht, W. S. (1993). Check kiting: Detection, prosecution, and prevention. *FBI Law Enforcement Bulletin*, 62(11), 12–16.

# Practice Test Solutions

APPENDIX A

## Chapter 1

| | | | |
|---|---|---|---|
| 1. a | 6. b | 11. c | 16. b |
| 2. c | 7. d | 12. a | 17. c |
| 3. e | 8. e | 13. d | 18. a |
| 4. b | 9. a | 14. c | 19. b |
| 5. d | 10. e | 15. e | 20. d |

## Chapter 2

| | | | |
|---|---|---|---|
| 1. c | 6. e | 11. b | 16. b |
| 2. e | 7. d | 12. e | 17. a |
| 3. b | 8. a | 13. b | 18. c |
| 4. c | 9. b | 14. d | 19. d |
| 5. a | 10. c | 15. a | 20. d |

## Chapter 3

| | | | |
|---|---|---|---|
| 1. a | 6. a | 11. d | 16. b |
| 2. d | 7. d | 12. b | 17. c |
| 3. b | 8. e | 13. c | 18. d |
| 4. c | 9. b | 14. d | 19. a |
| 5. c | 10. a | 15. e | 20. b |

## Chapter 4

| | | | |
|---|---|---|---|
| 1. c | 6. b | 11. a | 16. a |
| 2. a | 7. b | 12. e | 17. a |
| 3. d | 8. a | 13. d | 18. e |
| 4. e | 9. c | 14. c | 19. b |
| 5. c | 10. d | 15. b | 20. d |

## Chapter 5

| | | | |
|---|---|---|---|
| 1. b | 6. c | 11. d | 16. e |
| 2. d | 7. a | 12. c | 17. d |
| 3. c | 8. e | 13. a | 18. b |
| 4. e | 9. d | 14. e | 19. c |
| 5. a | 10. b | 15. d | 20. a |

## Chapter 6

| | | | |
|---|---|---|---|
| 1. e | 6. b | 11. e | 16. c |
| 2. b | 7. a | 12. d | 17. a |
| 3. c | 8. b | 13. e | 18. d |
| 4. d | 9. e | 14. a | 19. b |
| 5. a | 10. b | 15. c | 20. c |

## Chapter 7

| | | | |
|---|---|---|---|
| 1. e | 6. a | 11. e | 16. d |
| 2. d | 7. e | 12. c | 17. e |
| 3. a | 8. d | 13. a | 18. c |
| 4. c | 9. b | 14. c | 19. d |
| 5. b | 10. a | 15. b | 20. d |

## Chapter 8

| | | | |
|---|---|---|---|
| 1. d | 6. b | 11. c | 16. e |
| 2. b | 7. c | 12. a | 17. b |
| 3. e | 8. a | 13. e | 18. d |
| 4. a | 9. d | 14. b | 19. a |
| 5. c | 10. e | 15. c | 20. c |

## Chapter 9

| | | | |
|---|---|---|---|
| 1. c | 6. b | 11. b | 16. c |
| 2. b | 7. d | 12. a | 17. e |
| 3. c | 8. c | 13. d | 18. c |
| 4. d | 9. a | 14. e | 19. a |
| 5. e | 10. e | 15. d | 20. b |

## Chapter 10

| | | | |
|---|---|---|---|
| 1. b | 6. d | 11. b | 16. a |
| 2. a | 7. a | 12. a | 17. e |
| 3. d | 8. c | 13. e | 18. d |
| 4. c | 9. d | 14. d | 19. a |
| 5. e | 10. a | 15. c | 20. c |

## Chapter 11

| | | | |
|---|---|---|---|
| 1. d | 6. b | 11. d | 16. a |
| 2. a | 7. a | 12. b | 17. b |
| 3. e | 8. a | 13. c | 18. d |
| 4. b | 9. c | 14. b | 19. e |
| 5. d | 10. d | 15. d | 20. a |

## Chapter 12

| | | | |
|---|---|---|---|
| 1. c | 6. b | 11. a | 16. b |
| 2. b | 7. c | 12. c | 17. c |
| 3. a | 8. a | 13. d | 18. a |
| 4. d | 9. e | 14. e | 19. e |
| 5. e | 10. c | 15. a | 20. c |

# APPENDIX B

# Declaration of Independence

## IN CONGRESS, July 4, 1776

The unanimous Declaration of the thirteen united States of America,

When in the Course of human events, it becomes necessary for one people to dissolve the political bands which have connected them with another, and to assume among the powers of the earth, the separate and equal station to which the Laws of Nature and of Nature's God entitle them, a decent respect to the opinions of mankind requires that they should declare the causes which impel them to the separation.

We hold these truths to be self-evident, that all men are created equal, that they are endowed by their Creator with certain unalienable Rights, that among these are Life, Liberty and the pursuit of Happiness.—That to secure these rights, Governments are instituted among Men, deriving their just powers from the consent of the governed,—That whenever any Form of Government becomes destructive of these ends, it is the Right of the People to alter or to abolish it, and to institute new Government, laying its foundation on such principles and organizing its powers in such form, as to them shall seem most likely to effect their Safety and Happiness. Prudence, indeed, will dictate that Governments long established should not be changed for light and transient causes; and accordingly all experience hath shewn, that mankind are more disposed to suffer, while evils are sufferable, than to right themselves by abolishing the forms to which they are accustomed. But when a long train of abuses and usurpations, pursuing invariably the same Object evinces a design to reduce them under absolute Despotism, it is their right, it is their duty, to throw off such Government, and to provide new Guards for their future security.—Such has been the patient sufferance of these Colonies; and such is now the necessity which constrains them to alter their former Systems of Government. The history of the present King of Great Britain is a history of repeated injuries and usurpations, all having in direct object the establishment of an absolute Tyranny over these States. To prove this, let Facts be submitted to a candid world.

He has refused his Assent to Laws, the most wholesome and necessary for the public good.

He has forbidden his Governors to pass Laws of immediate and pressing importance, unless suspended in their operation till his Assent should be obtained; and when so suspended, he has utterly neglected to attend to them.

He has refused to pass other Laws for the accommodation of large districts of people, unless those people would relinquish the right of Representation in the Legislature, a right inestimable to them and formidable to tyrants only.

He has called together legislative bodies at places unusual, uncomfortable, and distant from the depository of their public Records, for the sole purpose of fatiguing them into compliance with his measures.

He has dissolved Representative Houses repeatedly, for opposing with manly firmness his invasions on the rights of the people.

He has refused for a long time, after such dissolutions, to cause others to be elected; whereby the Legislative powers, incapable of Annihilation, have returned to the People at large for their exercise; the State remaining in the mean time exposed to all the dangers of invasion from without, and convulsions within.

He has endeavoured to prevent the population of these States; for that purpose obstructing the Laws for Naturalization of Foreigners; refusing to pass others to encourage their migrations hither, and raising the conditions of new Appropriations of Lands.

He has obstructed the Administration of Justice, by refusing his Assent to Laws for establishing Judiciary powers.

He has made Judges dependent on his Will alone, for the tenure of their offices, and the amount and payment of their salaries.

He has erected a multitude of New Offices, and sent hither swarms of Officers to harrass our people, and eat out their substance.

He has kept among us, in times of peace, Standing Armies without the Consent of our legislatures.

He has affected to render the Military independent of and superior to the Civil power.

He has combined with others to subject us to a jurisdiction foreign to our constitution, and unacknowledged by our laws; giving his Assent to their Acts of pretended Legislation:

For Quartering large bodies of armed troops among us:

For protecting them, by a mock Trial, from punishment for any Murders which they should commit on the Inhabitants of these States:

For cutting off our Trade with all parts of the world:

For imposing Taxes on us without our Consent:

For depriving us in many cases, of the benefits of Trial by Jury:

For transporting us beyond Seas to be tried for pretended offences:

For abolishing the free System of English Laws in a neighbouring Province, establishing therein an Arbitrary government, and enlarging its Boundaries so as to render it at once an example and fit instrument for introducing the same absolute rule into these Colonies:

For taking away our Charters, abolishing our most valuable Laws, and altering fundamentally the Forms of our Governments:

For suspending our own Legislatures, and declaring themselves invested with power to legislate for us in all cases whatsoever.

He has abdicated Government here, by declaring us out of his Protection and waging War against us.

He has plundered our seas, ravaged our Coasts, burnt our towns, and destroyed the lives of our people.

He is at this time transporting large Armies of foreign Mercenaries to compleat the works of death, desolation and tyranny, already begun with circumstances of Cruelty & perfidy scarcely paralleled in the most barbarous ages, and totally unworthy of the Head of a civilized nation.

He has constrained our fellow Citizens taken Captive on the high Seas to bear Arms against their Country, to become the executioners of their friends and Brethren, or to fall themselves by their Hands.

He has excited domestic insurrections amongst us, and has endeavoured to bring on the inhabitants of our frontiers, the merciless Indian Savages, whose known rule of warfare, is an undistinguished destruction of all ages, sexes and conditions.

In every stage of these Oppressions We have Petitioned for Redress in the most humble terms: Our repeated Petitions have been answered only by repeated injury. A Prince whose character is thus marked by every act which may define a Tyrant, is unfit to be the ruler of a free people.

Nor have We been wanting in attentions to our British brethren. We have warned them from time to time of attempts by their legislature to extend an unwarrantable jurisdiction over us. We have reminded them of the circumstances of our emigration and settlement here. We have appealed to their native justice and magnanimity, and we have conjured them by the ties of our common kindred to disavow these usurpations, which, would inevitably interrupt our connections and correspondence. They too have been deaf to the voice of justice and of consanguinity. We must, therefore, acquiesce in the necessity, which denounces our Separation, and hold them, as we hold the rest of mankind, Enemies in War, in Peace Friends.

We, therefore, the Representatives of the united States of America, in General Congress, Assembled, appealing to the Supreme Judge of the world for the rectitude of our intentions, do, in the Name, and by Authority of the good People of these Colonies, solemnly publish and declare, That these United Colonies are, and of Right ought to be Free and Independent States; that they are Absolved from all Allegiance to the British Crown, and that all political connection between them and the State of Great Britain, is and ought to be totally dissolved; and that as Free and Independent States, they have full Power to levy War, conclude Peace, contract Alliances, establish Commerce, and to do all other Acts and Things which Independent States may of right do. And for the support of this Declaration, with a firm reliance on the protection of divine Providence, we mutually pledge to each other our Lives, our Fortunes and our sacred Honor.

Courtesy of the Constitution Society (http://constitution.org/us_doi.htm)

# APPENDIX C

# Constitution for the United States of America

We the People of the United States, in Order to form a more perfect Union, establish Justice, insure domestic Tranquility, provide for the common defence, promote the general Welfare, and secure the Blessings of Liberty to ourselves and our Posterity, do ordain and establish this Constitution for the United States of America.

## Article I

### Section 1

All legislative Powers herein granted shall be vested in a Congress of the United States, which shall consist of a Senate and House of Representatives.

### Section 2

The House of Representatives shall be composed of Members chosen every second Year by the People of the several States, and the Electors in each State shall have the Qualifications requisite for Electors of the most numerous Branch of the State Legislature.

No Person shall be a Representative who shall not have attained to the Age of twenty five Years, and been seven Years a Citizen of the United States, and who shall not, when elected, be an Inhabitant of that State in which he shall be chosen.

*Representatives and direct Taxes shall be apportioned among the several States which may be included within this Union, according to their respective Numbers, which shall be determined by adding to the whole Number of free Persons, including those bound to Service for a Term of Years, and excluding Indians not taxed, three fifths of all other Persons* [Modified by Amendment XIV]. The actual Enumeration shall be made within three Years after the first Meeting of the Congress of the United States, and within every subsequent Term of ten Years, in such Manner as they shall by Law direct. The Number of Representatives shall not exceed one for every thirty Thousand, but each State shall have at Least one Representative; and until such enumeration shall be made, the State of New Hampshire shall be entitled to chuse three, Massachusetts eight, Rhode-Island and Providence Plantations one, Connecticut five, New-York six, New Jersey four, Pennsylvania eight, Delaware one, Maryland six, Virginia ten, North Carolina five, South Carolina five, and Georgia three.

When vacancies happen in the Representation from any State, the Executive Authority thereof shall issue Writs of Election to fill such Vacancies.

The House of Representatives shall chuse their Speaker and other Officers; and shall have the sole Power of Impeachment.

### Section 3

The Senate of the United States shall be composed of two Senators from each State, *chosen by the Legislature thereof* [Modified by Amendment XVII], for six Years; and each Senator shall have one Vote.

Immediately after they shall be assembled in Consequence of the first Election, they shall be divided as equally as may be into three Classes. The Seats of the Senators of the first Class shall be vacated at the Expiration of the second Year, of the second Class at the Expiration of the fourth Year, and of the third Class at the Expiration of the sixth Year, so that one third may be chosen every second Year; *and if Vacancies happen by Resignation, or otherwise, during the Recess of the Legislature of any State, the Executive thereof may make temporary Appointments until the next Meeting of the Legislature, which shall then fill such Vacancies* [Modified by Amendment XVII].

No Person shall be a Senator who shall not have attained to the Age of thirty Years, and been nine Years a Citizen of the United States, and who shall not, when elected, be an Inhabitant of that State for which he shall be chosen.

The Vice President of the United States shall be President of the Senate, but shall have no Vote, unless they be equally divided.

The Senate shall chuse their other Officers, and also a President pro tempore, in the Absence of the Vice President, or when he shall exercise the Office of President of the United States.

The Senate shall have the sole Power to try all Impeachments. When sitting for that Purpose, they shall be on Oath or Affirmation. When the President of the United States is tried, the Chief Justice shall preside: And no Person shall be convicted without the Concurrence of two thirds of the Members present.

Judgment in Cases of Impeachment shall not extend further than to removal from Office, and disqualification to hold and enjoy any Office of honor, Trust or Profit under the United States: but the Party convicted shall nevertheless be liable and subject to Indictment, Trial, Judgment and Punishment, according to Law.

## Section 4

The Times, Places and Manner of holding Elections for Senators and Representatives, shall be prescribed in each State by the Legislature thereof; but the Congress may at any time by Law make or alter such Regulations, except as to the Places of chusing Senators.

The Congress shall assemble at least once in every Year, *and such Meeting shall be on the first Monday in December* [Modified by Amendment XX], unless they shall by Law appoint a different Day.

## Section 5

Each House shall be the Judge of the Elections, Returns and Qualifications of its own Members, and a Majority of each shall constitute a Quorum to do Business; but a smaller Number may adjourn from day to day, and may be authorized to compel the Attendance of absent Members, in such Manner, and under such Penalties as each House may provide.

Each House may determine the Rules of its Proceedings, punish its Members for disorderly Behaviour, and, with the Concurrence of two thirds, expel a Member.

Each House shall keep a Journal of its Proceedings, and from time to time publish the same, excepting such Parts as may in their Judgment require Secrecy; and the Yeas and Nays of the Members of either House on any question shall, at the Desire of one fifth of those Present, be entered on the Journal.

Neither House, during the Session of Congress, shall, without the Consent of the other, adjourn for more than three days, nor to any other Place than that in which the two Houses shall be sitting.

## Section 6

The Senators and Representatives shall receive a Compensation for their Services, to be ascertained by Law, and paid out of the Treasury of the United States. They shall in all Cases, except Treason, Felony and Breach of the Peace, be privileged from Arrest during their Attendance at the Session of their respective Houses, and in going to and returning from the same; and for any Speech or Debate in either House, they shall not be questioned in any other Place.

No Senator or Representative shall, during the Time for which he was elected, be appointed to any civil Office under the Authority of the United States, which shall have been created, or the Emoluments whereof shall have been encreased during such time; and no Person holding any Office under the United States, shall be a Member of either House during his Continuance in Office.

## Section 7

All Bills for raising Revenue shall originate in the House of Representatives; but the Senate may propose or concur with Amendments as on other Bills.

Every Bill which shall have passed the House of Representatives and the Senate, shall, before it become a Law, be presented to the President of the United States; If he approve he shall sign it, but if not he shall return it, with his Objections to that House in which it shall have originated, who shall enter the Objections at large on their Journal, and proceed to reconsider it. If after such Reconsideration two thirds of that House shall agree to pass the Bill, it shall be sent, together with the Objections, to the other House, by which it shall likewise be reconsidered, and if approved by two thirds of that House, it shall become a Law. But in all such Cases the Votes of both Houses shall be determined by yeas and Nays, and the Names of the Persons voting for and against the Bill shall be entered on the Journal of each House respectively. If any Bill shall not be returned by the President within ten Days (Sundays excepted) after it shall have been presented to him, the Same shall be a Law, in like Manner as if he had signed it, unless the Congress by their Adjournment prevent its Return, in which Case it shall not be a Law.

Every Order, Resolution, or Vote to which the Concurrence of the Senate and House of Representatives may be necessary (except on a question of Adjournment) shall be presented to the President of the United States; and before the Same shall take Effect, shall be approved by him, or being disapproved by him, shall be repassed by two thirds of the Senate and

House of Representatives, according to the Rules and Limitations prescribed in the Case of a Bill.

## Section 8

The Congress shall have Power To lay and collect Taxes, Duties, Imposts and Excises, to pay the Debts and provide for the common Defence and general Welfare of the United States; but all Duties, Imposts and Excises shall be uniform throughout the United States;

To borrow Money on the credit of the United States;

To regulate Commerce with foreign Nations, and among the several States, and with the Indian Tribes;

To establish an uniform Rule of Naturalization, and uniform Laws on the subject of Bankruptcies throughout the United States;

To coin Money, regulate the Value thereof, and of foreign Coin, and fix the Standard of Weights and Measures;

To provide for the Punishment of counterfeiting the Securities and current Coin of the United States;

To establish Post Offices and post Roads;

To promote the Progress of Science and useful Arts, by securing for limited Times to Authors and Inventors the exclusive Right to their respective Writings and Discoveries;

To constitute Tribunals inferior to the supreme Court;

To define and punish Piracies and Felonies committed on the high Seas, and Offences against the Law of Nations;

To declare War, grant Letters of Marque and Reprisal, and make Rules concerning Captures on Land and Water;

To raise and support Armies, but no Appropriation of Money to that Use shall be for a longer Term than two Years;

To provide and maintain a Navy;

To make Rules for the Government and Regulation of the land and naval Forces;

To provide for calling forth the Militia to execute the Laws of the Union, suppress Insurrections and repel Invasions;

To provide for organizing, arming, and disciplining, the Militia, and for governing such Part of them as may be employed in the Service of the United States, reserving to the States respectively, the Appointment of the Officers, and the Authority of training the Militia according to the discipline prescribed by Congress;

To exercise exclusive Legislation in all Cases whatsoever, over such District (not exceeding ten Miles square) as may, by Cession of particular States, and the Acceptance of Congress, become the Seat of the Government of the United States, and to exercise like Authority over all Places purchased by the Consent of the Legislature of the State in which the Same shall be, for the Erection of Forts, Magazines, Arsenals, dock-Yards, and other needful Buildings;—And

To make all Laws which shall be necessary and proper for carrying into Execution the foregoing Powers, and all other Powers vested by this Constitution in the Government of the United States, or in any Department or Officer thereof.

## Section 9

The Migration or Importation of such Persons as any of the States now existing shall think proper to admit, shall not be prohibited by the Congress prior to the Year one thousand eight hundred and eight, but a Tax or duty may be imposed on such Importation, not exceeding ten dollars for each Person.

The Privilege of the Writ of Habeas Corpus shall not be suspended, unless when in Cases of Rebellion or Invasion the public Safety may require it.

No Bill of Attainder or ex post facto Law shall be passed.

No Capitation, or other direct, Tax shall be laid, unless in Proportion to the Census or Enumeration herein before directed to be taken.

No Tax or Duty shall be laid on Articles exported from any State.

No Preference shall be given by any Regulation of Commerce or Revenue to the Ports of one State over those of another; nor shall Vessels bound to, or from, one State, be obliged to enter, clear, or pay Duties in another.

No Money shall be drawn from the Treasury, but in Consequence of Appropriations made by Law; and a regular Statement and Account of the Receipts and Expenditures of all public Money shall be published from time to time.

No Title of Nobility shall be granted by the United States: And no Person holding any Office of Profit or Trust under them, shall, without the Consent of the Congress, accept of any present, Emolument, Office, or Title, of any kind whatever, from any King, Prince, or foreign State.

## Section 10

No State shall enter into any Treaty, Alliance, or Confederation; grant Letters of Marque and Reprisal; coin Money; emit Bills of Credit; make any Thing but gold and silver Coin a Tender in Payment of Debts; pass any Bill of Attainder, ex post facto Law, or Law impairing the Obligation of Contracts, or grant any Title of Nobility.

No State shall, without the Consent of the Congress, lay any Imposts or Duties on Imports or Exports, except what may be absolutely necessary for executing it's inspection Laws; and the net Produce of all Duties and Imposts, laid by any State on Imports or Exports, shall be for the Use of the Treasury of the United States; and all such Laws shall be subject to the Revision and Controul of the Congress.

No State shall, without the Consent of Congress, lay any Duty of Tonnage, keep Troops, or Ships of War in time of Peace, enter into any Agreement or Compact with another State, or with a foreign Power, or engage in War, unless actually invaded, or in such imminent Danger as will not admit of delay.

# Article II

## Section 1

The executive Power shall be vested in a President of the United States of America. He shall hold his Office during the Term of four Years, and, together with the Vice President, chosen for the same Term, be elected, as follows:

Each State shall appoint, in such Manner as the Legislature thereof may direct, a Number of Electors, equal to the whole Number of Senators and Representatives to which the State may be entitled in the Congress: but no Senator or Representative, or Person holding an Office of Trust or Profit under the United States, shall be appointed an Elector.

*The Electors shall meet in their respective States, and vote by Ballot for two Persons, of whom one at least shall not be an Inhabitant of the same State with themselves. And they shall make a List of all the Persons voted for, and of the Number of Votes for each; which List they shall sign and certify, and transmit sealed to the Seat of the Government of the United States, directed to the President of the Senate. The President of the Senate shall, in the Presence of the Senate and House of Representatives, open all the Certificates, and the Votes shall then be counted. The Person having the greatest Number of Votes shall be the President, if such Number be a Majority of the whole Number of Electors appointed; and if there be more than one who have such Majority, and have an equal Number of Votes, then the House of Representatives shall immediately chuse by Ballot one of them for President; and if no Person have a Majority, then from the five highest on the List the said House shall in like Manner chuse the President. But in chusing the President, the Votes shall be taken by States, the Representation from each State having one Vote; a quorum for this Purpose shall consist of a Member or Members from two thirds of the States, and a Majority of all the States shall be necessary to a Choice. In every Case, after the Choice of the President, the Person having the greatest Number of Votes of the Electors shall be the Vice President. But if there should remain two or more who have equal Votes, the Senate shall chuse from them by Ballot the Vice President* [Modified by Amendment XII].

The Congress may determine the Time of chusing the Electors, and the Day on which they shall give their Votes; which Day shall be the same throughout the United States.

No Person except a natural born Citizen, or a Citizen of the United States, at the time of the Adoption of this Constitution, shall be eligible to the Office of President; neither shall any Person be eligible to that Office who shall not have attained to the Age of thirty five Years, and been fourteen Years a Resident within the United States.

*In Case of the Removal of the President from Office, or of his Death, Resignation, or Inability to discharge the Powers and Duties of the said Office, the Same shall devolve on the Vice President, and the Congress may by Law provide for the Case of Removal, Death, Resignation or Inability, both of the President and Vice President, declaring what Officer shall then act as President, and such Officer shall act accordingly, until the Disability be removed, or a President shall be elected* [Modified by Amendment XXV].

The President shall, at stated Times, receive for his Services, a Compensation, which shall neither be increased nor diminished during the Period for which he shall have been elected, and he shall not receive within that Period any other Emolument from the United States, or any of them.

Before he enter on the Execution of his Office, he shall take the following Oath or Affirmation:—"I do solemnly swear (or affirm) that I will faithfully execute the Office of President of the United States, and will to the best of my Ability, preserve, protect and defend the Constitution of the United States."

## Section 2

The President shall be Commander in Chief of the Army and Navy of the United States, and of the Militia of the several States, when called into the actual Service of the United States; he may require the Opinion, in writing, of the principal Officer in each of the executive Departments, upon any Subject relating to the Duties of their respective Offices, and he shall have Power to grant Reprieves and Pardons for Offences against the United States, except in Cases of Impeachment.

He shall have Power, by and with the Advice and Consent of the Senate, to make Treaties, provided two thirds of the Senators present concur; and he shall nominate, and by and with the Advice and

Consent of the Senate, shall appoint Ambassadors, other public Ministers and Consuls, Judges of the supreme Court, and all other Officers of the United States, whose Appointments are not herein otherwise provided for, and which shall be established by Law: but the Congress may by Law vest the Appointment of such inferior Officers, as they think proper, in the President alone, in the Courts of Law, or in the Heads of Departments.

The President shall have Power to fill up all Vacancies that may happen during the Recess of the Senate, by granting Commissions which shall expire at the End of their next Session.

## Section 3

He shall from time to time give to the Congress Information of the State of the Union, and recommend to their Consideration such Measures as he shall judge necessary and expedient; he may, on extraordinary Occasions, convene both Houses, or either of them, and in Case of Disagreement between them, with Respect to the Time of Adjournment, he may adjourn them to such Time as he shall think proper; he shall receive Ambassadors and other public Ministers; he shall take Care that the Laws be faithfully executed, and shall Commission all the Officers of the United States.

## Section 4

The President, Vice President and all civil Officers of the United States, shall be removed from Office on Impeachment for, and Conviction of, Treason, Bribery, or other high Crimes and Misdemeanors.

# Article III

## Section 1

The judicial Power of the United States shall be vested in one supreme Court, and in such inferior Courts as the Congress may from time to time ordain and establish. The Judges, both of the supreme and inferior Courts, shall hold their Offices during good Behaviour, and shall, at stated Times, receive for their Services a Compensation, which shall not be diminished during their Continuance in Office.

## Section 2

The judicial Power shall extend to all Cases, in Law and Equity, arising under this Constitution, the Laws of the United States, and Treaties made, or which shall be made, under their Authority;—to all Cases affecting Ambassadors, other public Ministers and Consuls;—to all Cases of admiralty and maritime Jurisdiction;—to Controversies to which the United States shall be a Party;—to Controversies between two or more States;—*between a State and Citizens of another State* [Modified by Amendment XI];—between Citizens of different States;—between Citizens of the same State claiming Lands under Grants of different States, and between a State, or the Citizens thereof, and foreign States, Citizens or Subjects.

In all Cases affecting Ambassadors, other public Ministers and Consuls, and those in which a State shall be Party, the supreme Court shall have original Jurisdiction. In all the other Cases before mentioned, the supreme Court shall have appellate Jurisdiction, both as to Law and Fact, with such Exceptions, and under such Regulations as the Congress shall make.

The Trial of all Crimes, except in Cases of Impeachment, shall be by Jury; and such Trial shall be held in the State where the said Crimes shall have been committed; but when not committed within any State, the Trial shall be at such Place or Places as the Congress may by Law have directed.

## Section 3

Treason against the United States shall consist only in levying War against them, or in adhering to their Enemies, giving them Aid and Comfort. No Person shall be convicted of Treason unless on the Testimony of two Witnesses to the same overt Act, or on Confession in open Court.

The Congress shall have Power to declare the Punishment of Treason, but no Attainder of Treason shall work Corruption of Blood, or Forfeiture except during the Life of the Person attainted.

# Article IV

## Section 1

Full Faith and Credit shall be given in each State to the public Acts, Records, and judicial Proceedings of every other State. And the Congress may by general Laws prescribe the Manner in which such Acts, Records and Proceedings shall be proved, and the Effect thereof.

## Section 2

The Citizens of each State shall be entitled to all Privileges and Immunities of Citizens in the several States.

A Person charged in any State with Treason, Felony, or other Crime, who shall flee from Justice, and be found in another State, shall on Demand of the executive Authority of the State from which he fled, be delivered up, to be removed to the State having Jurisdiction of the Crime.

*No Person held to Service or Labour in one State, under the Laws thereof, escaping into another, shall, in*

*Consequence of any Law or Regulation therein, be discharged from such Service or Labour, but shall be delivered up on Claim of the Party to whom such Service or Labour may be due* [Modified by Amendment XIII].

## Section 3

New States may be admitted by the Congress into this Union; but no new State shall be formed or erected within the Jurisdiction of any other State; nor any State be formed by the Junction of two or more States, or Parts of States, without the Consent of the Legislatures of the States concerned as well as of the Congress.

The Congress shall have Power to dispose of and make all needful Rules and Regulations respecting the Territory or other Property belonging to the United States; and nothing in this Constitution shall be so construed as to Prejudice any Claims of the United States, or of any particular State.

## Section 4

The United States shall guarantee to every State in this Union a Republican Form of Government, and shall protect each of them against Invasion; and on Application of the Legislature, or of the Executive (when the Legislature cannot be convened), against domestic Violence.

## Article V

The Congress, whenever two thirds of both Houses shall deem it necessary, shall propose Amendments to this Constitution, or, on the Application of the Legislatures of two thirds of the several States, shall call a Convention for proposing Amendments, which, in either Case, shall be valid to all Intents and Purposes, as Part of this Constitution, when ratified by the Legislatures of three fourths of the several States, or by Conventions in three fourths thereof, as the one or the other Mode of Ratification may be proposed by the Congress; Provided that no Amendment which may be made prior to the Year One thousand eight hundred and eight shall in any Manner affect the first and fourth Clauses in the Ninth Section of the first Article; and that no State, without its Consent, shall be deprived of its equal Suffrage in the Senate.

## Article VI

All Debts contracted and Engagements entered into, before the Adoption of this Constitution, shall be as valid against the United States under this Constitution, as under the Confederation.

This Constitution, and the Laws of the United States which shall be made in Pursuance thereof; and all Treaties made, or which shall be made, under the Authority of the United States, shall be the supreme Law of the Land; and the Judges in every State shall be bound thereby, any Thing in the Constitution or Laws of any State to the Contrary notwithstanding.

The Senators and Representatives before mentioned, and the Members of the several State Legislatures, and all executive and judicial Officers, both of the United States and of the several States, shall be bound by Oath or Affirmation, to support this Constitution; but no religious Test shall ever be required as a Qualification to any Office or public Trust under the United States.

## Article VII

The Ratification of the Conventions of nine States, shall be sufficient for the Establishment of this Constitution between the States so ratifying the Same.

Courtesy of the Constitution Society (http://constitution.org/cons/constitu.htm)

# Bill of Rights

**APPENDIX D**

## Amendment I

Congress shall make no law respecting an establishment of religion, or prohibiting the free exercise thereof; or abridging the freedom of speech, or of the press; or the right of the people peaceably to assemble, and to petition the Government for a redress of grievances.

## Amendment II

A well regulated Militia, being necessary to the security of a free State, the right of the people to keep and bear Arms, shall not be infringed.

## Amendment III

No Soldier shall, in time of peace be quartered in any house, without the consent of the Owner, nor in time of war, but in a manner to be prescribed by law.

## Amendment IV

The right of the people to be secure in their persons, houses, papers, and effects, against unreasonable searches and seizures, shall not be violated, and no Warrants shall issue, but upon probable cause, supported by Oath or affirmation, and particularly describing the place to be searched, and the persons or things to be seized.

## Amendment V

No person shall be held to answer for a capital, or otherwise infamous crime, unless on a presentment or indictment of a Grand Jury, except in cases arising in the land or naval forces, or in the Militia, when in actual service in time of War or public danger; nor shall any person be subject for the same offence to be twice put in jeopardy of life or limb; nor shall be compelled in any criminal case to be a witness against himself, nor be deprived of life, liberty, or property, without due process of law; nor shall private property be taken for public use, without just compensation.

## Amendment VI

In all criminal prosecutions, the accused shall enjoy the right to a speedy and public trial, by an impartial jury of the State and district wherein the crime shall have been committed, which district shall have been previously ascertained by law, and to be informed of the nature and cause of the accusation; to be confronted with the witnesses against him; to have compulsory process for obtaining witnesses in his favor, and to have the Assistance of Counsel for his defence.

## Amendment VII

In Suits at common law, where the value in controversy shall exceed twenty dollars, the right of trial by jury shall be preserved, and no fact tried by a jury, shall be otherwise re-examined in any Court of the United States, than according to the rules of the common law.

## Amendment VIII

Excessive bail shall not be required, nor excessive fines imposed, nor cruel and unusual punishments inflicted.

## Amendment IX

The enumeration in the Constitution, of certain rights, shall not be construed to deny or disparage others retained by the people.

## Amendment X

The powers not delegated to the United States by the Constitution, nor prohibited by it to the States, are reserved to the States respectively, or to the people.

Courtesy of the Constitution Society (http://constitution.org/cons/constitu.htm)

# Glossary

## A

**Abandoned property** Intentionally discarded property in which the owner has no apparent intent to retrieve the property in question or claim ownership.

**Abatement** Court order to cease or eliminate conditions or behavior causing a nuisance.

**Accessory after the fact** One who assists, aids, or abets an accused after commission of a crime.

**Accessory before the fact** Person who was not at or near the scene of a crime but who did assist, aid, or abet a perpetrator prior to the commission of a crime.

**Accident** Incident or transaction that occurs without intent to cause that event.

**Accomplice** Person who assists, aids, or abets a perpetrator before the commission of a crime or who fails to prevent the commission of a crime when possessing a legal duty.

**Actual breaking** Use of physical force to enter a building or structure.

**Actual cause** Harm caused by the actual conduct of an accused person.

**Actual entry** Physical insertion of a body part to enter a building or structure.

**Actual possession** Physical possession of money or property.

*Actus reus* Latin for "guilty act."

**Adequate provocation** Actions intended and calculated to interfere with rational thinking skills; unlawful conduct sufficient to provoke a reasonable person to inflict harm.

**Administrative law** Policies and regulations that govern and restrict behavior within government agencies.

**Adulteration** Adding or removing a substance, compound, or other ingredient from a product.

**Adultery** Sexual intercourse with someone other than a lawful spouse while married to another.

**Affinity** Familial relation by marriage (not blood).

**Affirmative defenses** Accused acknowledges commission of a crime while concurrently offering some justification or excuse to negate culpability.

**Aforethought** Advance planning or design.

**Age of consent** Age at which a minor may engage in legal decisions such as the decision to marry, contract, or engage in sexual activity.

**Aggravated assault** Crime that causes or intends to cause serious bodily injury, mayhem, or permanent disfigurement.

**Aggravating circumstance** Factor that heightens the severity of a crime.

**Alibi** Defense to establish that an accused could not have committed the crime in question because he or she was in another physical location.

**Alteration** Addition, deletion, or manipulation of a document or instrument.

**Anti-Car Theft Act of 1992** Legislation that made it a federal crime to commit a carjacking.

**Antiterrorism and Effective Death Penalty Act of 1996** Legislation that granted federal authority to prosecute crimes normally pursued in state courts when their motivation was to coerce, intimidate, or retaliate against a government or civilian population.

**Armed robbery** Taking of money or property from the person or presence of another through use or threat of force with a deadly weapon.

**Arson** At common law, the malicious burning of the dwelling of another.

**Assault** At common law, an attempted battery or threatened battery.

**Attempt** Separate offenses which render illegal certain steps taken in furtherance of an intended crime.

**Attempted battery** Overt act to cause harm through physical contact.

**Attendant circumstances** Event that must accompany certain crime definitions.

**Automatism** Involuntary action(s) committed during a mental state of incapacity.

**Aviation and Transportation Security Act of 2001** Legislation which created the Transportation Security Administration (TSA) and vested certain TSA employees with federal law enforcement powers.

## B

**Bail** Money or property that must be deposited before a defendant may be released pending trial.

**Bank Secrecy Act of 1970** Legislation that mandated financial institutions to report transactions exceeding $10,000 and implement a reporting system for suspicious activity.

**Battered wife syndrome** Extreme emotional state caused by a cycle of domestic violence.

**Battery** Nonlethal culmination of an assault.

**Bestiality** Sex with an animal.

**Beyond a reasonable doubt** Proof of moral certainty; standard for criminal conviction.

**Bifurcated proceeding** Refers to a legal proceeding delineated into two phases; for example, trials involving the death penalty consist of a guilt phase and a sentencing phase.

**Bigamy** Entering a purported marriage while legally wed to another; constitutes a crime and grounds for divorce and/or annulment.

**Bilateral theory** Two or more parties collaborate to constitute a conspiracy.

**Blackmail** Form of extortion where the threat is to expose secrets or damaging information.

**Blockburger test** Stipulates that one criminal act can constitute two or more separate offenses only if each requires proof of an additional fact which the other does not.

**Blood alcohol level** Milligrams of alcohol per milliliter of blood.

**Born alive standard** Fetus must achieve independent circulation to be considered a human being.

**Brain death** Complete cessation of electrical impulses in the brain.

**Breach of the peace** Disturbance of the peace and tranquility of a community.

**Breathalyzer** Instrument that detects and records blood–alcohol content levels.

**Bribery** Agreement to do (or refrain from) required acts in exchange for money or property.

**Buggery** Anal intercourse (penetration of the rectum).

**Burden of proof** Legal standard required to hold one accountable for criminal and civil harm.

**Burglary** At common law, the breaking and entering of the dwelling of another in the nighttime with intent to commit a felony therein.

**Burning** Structural degradation caused by fire.

**But-for test** Regarded as actual cause when harm would not have occurred but for the conduct.

## C

**Canon law** Laws of the Catholic Church.

**Capital felony** Crimes eligible for the punishment of death or life imprisonment.

**Carjacking** Taking of a motor vehicle from an occupant through use or threat of force.

**Carnal knowledge** Penile–vaginal intercourse.

**Case law** Body of law derived over time by judicial opinion.

**Castle doctrine** Removes the duty to retreat during home invasions when it is reasonable to believe their lives (or lives of others) are in immediate danger.

**Caveat emptor** Latin for "let the buyer beware."

**Chaste character** Never engaged in sexual intercourse—a virgin.

**Chattel** Common law rule that wives were the personal property of their husbands.

**Check kiting** Use of multiple businesses, banks, or other institutions in conjunction with issuing worthless checks.

**Checks and balances** Government system designed to prevent tyranny by any singular branch.

**Child exploitation** Variety of acts calculated to derive financial, sexual, or other benefits from the manipulation of children.

**Churning** Excessive trade transactions to generate commissions for a broker at an investor's expense.

**Civil contempt** Court effort to obtain compliance (not punish) from those not obeying judicial orders, decrees, and judgments.

**Civil forfeiture** Loss of property resulting from legal proceedings.

**Civil law** A body of law that regulates claims of private wrongs.

**Claim of right** A party with the singular right to possess can legally retrieve property through stealth and not be guilty of larceny.

**Clean Air Act of 1970** Legislation that granted federal authority to regulate air pollution.

**Clean Water Act of 1972** Legislation that granted federal authority to regulate discharge of material into navigable waters.

**Code of Hammurabi** Widely regarded as the first set of written laws; contains some 300 criminal and civil laws developed by Babylon's King Hammurabi between 1792 and 1750 B.C.

**Cognition** Did the accused possess substantial mental capacity to distinguish right from wrong?

**Commercial bribery** Illegal influence on business officials and transactions by getting them to violate duties in exchange for money or other value.

**Common law** Laws common to the circuits of Old England; consisted of judicial rulings regarding the application and interpretation of laws, customs, and prior case decisions.

**Compensatory damage** Actual expenses associated with wrongful conduct.

**Compounding** Accepting things of value in exchange for failing to report a crime.

**Compulsory process** Sixth Amendment constitutional right to secure presence of witnesses.

**Consanguinity** Familial relation by blood.

**Consent of the victim** Defense that negates culpability when the victim, in advance, voluntarily acquiesced to nonserious bodily harm.

**Consideration** Something of value exchanged or proposed for exchange.

**Conspiracy** Multiple parties agree to commit a crime; concert in criminal purpose.

**Conspirators** The parties to a criminal agreement (or conspiracy).

**Constitutional law** Substantive and procedural dictates contained within the constitutions of the United States and its independent states.

**Constructive asportation** Causing an innocent third party to move money or property that an accused never touched.

**Constructive breaking** Causing an opening to effect entrance without physical contact.

**Constructive contempt** Unruly behavior occurring outside the presence of the court but nonetheless disrespectful.

**Constructive entry** Causing entry without physical insertion of a body part but through the use of an instrument or tool.

**Constructive intent** Actions committed with recklessness or negligence.

**Constructive possession** Causing money or property to be possessed without physical interaction.

**Constructive taking** Causing an innocent third party to gain possession of money or property that the accused never possessed.

**Continuing offense** Concealing assets, ongoing fraud, and other similar crimes.

**Continuing trespass** Intent to permanently deprive formed at some point beyond a property's being taken and carried away.

**Contractual theory** Common law standard that wives consent to all sexual intercourse, consensual and forcible, when entering marital contracts.
**Conversion** Transforming property into something other than its original status.
***Corpus delicti*** Latin for "body of the crime," meaning good reason to believe that a crime was committed and that the accused committed the crime.
**Counterfeiting** Making or possessing forged obligations or securities of the United States.
**Courtroom decorum** Orderly and professional atmosphere required in courts of law.
**Courts of equity** Common law courts that possessed broad authority to award relief with the goal of achieving fairness.
**Creation** Manufacture of a document or instrument.
**Crime** Public wrong committed against the welfare of society; commission of a prohibited act or omission of a required act, without defense, and codified as a felony or misdemeanor.
**Crimes against habitation** Crimes that seek to deter violations against a person's residence—the home (or dwelling): arson and burglary.
**Criminal contempt** Punishes one who violates a court order.
**Criminal enterprise** Individual, partnership, corporation, association or other legal entity, or any union or group of individuals associated in crime though not a legal entity.
**Criminal forfeiture** Loss of property as a penalty for committing a crime.
**Criminal law** Body of law comprised of substantive and procedural rules of conduct.
**Criminal Lunatics Act of 1800** Legislation that created the verdict "not guilty on account of insanity."
**Culpable** Worthy of blame.
**Cunnilingus** Oral stimulation of the female sexual organ (vagina).
**Custody** Limited right to use property within one's care but little real discretion regarding how the property is exercised.

# D

**Dangerous proximity test** Examines whether a defendant was dangerously close to committing an intended crime.
**Deadly weapon doctrine** Infers malice from the use of a deadly weapon.
**Death-qualified jury** *Voir dire* has established the ability of a jury to consider the death penalty as a possible punishment.
**Declaratory relief** Civil determination of a person's rights under a contract or statute.
**Deliberation** Careful reflection upon the wisdom of putting into action premeditated thoughts.
**Democracy** Form of government where elected leaders make decisions for the populous with no legal safeguards.
**Depraved-heart murder** Infers malice from actions that exhibit signs of an abandoned and malignant heart.
**Determinate sentence** Legislature proscribes specific terms of incarceration.
**Deterrence** Theory that suggests authorized punishments will prevent individuals from engaging in illegal acts.
**Deviance** Behavior that breaches or deviates from social norms; or a statistical anomaly.
**Diminished capacity** Does not possess the *mens rea* required of an intent crime.

**Direct asportation** Physically moving money or property.
**Direct contempt** Unruly behavior that occurs in the presence of the court.
**Direct taking** Physical custody of money or property.
**Disablement** Loss of the use of a body part or organ.
**Dismemberment** Loss of some portion of a body part or organ.
**Disorderly conduct** Acts causing a public disturbance, or otherwise threatening or menacing.
**Doctrine of overbreadth** Laws so general that they could criminalize both legal and illegal behavior; typically raised in cases involving First Amendment protections such as freedom of assembly and freedom of speech.
**Document** Anything with writing on its surface.
**Domestic terrorism** Acts of terrorism occurring within the United States or its territories.
**Domestic violence** Assault against spouses or intimate partners.
**Drug possession** Dominion or control of drugs known to be illegal.
**Dual sovereignty** Multiple prosecutions by different governments with lawful authority.
**Due process** Fifth Amendment constitutional right requiring government to follow certain procedures when infringing on life, liberty, or property.
**Duress** Defense that argues an accused was coerced to involuntarily commit a crime
**Durham Rule** Broadened the standard for insanity by declaring that an accused is not criminally responsible when an unlawful act was the product of mental disease or mental defect.
**Dwelling** Primary safe haven where one sleeps and eats.

# E

**Ecclesiastical courts** English court of equity charged with enforcing canon law.
**Embezzlement** Unlawful conversion or misappropriation of another's property by one to whom property was entrusted.
**Embracery** Unlawful attempt to influence a jury or juror.
**Eminent domain** Fifth Amendment constitutional right requiring citizens be given just compensation when government seizes private property for personal use.
**Entrapment** Defense that argues that police were responsible for making an accused commit a crime that otherwise would not have been contemplated.
**Environmental Protection Agency** Federal agency with jurisdiction over most cases involving environmental offenses and enforcing laws regarding pollution, facility permits, and reporting.
**Equal protection clause** Prohibits states from making arbitrary and unreasonable distinctions in terms of rights and freedoms.
**Equivocality test** Determines whether a defendant's actions were indicative of criminal intent.
**Escape** Lawful detainee who leaves or fails to return without official permission.
**Espionage** Spying; to gather (or attempt to gather), transmit, or deliver intelligence information to unauthorized parties.
**Euthanasia** Intentional mercy killing.
**Excessive bail** Exceeds an amount that would assure the presence of a person in court.
**Exclusionary rule** Prohibits introduction of evidence seized in violation of the Fourth Amendment into criminal trials.

**Excusable homicide** Noncriminal homicide due to mitigating circumstances.

**Excuse defense** *Mens rea* associated with criminal wrongdoing is negated (or mitigated) because of mental incapacity.

**Executive branch** Government entity vested with power to enforce law.

**Extortion** Demands things of value (primarily money) in exchange for not causing harm.

# F

**Factual impossibility** Inability to commit a crime due because certain facts were unknown or beyond control of the defendant.

**False advertising** Luring consumers into purchasing goods and services at excessive costs through dissemination of false information.

**False imprisonment** At common law, unlawfully restricting the freedom of another but where the victim is not moved to another location.

**False pretenses** Acquiring ownership of another's property through fraudulent means.

**Federal Anti-Riot Act of 1968** Criminalized riots involving interstate travel or communication.

**Federal Sentencing Guidelines** Designed to reduce sentencing disparity in federal courts through reducing judicial discretion and providing an objective standard for sentencing.

**Federal Trade Commission** Federal agency with jurisdiction over false advertising.

**Federalism** Nationalized strong central government that recognizes state sovereignty.

**Fellatio** Oral stimulation of the male sexual organ (penis).

**Felony** Crime for which punishment is death or 1 year or more in a federal or state prison.

**Felony murder** Classifies as murder any death resulting from reckless or negligent actions committed during the perpetration of designated felonies.

**Feticide** The killing of an unborn child.

**Field sobriety test** Observations designed to detect mental impairment of drivers.

**Fighting words** Speech that inflicts injury or creates a breach of the peace and is not central to the exposition of an idea.

**Fine** Fixed sum of money paid as a penalty for committing a crime.

**First-degree murder** Malicious killing of a human being with premeditation and deliberation.

**Float** Time interval between issuance of a check and its presentation to a bank.

**Fondling** Adult handles, touches, or rubs a child under a specified age for the purpose of gratification of lust.

**Food and Drug Administration** Regulates food and drug consumption in the United States.

**Food, Drug, and Cosmetic Act** Defines prohibited acts and penalties for federal food and drug violations.

**Forcible rape** At common law, carnal knowledge of a female against her will and through use or threat of force.

**Foreign Intelligence and Surveillance Act of 1978** Federal authority to exercise warrantless (mainly electronic) searches of foreigners within the United States.

**Foreign Intelligence and Surveillance Court** Courts created to issue warrants subject to FISA authorization.

**Forfeiture** Taking or seizing property that was used to commit or facilitate a crime.

**Forgery** Unlawful creation or alteration of a document possessing apparent legal significance and with the intent to defraud.

**Fornication** Consensual sexual intercourse by an unmarried person.

# G

**Gambling** Risking something of value to accumulate greater value.

**Gaming** Legal participation in games of chance.

**General deterrence** Goal of punishment whereby society is deterred from engaging in crime as a result of viewing punishment meted out on others.

**General intent** Malevolent or wrongful design committed with no particularized objective.

**Genocide** Actions that intend to destroy a national, ethnic, racial, or religious group.

**Grand jury** Body of citizens authorized to determine whether sufficient proof exists to move forward with criminal charges and trial.

**Grand larceny** Theft of property valued at or exceeding a predetermined amount.

**Gross misdemeanor** Crime for which incarceration ranges from 6 to 12 months.

**Gross negligence** Negligence so extreme that it carries penalties associated with recklessness.

**Guilty but mentally ill** Legal standard that negates punishment when an accused is mentally ill at the time of the offense, yet not to the extent required to plead insanity.

# H

**Habitual offender statute** Imposes mandatory sentence upon conviction of third felony.

**Harassment** Statutes that adjudicate assaults of a minor nature—less than bodily injury from mere pushing or shoving.

**Hate Crimes Statistics Act of 1990** Legislation that requires the United States Attorney General to collect and publish data regarding the extent to which hate crimes occur in America.

**Hearsay** Oral or written statement made out of court that is offered in court to prove the truth of the matter asserted in the statement.

**Heat of passion** Significant impairment with one's ability to deliberate on pending actions.

**Homeland Security Act of 2002** Legislation that created the Department of Homeland Security, placing more than 20 federal agencies under its authority.

**Homicide** Killing of one human being by another.

# I

**Identity theft** Series of offenses wherein a person represents himself or herself as another to obtain a line of credit, loan, property, money, or any other good or service from financial institutions.

**Identity Theft and Assumption Deterrence Act of 1998** Regulates production, transfer, or possession of false identification issued by federal, foreign, state, or local governments.

**Identity Theft Penalty Enhancement Act of 2004** Amended federal law to authorize harsher penalties for identity theft associated with serious crimes such as illegal immigration.

**Imperfect self-defense** Subjective belief that deadly force is necessary, but where objective circumstances do not actually warrant such action.

**Implied consent statute** When obtaining a driver's license, drivers give advance consent to field sobriety and breathalyzer tests in future police interaction.

**Implied malice** Presence of malice is inferred without express evidence of its existence.

**Incapacitation** Removal of offenders from society to avoid future harm.

**Incest** Intercourse or marriage between individuals not too closely related by blood or marriage.

**Inchoate** Incipient crime that generally leads to another crime; uncompleted crime.

**Incitement of a riot** Crime committed by those who organize, promote, encourage, participate in, or carry on a riot.

**Incorporation** Process wherein Bill of Rights provisions are applied to states.

**Indecent exposure** Intentionally exposing private parts in a manner that others are likely to view and for the purpose of gratifying licentious desire.

**Indeterminate sentence** Legislature proscribes minimum and maximum incarceration period, but trial judge, correctional authorities, or parole boards determine actual moment of release.

**Indispensable element test** Assesses whether a defendant has the ability to carry out a crime.

**Inducement** Actions that present a person with an opportunity to engage in certain conduct.

**Infancy** English common law standard that absolved children of criminal responsibility.

**Information** Formal charging document filed by the prosecutor with the court.

**Injunctive relief** Court order requiring one to do or stop doing harm.

**Insanity defense** State of mind rendering one not culpable for criminal action due to mental defect or disease.

**Insanity Defense Reform Act of 1984** Federal insanity guideline requiring that persons suffer severe mental disease or defect and be unable to appreciate the nature and quality of wrongful acts.

**Insider trading** Securities bought or sold based on knowledge not yet available to the public.

**Instrument** Written legal document (such as a contract, deed or will).

**Intangible property** Items with value but no actual concrete qualities.

**Intensive supervision probation** Offenders remain in the community to serve sentence but under supervision and conditions more stringent than customary.

**Intermediary** Third party used to solicit others.

**International terrorism** Acts of terrorism occurring outside territories of the United States.

**Interstate Wire Act of 1961** Regulates online sports gambling as illegal without state intervention.

**Intervening cause** Event that severs (or breaks) the connection between conduct and harmful consequences, thereby negating legal causation.

**Intoxication** Refers to diminished mental capacity by way of alcohol or drug use.

**Involuntary intoxication** Unknowing ingestion of alcohol or drugs.

**Involuntary manslaughter** Death resulting from reckless or negligent conduct, or during the commission of a misdemeanor.

**Involuntary renunciation** Reluctantly abandoning intent to commit a crime due to intervening causes.

**Irresistible impulse test** Declares insane individuals incapable of controlling conduct due to a mental disease or defect.

## J

**Jihad** Islamic concept of "Holy War" used by extremists to justify terrorist acts.

**Johns** Customers of prostitution transactions.

**Jostling** Statute that addresses bumping and pushing for the purpose of committing theft.

**Judicial activism** Occurs when a judge relies on personal ideology to guide decisions as opposed to the facts of the case and rule of law.

**Judicial branch** Government entity vested with the power to interpret law.

**Jurisdiction** Court authority to hear and decide a case.

**Justifiable homicide** Noncriminal homicide due to the exercise of a right or duty.

**Justification defense** Accused had a right or duty to commit what normally is a crime.

## K

**Kidnapping** At common law, unlawfully restricting the freedom of another and moving that person to another location against their will.

**Kings courts** English law courts operated under the rule of the monarchy.

**Knowing endangerment** Subject to the Clean Water Act, knowledge that a clean water offense has placed others in danger of imminent death or serious injury.

## L

**La Cosa Nostra** Italian for "our thing;" Italian-American Mafia.

**Larceny** Taking and carrying away personal property of another with intent to deprive permanently.

**Larceny by trick** Taking and carrying away property of another with consent that is invalid because it was obtained through fraud (trickery or deceit).

**Last act test** Evaluates whether a defendant's actions constitute an attempt.

**Law courts** Common law court restricted to awarding monetary damages.

**Law Reform (Year and a Day Rule) Act of 1996** Legislation passed in England that abolished its common law year-and-a day rule.

**Least restrictive mechanism** Binding promise that government action will be implemented with every effort to minimize intrusion into the lives of its people.

**Legal cause** Recognizes the unfairness of imposing criminal penalties when the person who caused the harm did not intend to do so or was unable to reasonably anticipate danger.

**Legal efficacy** Apparent legal significance.

**Legal impossibility** Cannot be guilty of a certain crime when the required intent is absent, or the wrong crime has been charged.

**Legislative branch** Government entity vested with power to make law.

**Lesser included offense** Crime possessing most elements of a more serious crime but missing some key component.

**Loitering** Wandering about with no apparent lawful purpose.

# M

**M'Naghten Rule** While laboring under a defect of reason or disease of the mind, an accused did not to know the nature and quality of the act, or the difference between right and wrong.

**Mail fraud** Deceptive, fraudulent, or otherwise illegal activity regarding the delivery or receipt of the United States mail.

*Mala in se* Latin for "wrong in itself."

*Mala prohibita* Wrong merely because it is legally regulated.

**Malice** Intent to cause harm; usually accompanied with ill will, hate, or revenge.

**Malicious intent** Voluntary (or willful) harm without justification or excuse.

**Malicious mischief** Willful and intentional damage or destruction to the property of another.

**Mann Act** Prohibits interstate transportation of women (and girls) for the purpose of prostitution or other immoral behavior.

**Manslaughter** Unlawful killing of a human being without malice.

**Marital rape exemption** Common law rule that a man could not legally rape his wife.

**Materiality** Evidence germane to the outcome of a judicial proceeding.

**Mayhem** Dismemberment or disablement of a body part or organ.

**Megan's Law** States must provide information to the public about registered sex offenders.

*Mens rea* Latin for "guilty mind."

**Menacing** Serious bodily threats that do not rise to physical action.

**Mental defect** Mental illness that is permanent and unchanging.

**Mental disease** Mental illness that could improve or worsen over time.

**Misadventure** Intentional but misdirected conduct where chain of events were justifiably set into motion.

**Misappropriation** Unauthorized use of unconverted property.

**Misbranding** Manipulating, destroying, or removing labels from products.

**Misdemeanor** Crime for which punishment is less than 1 year in jail and/or a fine.

**Misdemeanor-manslaughter rule** Unlawful conduct that causes the death of another is manslaughter regardless of one's awareness of pending danger.

**Misprision of felony** Concealment or suppression of a felony committed by another.

**Misprision of treason** Concealment or suppression of the treasonous activities of others.

**Mistake of age** Defense that argues an accused was not aware of a victim's youthfulness.

**Mistake of fact** Defense that argues that an accused made an honest error.

**Mistake of law** Defense that relies on the genuine and honest belief that an accused acted in accordance with the law.

**Mitigating circumstance** Factor that diminishes the severity of a crime.

**Money laundering** Disguising (or washing) illegal income to create appearance of authenticity.

**Money Laundering Control Act of 1986** Augmented the Bank Secrecy Act to criminalize use of structured transactions to avoid detection.

**Murder** Unlawful killing of a human being with malice aforethought.

**Mutual affray** Exempts from criminal classification what otherwise would constitute a battery (or assault) were it not for the mutual consent of the engaging parties.

# N

**Narco terrorism** Acts of terrorism by drug cartels to eliminate competitors and intimidate government officials.

**National Stolen Property Act** Federal authority to prosecute offenses in which fraudulent or forged securities instruments worth $5,000 or more are transported, transmitted, bought, or sold within the confines of interstate commerce.

**Natural law** Rules of conduct established by the author of human nature.

**Necessity** Defense that assumes existing conditions (in nature or otherwise) caused the accused to commit a criminal act as the lesser of two evils.

**Necrophilia** Sex with a human corpse.

**Negligence** Failure to exercise a reasonable standard of care when unaware of pending danger.

**Nighttime** Period between dusk and dawn.

**No bill** Grand jury verdict stipulating insufficient evidence to go to trial.

**Noncriminal homicide** Homicide committed with legal justification or excuse.

**Nuisance** Excessive noise, offensive conditions, or interference with the lawful use of property resulting in annoying or harmful effects.

*Nulla poena sine lege* Latin for "no penalty without a law."

**Nystagmus gaze** Automatic tracking mechanism of eyes when responding to moving objects.

# O

**Obstruction of justice** Behavior that impedes or hinders the administration of justice.

**Of another** Refers to rightful possession (not ownership).

**Omnibus Crime Control and Safe Streets Act of 1968** Created and funded agencies within the Justice Department to combat organized crime activities.

**Ordinance** Regulation of behavior at the county and municipal level.

**Ordinary misdemeanor** Crime for which incarceration ranges from 3 to 6 months in jail.

**Ordinary negligence** Form of negligence that does not rise to culpable levels, meaning that the law does not regard such action as criminal.

**Organized crime** Unlawful activities of a highly organized and disciplined association engaged in supplying illegal goods and services.

**Organized Crime Control Act of 1970** Legislation that enhanced the availability of law enforcement tools to combat organized crime; known for creating RICO.

**Overt act** Action beyond mere preparation.

**Ownership** Possessing title to property.

## P

***Parens patriae*** Latin for "king is the father," and refers to the ability of the state to serve as the ultimate parent or guardian of persons with certain limitations.

**Parental Kidnapping Prevention Act of 1980** Legislation that eliminated jurisdictional disputes in child custody cases by usurping all authority over such matters.

**Pederasty** Unnatural intercourse between man and boy.

**Penetration** Insertion of the penis or other body part into a vagina or other body opening.

**Perfect self-defense** Necessary and reasonable force used in defense of self or others.

**Perjury** False statement made during a judicial proceeding while under oath or affirmation and without belief in the truth of the statement.

**Personal property** Items with value not affixed to land or real estate.

**Petit larceny** Theft of property with value less than that associated with grand larceny.

**Petty misdemeanor** Crime for which incarceration ranges from ten to thirty days in jail.

**Physical proximity test** Assesses what remains for a person to commit an intended crime.

**Pinto Papers** Late-1960s memo that indicated Ford was aware of a design flaw concerning its Pinto model but determined it was cheaper to settle lawsuits than modify the design.

**Policeman at the elbow test** Benchmark for assessing irresistible impulse through testing whether an accused would have committed the same offense in the presence of a police officer.

**Political-dissident terrorism** Acts of terrorism committed when citizens of a nation or state attack the government or society.

**Polygamy** Common law crime which sought to deter multiple (two or more) extramarital unions.

**Positive law** Man-made law to protect societal members.

**Possession** Discretion regarding the use of property within one's control or care.

**Possession of burglar tools** Crime that allows the prosecution to infer intent to commit burglary when persons possess instruments used in burglaries without a legitimate reason.

**Possession with intent to distribute** Intent to distribute illegal drugs.

**Precedent** Judicial practice where inferior (lower) courts evaluate higher court decisions when addressing legal issues.

**Precursors** Ingredients used to manufacture illegal drugs.

**Predicate crime** Previous offenses for which a defendant has been convicted.

**Predisposition** Refers to past or present behavior of an accused that indicates the person was inclined to commit certain acts.

**Premeditation** Advance planning (even for one second).

**Premenstrual syndrome** Diminished capacity defense that unsuccessfully set forth that some females are incapable of controlling their actions due to severe menstrual symptoms.

**Preponderance of the evidence** Standard used in civil court where one need only establish a greater likelihood that harm occurred.

**Principal at the fact** Accomplice present at the scene of the crime.

**Prison break** Form of common law escape during which force was used during departure.

**Privilege against self-incrimination** Fifth Amendment constitutional right prohibiting government from forcing a witness to testify against self.

**Probable cause** Good reason to believe that a particular person committed a crime (also applied to other judicial matters, such as search warrants).

**Probable desistance test** Evaluates the likelihood one will desist from committing a crime.

**Probation** Offender remains in the community while serving sentence under supervision.

**Procedural due process** Fair process afforded to accused persons before permitting the deprivation of life, liberty, or property.

**Procedural law** Branch of law that outlines the procedures to be following by those empowered with criminal justice duties.

**Property crime** Designated by the Federal Bureau of Investigation as the most serious property crimes in America: burglary, arson, larceny/theft, and motor vehicle theft.

**Proportionality of punishment** Constitutional principle that punishment must be graduated and proportioned to the criminal offense.

**Prostitution** Sexual favors for hire.

**Protect America Act of 2007** FISA amendment permitting warrantless electronic monitoring of American citizens provided no singular citizen is the focus of investigation.

**Protection of Children from Child Exploitation Act of 1977** Prohibits the use of children under the age of 16 in sexually explicit materials.

**Proximate cause** Proof that the defendant was, in fact, the one who caused the harm in question.

**Prurient interest** Shameful or morbid interest in nudity, sex, or excretion.

**Public intoxication** Prohibits being drunk in a public place.

**Punitive damage** Court damages calculated to punish people with the aim of deterring similar harmful acts in the future; goal is to teach wrongdoers a lesson that will not be forgotten.

**Purported marriage** Second or subsequent illegitimate marital union.

## Q

**Quasi-political terrorism** Acts of terrorism by individuals with no intrinsic desire for political or religious change but with a strong desire to instill fear.

**Quick fetus** Mother can detect fetal movement in the womb.

## R

**Racketeering** Individuals associated with a criminal enterprise who commit certain crimes more than once during a 10-year period.

**Rape by instrumentation** Unlawful penetration of the genitals, anus, or perineum with an object other than the penis.

**Rape shield laws** Prohibits introduction of evidence in a criminal proceeding that examines a victim's sexual history and reputation.

**Real property** Items not attached to real estate (or the ground).

**Reasonable-person standard** Objective standard that requires jurors to consider what a reasonable person would do in a like situation.

**Receiving stolen property** Acquiring property of another with knowledge of its stolen origin and with no intent of returning it to its rightful owner.

**Recklessness** Failure to exercise a reasonable standard of care when danger was foreseeable.

**Rehabilitation** Punishment or sanctions designed to correct or reform.

**Religious-extremist terrorism** Acts of terrorism based on religious moral justification.

**Renunciation** Defendant voluntarily abandons criminal endeavor.

**Republic** Form of government where elected leaders operate under the dictates of a Constitution to safeguard the best interest of the nation.

**Rescue** Common law crime where one assists inmates with an escape or prison break.

**Resisting arrest** Attempt to thwart or avoid being taken into lawful custody.

**Restitution** Service or payment from an offender to a victim as compensation for wrongdoing.

**Restorative justice** Healing process whereby victims and offenders, with the assistance of a trained mediator, identify and address consequences of crime.

**Retardation** Delayed mental development, cognitive abilities, communication skills, or limited comprehension of health and safety.

**Retribution** Referred to as "just desserts"—offenders deserve punishment for their wrongs.

**RICO Act of 1961** Racketeer Influenced and Corrupt Organizations Act; provides law enforcement and prosecution with great flexibility to investigate and convict organized crime.

**Right of locomotion** Legal maxim stipulating that citizens have the right to freely come and go.

**Riot** Unlawful gathering with intent to create a public disturbance that poses a significant risk of personal injury or property damage.

**Robbery** At common law, the felonious taking of the money or property of another from his or her person or immediate presence and through the use or threat of force or violence.

**Rout** Intermediate stage between unlawful assembly and riot.

**Rule of Consistency** Person may not be convicted of conspiracy when all co-conspirators have been acquitted.

# S

**Sabotage** Purposeful acts that seek to hinder defense capabilities of the U.S. government.

**Second-degree murder** Malicious killing of another absent premeditation and/or deliberation.

**Secondary traumatization** Occurs when interaction between victims and the criminal justice system is unproductive; examples include delays in court proceedings, repeated interviews, and stigmatization of sexual abuse victims.

**Securities and Exchange Commission** Agency vested with federal authority to monitor and regulate securities transactions in the United States.

**Securities fraud** Deceptive activities to affect the welfare of securities.

**Sedition** Communication that advocates overthrow of the federal government.

**Seduction** Adult male entices an unmarried woman of chaste character to engage in sexual intercourse through a false promise of marriage.

**Selective incorporation** Process of applying individual constitutional rights to the states.

**Sentencing disparity** Markedly different sentences for similar offenses.

**Sentencing Reform Act of 1984** Created the Federal Sentencing Commission and its Federal Sentencing Guidelines.

**Separation of powers** Distribution of authority among three branches of government: legislative, executive, and judicial.

**Serious bodily injury** High probability of death.

**Service** Paid work performed by others.

**Sexual battery** Penetration of the genital, oral, or anal cavities of another without consent.

**Shoplifting** Theft of merchandise from stores.

**Short-sale orders** Selling securities at high prices without actually owning them; illegal when security is not properly borrowed or the broker never returns the borrowed security.

**Simple assault** Causes, intends to cause, or threatens to cause less-than-serious bodily harm.

**Single legal entity theory** Two become one at time of marriage (unity in marriage).

**Smith Act of 1940** Known as the Alien Registration Act; codified sedition into U.S. law.

**Social contract theory** American citizens voluntarily waive rights, privileges, and liberties guaranteed in the United States Constitution in exchange for government protection.

**Sodomy** Oral or anal intercourses; considered an abomination against nature at common law.

**Solicitation** Commands, encourages, or requests another to commit crime.

**Specific deterrence** Goal of punishment whereby offenders will be deterred or prevented from committing crime because of the severity of punishment.

**Specific intent** Willful, intentional, and stubborn purpose.

**Speedy trial** Sixth Amendment constitutional right to a trial without unnecessary delay.

**Sports bribery** Illegal influence on sports officials and athletes by getting them to violate duties in exchange for money or other value.

**Stalking** Intentionally and repeatedly scaring another through watching and/or following.

**Stand your ground laws** Permits persons to defend themselves in their homes regardless of immediate danger.

*Stare decisis* Latin for "let the decision stand."

**State-sponsored terrorism** Acts of terrorism perpetrated by government.

**Statute of limitations** Proscribes the time frame between the commission or discovery of a crime and the validity of any subsequent arrest or indictment.

**Statutory law** Laws which are enacted by legislatures.

**Statutory rape** At common law, carnal knowledge of a chaste female under a designated age by an older male to whom she is not married.

**Stealth** Sneaking away with property without permission.

**Stop-loss orders** Used to determine thresholds for buying (at a low price) or selling (at a high price) securities; often used to illegally manipulate stock prices.

**Strict liability** Presumed guilty without regard to mental fault.

**Strong-armed robbery** Taking of money or property from the person or presence of another through use or threat of force that does not rely on a deadly weapon.

**Structural degradation** Permanent change in the composition of material.

**Structured transactions** Breaking large amounts of money into multiple smaller deposits.

**Subornation of perjury** Willful and corrupt procurement of false testimony from another.

**Substantial capacity test** Person is insane when conduct resulted from mental disease or defect, and accused lacked substantial capacity to appreciate its criminality or conform conduct to the requirements of the law.

**Substantial factor test** Assesses whether actions contributed significantly to resulting harm.

**Substantial product hazard** A failure to comply with consumer product safety rules that creates substantial risk of injury to the public.

**Substantial step test** Assesses whether a substantial step has been taken toward committing an intended crime.

**Substantive due process** Freedoms and protections inherent in the pursuit of liberty.

**Substantive law** Branch of law proscribing behavioral mandates placed on people.

**Superior right of possession** First-order right to possess among multiple possessors.

**Surety** Third party to whom property has been entrusted.

# T

**Tangible property** Items that possess concrete qualities and can be moved.

**Terrorism** Political acts of violence by subnational groups against noncombatant members of other groups.

**Threatened battery** Imminent threat to batter a person with some degree of mental fault.

**Till skimming** Taking money from a cash register, overcharging customers, or under ringing sales to compensate for cash losses.

**Tolling** A pause (or stoppage) in the time limit imposed by a statute of limitation.

**Tort** Civil cause of action wherein the plaintiff seeks monetary or injunctive relief.

**Tortfeasor** Person accused in civil court of causing private harm.

**Toxic Substances Control Act of 1976** Regulates the manufacture, distribution, and use of toxic chemicals in the United States.

**Transferred intent** General intent to harm is transferred from an intended target to an unintended target.

**Treason** Overt act of levying war, or giving aid or comfort to enemies.

**Treble damages** Restitution requiring three times the worth of property fraudulently acquired.

**Trespass** Unlawful interference with the person or property of another.

**True bill** Grand jury verdict stipulating sufficient evidence to go to trial.

**Tumultuous** Significant risk of personal injury or damage to property.

**Twinkie defense** Diminished capacity defense that unsuccessfully set forth that certain persons are rendered incapable of controlling actions because of excess consumption of junk food.

**Two-witness rule** Prosecution must present two witnesses to establish crime of perjury.

# U

**Unauthorized use** Taking and carrying away the personal property of another without permission (or consent) but with the intent only to temporarily deprive.

**Uniform Child Custody Jurisdiction Act of 1968** Custody always remains with the home custodial state where the court rendered the decision.

**Uniform Controlled Substances Act of 1970** Regulates controlled substances through creation of five classification schedules including all narcotics, marijuana, and dangerous drugs.

**Uniform Crime Reports** Annual statistical portrait of crime in America compiled by the Federal Bureau of Investigation.

**Uniform Determination of Death Act** Formal determination standard for brain death.

**Uniform Vehicle Code** Collection of traffic laws entitled Rules of the Road, compiled by the private organization National Committee on Uniform Traffic Laws and Ordinances.

**Unilateral theory** Permits conspiracy conviction of one person if he or she believes an agreement existed and intended to commit a crime.

**Unlawful assembly** Group who gathers (usually more than three) to commit an unlawful act, or a lawful act in an unlawful manner.

**Unlawful fleeing** Flight from law enforcement by vehicle.

**USA PATRIOT Act of 2001** Uniting and Strengthening America by Providing Appropriate Tools Required to Intercept and Obstruct Terrorism Act; broadened federal powers regarding surveillance.

**Usury** Issuance of loans with excessive interest rates; "loan sharking."

**Uttering** Unlawful passing of a forged document with intent to defraud.

# V

**Vagrancy** Wandering or loitering with no visible means of support.

**Vandalism** Willful or negligent damage to the property of another.

**Viable fetus** High probability that a fetus can maintain life outside the womb.

**Victim and Witness Protection Act of 1982** Protects witnesses by providing government protection through the course of judicial proceedings.

**Vigorish** Interest on usurious loans.

**Violation** State-sanctioned crime punished with fines only (not recorded as criminal).

**Violent crime** Crimes designated by the FBI as the most serious violent crime in America: murder, forcible rape, aggravated assault, and robbery.

**Violent Crime Control and Law Enforcement Act of 1994** Legislation that enhances penalties for federal crimes committed on the basis of race, color, religion, national origin, ethnicity, gender, disability, and sexual orientation.

**Void for vagueness** Law so unclear that the average person is not able to determine what conduct is legal or illegal.

**Voir dire** Questioning of prospective jurors to assess qualification to serve on a jury.

**Volition** Did the accused possess substantial mental capacity to act in accordance with the law?

**Volstead Act of 1919** Authorized federal government to regulate the manufacture, importation, exportation, and possession of alcohol in the United States.

**Voluntary intoxication** Purposeful ingestion of alcohol or drugs.

**Voluntary manslaughter** Deaths resulting from heat of passion in response to adequate provocation, or deaths resulting from imperfect self-defense.

**Voyeurism** Viewing or attempting to view others' naked bodies or sexual acts without their knowledge or consent.

## W

**Waiver** Transferring jurisdiction of a juvenile to the adult court system.

**Wergild** Required offenders at common law to pay compensation to the state and victim.

**Wharton's Rule** Conspiracy applies only to crimes that require participation from multiple persons (usually two or more).

**White collar crime** Crime committed within the scope of legitimate business.

**Wire fraud** Deceptive, fraudulent, or otherwise illegal activity committed through telephone, radio, television, cable (i.e., the Internet), or other electronic means of communication.

**Witness tampering** Unlawful attempt to influence, delay, or prevent witness testimony or production of evidence.

**Wobblers** Crime that can be a misdemeanor or felony.

## X

**XYY chromosome abnormality** Diminished capacity defense that unsuccessfully set forth that some males are unable to conform to the law because of an extra Y chromosome.

## Y

**Year-and-a-day rule** Prohibition against charging a person with criminal homicide when death does not occur within 1 year and a day from infliction of harm.

# INDEX

*Exhibits, figures, and tables are indicated by exh., f, and t following page numbers.*

## A

Abandoned property, 39–40
Abandonment. *See* Renunciation
Abatement, 128
Abortion, 6, 7 (exh.), 22, 73, 173
Accessory after the fact, 141
Accessory before the fact, 141–142
Accidents, 75, 75f
Accomplices, 141
Actual breaking, 54
Actual cause, 11, 14f
Actual entry, 55
Actual possession, 59
*Actus reus*, 11–12, 13, 14, 14f. *See also specific offenses*
Adequate provocation, 84
Administration of justice and public order, crimes against, 121–132
  corruption of judicial process by public officials, 126–127
  crimes affecting integrity of judicial process, 121–126
  overview, 121
  public order and safety, crimes against, 127–130
Administrative law as primary source of criminal law, 6, 7f
Adulteration, 185
Adultery, 109
Affinity, 101
Affirmative defenses, 146–160. *See also* Defenses; *specific defenses*
  consent of victim, 158
  duress, 158
  entrapment, 160, 161 (exh.)
  excuse defenses, 149–156
  jurisdiction defenses, 146–149
  mistake of fact or law, 156–158
  necessity, 158–160
Afghanistan, Taliban in, 173
Aforethought, 77
Age discrimination, 19
Age of consent, 96
Aggravated arson, 53–54
Aggravated assault, 14, 65–66
Aggravated circumstances, 30
Air Marshals, 175
Alcohol offenses, 115–116
  driving under influence (DUI), 115–116, 116f, 130
  public intoxication, 115, 115f
*Aleman v. Illinois* (1998), 162
*Aleman, United States ex rel. v. Illinois* (1997), 162
ALI (American Law Institute), 151, 152. *See also* Model Penal Code
Alibi, 160–161
Alien Registration Act of 1940, 177
al-Qaeda, 176

Alteration, defined, 45
American Law Institute (ALI), 151, 152. *See also* Model Penal Code
Ames, Aldrich, 178 (exh.)
Ancient origins of law, 3
Animals, harm to, 129
Anti-Car Theft Act of 1992, 63
Antiterrorism and Effective Death Penalty Act of 1996, 174
Armed robbery, 59–60
Arson, 51–54, 52f
  burning, 52
  common law, 52–53
  defined, 52
  dwelling of another, 52
  federal arson law, 54
  malicious intent, 52–53
  modern arson examined, 53–54
  as property crime, 14
Arthur Andersen (accounting firm), 182
Article III, Section 3 (U.S. Constitution), 176–177, 203
Asportation, 37
Assault, 63–68, 64f
  aggravated and simple assault distinguished, 65–66
  assault-related crimes, 66–67
  common law, 63–64, 64f
  defined, 63
  federal law, 65
  modern assault examined, 65
Assembly, freedom of, 22
*Atkins v. Virginia* (2002), 25, 26, 30
Attempt, 133–136, 133–134f
  act, 134
  defenses, 135–136
  defined, 133
  impossibility, 135
  indispensable element test, 135
  intent, 134
  preparation tests, 134, 134f
  renunciation, 135–136
Attempted battery, 63
Attendant circumstances, 13, 14f
Attorney, right to, 25, 31
Automatism, 155
Aviation and Transportation Security Act of 2001, 175
"Axis Sally," 176

## B

BAC (blood alcohol level), 116
Bail, 19, 25–26
*Bailey; United States v.* (1980), 125, 158, 159, 160
*Baker; United States v.* (1976), 156–157
*Baldwin v. New York* (1970), 124
Bank robbery, 62–63
Bank Secrecy Act of 1970, 171
*Barker v. Wingo* (1972), 23, 162–163
*Barnes v. Glen Theatre* (1991), 110
*Barron v. Baltimore* (1833), 18, 22

*Bartkus v. Illinois* (1959), 162
*Batson v. Kentucky* (1986), 24
*The Battered Woman* (Walker), 155
Battered woman syndrome, 155
Battered women, 85
Battery, 63–64, 64f
Beckwith, Byron De La, 163 (exh.)
Bestiality, 98
Beyond a reasonable doubt, 11
Bifurcated proceeding for capital punishment cases, 30
Bigamy, 108
Bilateral theory of conspiracy, 138
Bill of Rights, 17, 18, 19, 22, 25, 205–206. *See also specific constitutional amendments*
Blackmail, 63, 127
Blackouts as defense, 155
*Black's Law Dictionary*, 3, 19, 111, 133
*Blockburger v. United States* (1932), 162
Blockburger test, 162
Blood alcohol level (BAC), 116
Blood samples as evidence, 22
*Bloom v. Illinois* (1968), 124
Born alive standard, 72
*Bowers v. Hardwick* (1986), 99 (exh.)
Brady Bill, 152 (exh.)
Brady, James, 152 (exh.)
Brain death, 74
Branch Davidian compound raid, 173
Breach of the peace, 128
Breaking in. *See* Burglary
Breathalyzer, 116
Bribery, 126, 183
Buggery, 97
Burden of proof
  beyond a reasonable doubt, 11
  defined, 11
  preponderance of the evidence, 11
  strict liability, 12–13
Bureau of Alcohol, Tobacco, and Firearms, 175
Burglar tools, possession of, 58
Burglary, 54–58, 54f
  breaking and entering, 54–55
  common law, 54–56
  criminal trespass distinguished, 57–58
  defined, 54
  dwelling of another in the nighttime, 55–56
  federal law, 62–63
  intent to commit felony, 56
  modern burglary examined, 56–57
  possession of burglar tools, 58
  as property crime, 14
Burning, defined, 52
But-for test, 11, 12

## C

Camorra, 168
Canon law, 4
Capital felony, 9, 10f
Capital punishment, 26, 29–30, 94, 148–149, 174, 178 (exh.)
Capone, Al, 182

Carjacking, 63
Carnal knowledge, 92
Cartels, 170, 174
Case law
　defined, 6
　as primary source of criminal law, 6, 7f
Castellammarese War, 168
Castle doctrine, 146
Catholic Church, 3, 4
*Caveat emptor,* 36
Central Intelligence Agency (CIA), 178 (exh.)
Challenges for cause, jury trials, 24
Chancery Courts, 4
*Chaplinsky v. New Hampshire* (1942), 128
Charges, notice of, 24
Chaste character, 96
Chattels, 93
Check Kite Analysis System (CKAS), 190
Check kiting, 189–190
Checks and balances, 18
Checks, offenses involving, 189–190
*Chicago v. Morales* (1999), 22, 129
Chicago Black Sox scandal, 168
Child exploitation, 101, 102
Child molestation, 99–101, 100f
　children and vulnerable individuals, 101
　intent, 101
　touching, rubbing, or handling body part or member, 100–101
Child Online Protection Act of 1998, 112
Child pornography and exploitation, 101, 102 (exh.), 111, 160, 161 (exh.)
Child Protection Act of 1984, 161 (exh.)
Christianity, principles of, 3
Church of England, 3, 107
Churning, 187
CIA (Central Intelligence Agency), 178 (exh.)
Circumstantial evidence, 11
Civil contempt, 124
Civil forfeiture, 28
Civil law
　administrative law, 6
　burden of proof, 11
　origins of law, 3–4
　private wrongs, 6–8, 8f
Civil trespass, 128
CKAS (Check Kite Analysis System), 190
Claim of right, 38
Clean Air Act of 1970, 186
Clean Water Act of 1972, 186
"Club Fed," 181
Coast Guard, U.S., 175
*Coates v. City of Cincinnati* (1971), 22
Co-conspirators
　hearsay rule exemption, 141
　liability, 139–140
Code of Hammurabi, 3
Cognition, 151
*Coker v. Georgia* (1977), 94
Colombia, drug cartels in, 170
Colonial America and origin of law, 3, 4
*Colten v. Commonwealth* (1972), 128
Commercial bribery, 126
Commissions and omissions, 8–9
Common law
　definition of, 4
　felonies under, 9
　*mala in se,* 9, 10f
　misdemeanors under, 9
　modern law, study of, 5
　origins of, 2, 3–4
　primary source of law, 4, 7f
　*stare decisis,* 4, 6

Communications Decency Act of 1996, 111–112
Compensatory damage, 7, 8f
Compounding crime, 125
Compulsory process, 24–25
Confrontation of witnesses, right to, 24
Consanguinity, 101
Consent of the victim, 158
Conservative Vice Lords, 172 (exh.)
Consideration, 126
Conspiracy, 133f, 137–140, 138f
　agreement between parties, 137–138
　co-conspirators, 139–140
　conspirators defined, 138
　crime or unlawful objectives by unlawful means, 138–139
　defenses, 140
　defined, 137
　evidentiary considerations, 140
　federal law, 140–141, 183
　overt act, 139
　Rule of Consistency, 139
　special considerations, 139–140
　specific intent, 138
　Wharton's Rule, 139
Constitutional law. *See also specific amendments*
　Bill of Rights, 17, 18, 19, 22, 25, 205–206
　case law, 6
　criminal law under, 5, 17–18
　overview, 17
　as primary source of law, 4, 6, 7f
　principles and limitations, 6, 18–26
　punishment, 26–31
　text of Constitution, 199–204
　treason, 176–177
Constructive asportation, 37
Constructive breaking, 54–55
Constructive contempt, 124–125
Constructive entry, 55
Constructive intent, 13, 14f
Constructive possession, 59
Constructive taking, 37
Consumer product safety, 184, 185, 185f
Consumer Product Safety Commission, 184
Contempt of court, 124–125
Continuing offenses, 163
Continuing trespass, 39
Contraception, freedom of choice regarding, 22
Contractual theory, 93
Controlled substances. *See* Drug offenses
Conversion, 42
Corporate crime and liability, 182. *See also* White collar crime
*Corpus delicti,* 11, 14f
Correction officers, use of deadly force by, 148–149
Corruption of judicial process by public officials, 126–127
　bribery, 126
　ethical violations, 127
　extortion and blackmail, 126–127, 183
La Cosa Nostra, 168
Counsel, right to, 25, 31
Counterfeiting, 46, 47–48, 170–171
"Country club prisons," 181
Courtroom decorum, 124
Courts of equity, 4
*Crawford v. Washington* (2004), 24
Creation, defined, 45
Credit card theft/fraud, 190–191
Crime. *See also specific offenses*
　classification of, 9–10, 10f

　codified, 9, 9f
　commissions and omissions, 8–9
　defined, 8–9, 9f
　deviance distinguished, 10
　elements, 10–13
　felonies, misdemeanors, and violations, 2, 9, 10f
　historical development of law, 3
　*mala in se* and *mala prohibita,* 9–10, 10f
　overview, 2
　as public wrong, 8, 8f
　statistics, 14
　without legal defense, 9, 9f
*Crime in the United States* (FBI), 14
Crimes against administration of justice and public order. *See* Administration of justice and public order, crimes against
Crimes against habitation. *See* Habitation, crimes against
Crimes against moral values. *See* Moral values, crimes against
Crime statistics, 14. *See also specific offenses*
Criminal contempt, 124
Criminal enterprise, 170
Criminal forfeiture, 28
Criminal homicide, 71–89
　*corpus delicti,* 76
　defined, 71–77, 72f
　excusable homicide, 75, 75f
　federal law, 86–87
　genocide, 87
　human being requirement, 72–73
　justifiable homicide, 75, 75f
　legally dead requirement, 74, 74f
　living victim requirement, 73
　manslaughter, 82–86, 83f, 86f
　murder, 14, 71, 77–82, 79f, 83
　noncriminal homicide defined, 74–75, 75f
　overview, 71
　proximate cause, 76
　year-and-a-day rule, 76–77
Criminal law, defined, 8
Criminal Lunatics Act of 1800, 150
Criminal punishment. *See* Punishment
Criminal trespass, 57–58
Cruel and unusual punishment, 25, 26, 94
Cuba and state-sponsored terrorism, 173
Culpable
　defined, 11
　requirements for, 12, 13
Culpable negligence manslaughter, 86
Cunnilingus, 97
Curfew statutes, 22
Custodial thefts, 40–42
Custody, 36, 36f
Customs and Border Protection, Bureau of, 175
Cyberstalking, 67

## D

Dangerous proximity test, 134
Deadly force, use of, 146–149
Deadly weapon doctrine, 78
Death penalty. *See* Capital punishment
Death-qualified jury, 30
Declaration of Independence, text of, 197–198
Declaratory relief, 7, 8, 8f
Defenses, 145–166
　affirmative defenses, 146–160
　alibi, 160–161
　attempt, 135–136

consent of the victim, 158
conspiracy, 140
constitutional and statutory, 161–163
diminished capacity, 153–156
double jeopardy, 20–21, 161–162
duress, 158, 159
entrapment, 160, 161 (exh.)
excuse defenses, 146, 149–156
imperfect self-defense, 85, 129
insanity, 149–152
jurisdiction defenses, 146–149
legal and moral rationale for, 145
mistake of fact and law, 156–158
necessity, 158–160
overview, 145
party liability, 142
police officers and correction officers, use of deadly force by, 146, 148–149
self defense, defense of others, and defense of property, 146–148
solicitation, 137
speedy trial, 162–163
statute of limitations, 163, 164 (exh.)
Deliberation, 79
Delivery or sale of controlled substance, 114–115, 115f
Democracy, 2
Depraved-heart murder, 82, 83 (exh.)
Determinate sentences, 28
Deterrence, punishment as, 26–27
Deviance, distinguished from crime, 10
Diminished capacity, 153–156
    automatism, 155
    battered woman syndrome, 155
    infancy, 153
    intoxication, 153–154
    mental retardation, 154–155
    premenstrual syndrome (PMS), 155–156
    Twinkie defense, 156
    XYY chromosome abnormality, 155
Direct asportation, 37
Direct contempt, 124
Direct taking, 37
Disablement, 65–66
Dismemberment, 65–66
Disorderly conduct, 128
Disturbing the peace, 128
*Dixon; United States v.* (1993), 162
Doctrine of overbreadth, 22
Document, defined, 46
Domestic terrorism, 172
Domestic violence, 66–67, 85
Double jeopardy, 20–21, 161–162
*Downum v. United States* (1963), 162
Driving under the influence (DUI), 115–116, 116f, 130
*Drope v. Missouri* (1975), 154
Drug Abuse Prevention and Controlled Substances Act of 1970, 113
Drug cartels, 170, 174
Drug offenses, 113f
    delivery or sale of controlled substance, 114–115, 115f
    federal controlled substance law, 112–113
    manufacture, 114
    possession, 113–114, 113f
Drug possession, 113–114, 113f
Dual sovereignty, 21
Due process
    defined, 19, 22
    incorporation of, 19, 23
    jury trial, right to, 23
    parole and probation revocation, 31

peremptory challenges, use of, 24
punishment, imposition of, 27
vague laws, 18, 22, 111
witness, right to confront, 25
DUI. *See* Driving under the influence
*Duncan v. Louisiana* (1968), 23
Duress, 158, 159
*Durham v. United States* (1954), 150–151
Durham rule, 150–151
Dwelling, defined, 52

# E

Ecclesiastical courts, 4, 107, 109
Edward I (England), 3
Effective assistance of counsel, 25
Eighteenth Amendment, 115, 169
Eighth Amendment
    bail, excessive, 19, 25–26
    Bill of Rights, 18, 206
    cruel and unusual punishment, 25, 26, 94
    fines, excessive, 26
Elements of crime and liability, 10–13, 14t. *See also specific offenses*
    attendant circumstances, 13
    *corpus delicti* (phase I of *actus reus*), 11, 14f
    *mens rea*, 12–13, 14, 14f
    proximate cause (phase II of *actus reus*), 11–12, 14f
Elgin Prison Camp, 181
E-mail scams, 172
Embezzlement, 42–43, 42t
    conversion or misappropriation, 42
    defined, 42
    entrusted with property of another, 42–43
    intent to deprive possession, 36, 36f, 43
    till skimming, 190
Embracery, 123–124
Eminent domain, 23
English legal system, 3–4
Enron, 182
Entrapment, 160, 161 (exh.)
Environmental offenses, 185–187
Environmental Protection Agency (EPA), 6, 186, 192
Equal Protection Clause, 19, 95 (exh.)
Equivocality test, 134
Escape, 125–126, 159–160
Espionage, 177–178, 178 (exh.), 179 (exh.)
Ethical violations by public officials, 127
Euthanasia, 77
Evers, Medgar, 163 (exh.)
Evidence. *See also* Burden of proof
    circumstantial, 11
    conspiracy, 140
    hearsay, 140
    nontestimonial, 22
*Ewing v. California* (2003), 29
Excessive bail, 19, 25–26
Exclusionary rule, 20
Excusable homicide, 75, 75f
Excuse defenses, 146, 149–156
    diminished capacity, 153–156
    insanity, 149–152
    self-defense, 149
Executive branch, 18
Expression, freedom of, 110
Extortion, 61–63, 61f, 126–127, 183

# F

Factual impossibility, 135
False advertising, 183–184

False imprisonment, 67
False pretenses, 36, 36f, 43–44, 44f
FBI. *See* Federal Bureau of Investigation
FDA. *See* Food and Drug Administration
Federal Air Marshals, 175
Federal Anti-Riot Act of 1968, 127
Federal Bank Fraud statute, 190
Federal Bureau of Investigation (FBI), 14, 169, 176, 190 (exh.)
Federal Emergency Management Agency, 175
Federalism, 4, 18
Federalist Papers, 2
Federal law. *See also specific offenses and statutes*
    alcohol and drug offenses, 112–113
    arson, 54
    assault, 65
    blackmail, 127
    bribery, 127–128
    burglary, 62–63
    conspiracy and solicitation, 140–141
    criminal homicide, 86–87
    domestic violence, 66
    escape and rescue, 126
    fraudulent practices, 44–45
    gambling, 112
    insanity defense, 152
    jurisdiction, 18
    kidnapping, 67–68
    malicious mischief, 129
    misprision of felony, 125
    obscenity, 111
    obstruction of justice, 122
    party liability, 141
    perjury, 123
    resisting arrest, 122
    robbery, 62–63
    sex offenses, 103–104
    stalking, 67
    theft offenses, 44–45
    witnesses under, 123, 124
Federal Protection of Children Against Sexual Exploitation Act of 1977, 101, 111
Federal Sentencing Guidelines, 28–29
Federal Trade Commission (FTC), 182, 183, 184
Fellatio, 97
Felonies
    capital felony, 9
    common law, 9
    crime classification, 9, 10f
    statutes, 5 (exh.)
Felony murder, 80–82, 81f
    causal connection, 81
    defined, 80
    proximity, 81–82
    timing, 81
Female genital mutilation, 65
Feticide, 73
Field sobriety test, 116
Fifth Amendment, 20–23
    Bill of Rights, 18, 205
    double jeopardy, 20–21, 161–162
    due process, 18, 19, 22, 27, 111
    eminent domain, 23
    grand jury, 19, 20
    self-incrimination, privilege against, 21–22
Fighting, 128
Fighting words, 128
*Figueredo; United States v.* (1972), 139
Fines, 26, 27
Fingerprints, 22

First Amendment
    hate crimes, 117
    protections, 22, 110, 111–112, 127, 128
    sedition, 177
First-degree murder, 78–79
FISA (Foreign Intelligence and Surveillance Act of 1978), 174
FISA Amendments Act of 2008, 174
FISC (Foreign Intelligence and Surveillance Court), 174
*Fishman; United States v.* (1990), 153
Float, 189
Fondling, 100–101, 100*f*
Food and Drug Administration (FDA), 182, 185
Food, Drug, and Cosmetic Act of 1938, 185
Forcible rape
    carnal knowledge and gender of participants, 92
    consent, 93
    defined, 91–94, 92*f*
    force, 92–93
    marital rape, 93–94
    penalties, 94
    statistics, 91–94
    as violent crime, 14
Ford Motor Company, 186 (exh.)
Foreign Intelligence and Surveillance Act of 1978 (FISA), 174
Foreign Intelligence and Surveillance Court (FISC), 174
Forfeiture, 28
Forgery, 45–48, 45*f*
    creation and alteration, 45
    defined, 45
    documents with legal efficacy, 46, 47
    intent to defraud, 46
    modern law, 47–48
Fornication and adultery, 109
Foster, Jodie, 152 (exh.)
Fourteenth Amendment
    Bill of Rights, 18
    due process, 19, 22, 24, 27, 111
    Equal Protection Clause, 19, 95 (exh.)
    incorporation, 19, 23, 24, 26
    loitering statutes, 129
    overview, 19
Fourth Amendment
    Bill of Rights, 18, 205
    FISA, 174
    protections of, 19–20
    roving wiretaps, 174, 175 (exh.)
    sneak-and-peek warrants, 174, 175 (exh.)
    USA PATRIOT Act, 174–175
Fraud. *See also* Theft offenses and fraudulent practices
    embezzlement, 42–43
    false pretenses, 36, 36*f*, 43–44, 44*f*
    forgery and uttering, 45–48, 45*f*, 47*f*, 189
    securities fraud and insider trading, 187, 187*f*
*Freeman v. United States* (1966), 151
FTC. *See* Federal Trade Commission

## G

Gadahn, Adam, 177
*Gagnon v. Scarpelli* (1973), 31
Gambling, 112
Gaming, 112
Gangs and organized crime, 172, 172 (exh.)
GBMI (guilty but mentally ill), 152

Gender discrimination
    equal protection, 19, 95 (exh.)
    juries, 19
General deterrence, 27
General intent, 13, 14*f*
Genocide, 87
*Gideon v. Wainwright* (1963), 25
Good Samaritan laws, 86
*Grady v. Corbin* (1990), 162
Grand jury, 19, 20
Grand larceny, 38
Greek civilization and origin of law, 3
*Green v. United States* (1957), 21
*Griffin v. California* (1965), 22
*Griswold v. Connecticut* (1965), 22
Gross misdemeanor, 9, 10*f*
Gross negligence, 75, 82, 86
Guilty but mentally ill (GBMI), 152

## H

Habitation, crimes against
    arson, 51–54, 52*f*
    burglary, 54–58, 54*f*, 62–63
    defined, 51
Habitual offender statute, 29
Hanssen, Robert, 179 (exh.)
Harassment, 66
Harlem gangs, 168
Harmful products, 184–185, 185*f*, 186 (exh.)
Hate crimes, 116–117
Hate Crimes Statistics Act of 1990, 116
Hearsay, 140
*Heath v. Alabama* (1985), 21
Heat of passion, 71, 84
Hebrew civilization and origin of law, 3
Henry II (England), 3
Hinckley, John, Jr., 151–152, 152 (exh.)
Homeland Security Act of 2002, 175
Homeland Security Department, 170–171, 175
Homicide. *See* Criminal homicide
Homosexuality, 110
*Housand; United States v.* (1977), 158
*Hubert; People v.* (1897), 150
*Hurtado v. California* (1884), 20

## I

*Ibn-Tamas v. United States* (1979, 1983), 155
Identity theft, 191–192
Identity Theft and Assumption Deterrence Act of 1998, 191–192
Identity Theft Penalty Enhancement Act of 2004, 192
I-gaming, 112
*Illinois v. Vitale* (1980), 162
Immigration and Customs Enforcement, Bureau of, 175
Impartial jury, right to, 23–24
Imperfect self-defense, 85, 149
Implied consent statutes, 116
Implied malice, 78
Incapacitation, 27
Incarceration, 28–30
    alternatives to, 30–31
    determinate sentences, 28
    habitual offender statute, 29
    indeterminate sentences, 28
    mandatory sentences, 29
    sentencing guidelines, 28–29
    "three strikes" laws, 29
Incest, 13, 101–102
Inchoate offenses, 133–141

    attempt, 133–136, 133–134*f*
    common law, 133
    conspiracy, 133*f*, 137–141, 138*f*, 183
    defined, 133
    overview, 133, 133*f*
    party liability, 141–142
    solicitation, 133*f*, 136–137, 137*f*, 140–141
Incitement of riot, 127–128
Incorporation, 19, 23, 24, 26
Indecent exposure, 110, 110*f*
Indeterminate sentences, 28
Indispensable element test, 135
Inducement, 160
Infancy, 153
Information, 20, 24, 30
Injunctive relief, 8, 8*f*
Insanity, 149–152
    Durham rule, 150–151
    and excusable homicide, 75
    guilty but mentally ill (GBMI), 152
    irresistible impulse, 150, 151, 152
    post-Hinckley insanity defenses, 151–152
    substantial capacity, 150, 151, 152, 152 (exh.)
Insanity Defense Reform Act of 1984, 152, 153
Insider trading, 187, 187*f*
Instrument, defined, 46
Intangible property, 38
Intensive supervision probation (ISP), 30–31
Intent. *See* Mens rea
Intent-to-cause-serious-bodily-injury murder, 80
Intent-to-kill murder, 79–80
Intermediary, 137
Internal Revenue Service (IRS), 6, 182
International counterfeiting rings, 170–171
International terrorism, 172
Internet
    gambling, 112
    organized crime, 172
    pornography, 111–112
Interstate Wire Act of 1961, 112
Intervening cause, 12
Intoxication, 153–154
Involuntary intoxication, 154
Involuntary manslaughter, 83*f*, 85–86, 86*f*
    culpable negligence manslaughter, 86
    degree of manslaughter, 82
    unlawful act manslaughter, 85–86
Involuntary renunciation, 135–136
Iran and state-sponsored terrorism, 173
Iraq insurgency movement, 173
Irresistible impulse test, 150, 151, 152
IRS. *See* Internal Revenue Service
ISP (intensive supervision probation), 30–31

## J

*Jacobellis v. Ohio* (1964), 110
*Jacobson v. United States* (1992), 160, 161 (exh.)
*J.E.B. v. Alabama* (1994), 24
Jeffs, Warren, 108
Jewish organized crime, 168
Jihad, 173
"Johns," defined, 109
Jostling, 66
Judaism, principles of, 3
Judicial activism, 18
Judicial branch, 18
Judicial process, crimes affecting integrity of, 121–126
    contempt of court, 124–125
    embracery and witness tampering, 123–124

escape, 125–126
misprision of felony and compounding crime, 125
obstruction of justice, 121–122
perjury and subornation of perjury, 122–123
resisting arrest, 122
Juries
criminal contempt, right to trial for, 124
death-qualified jury, 30
embracery, 123–124
grand jury, 19, 20
impartial jury, right to, 23–24, 123
*voir dire,* 24, 30
Jurisdiction
civil law, 6
defined, 4
federal criminal law, 18
statutory law, 5
Jury tampering, 123–124
Justice Department, 169
Justifiable homicide, 75, 75*f*
Justification defenses, 146–149
police officers and correction officers, use of deadly force by, 146, 148–149
self-defense, defense of others, and defense of property, 146–148
Juvenile offenders, 26, 29–30, 153

## K

Kaczynski, Theodore "Ted," 173
*Kennedy v. Louisiana* (2008), 94 (exh.)
Kidnapping, 67–68
Kings courts, 4
*Kirby v. Illinois* (1972), 25
Knowing endangerment, 186
Ku Klux Klan, 163 (exh.), 174
*Kunak; United States v.* (1954), 150

## L

*Lambert v. California* (1957), 157–158
Lansky, Meyer, 168
*Lanza; United States v.* (1922), 21
*Lanzetta v. New Jersey* (1939), 22
Larceny, 37–40, 37*f*
defined, 36, 36*f*, 37
intent to permanently deprive, 38–39
personal property of another, 38
as property crime, 14
taking and carrying away, 37–38
by trick, 39
Last act test, 134
Law courts, 4
Law Reform (Year and a Day Rule) Act of 1996 (England), 77
*Lawrence v. Texas* (2003), 99 (exh.), 109–110
Lay, Kenneth, 182
Least restrictive mechanism, 2
Legal cause, 12, 14*t*
Legal efficacy, 46
Legal impossibility, 135
Legal wrongs, types, 6–8, 8*f*
private wrongs, 6–8
public wrongs, 8
Legislative branch, 18
Lesser included offenses, 14
Lewdness, 110
Loan sharking, 170, 171–172
*Lockyer v. Andrade* (2003), 29
Loitering, 22, 129
Lost, mislaid, and abandoned property, 39–40
Luciano, Charles (Lucky), 168

## M

Madison, James, 2
Mafia, 168
Mail fraud, 187–188
*Mala in se,* 9, 10*f. See also specific offenses*
*Mala prohibita,* 9–10, 10*f. See also specific offenses*
Malice, 77–78
Malicious intent, 52–53
Malicious mischief, 129
Mandatory sentences, 29
Mann Act of 1910, 109
Manslaughter, 82–86, 83*f*, 86*f*
federal law, 86–87
involuntary, 82, 83*f*, 85–86, 86*f*
as lesser included offense, 14
voluntary, 82–85
*Mapp v. Ohio* (1961), 19
Maranzano, Salvatore, 168
Marital rape exemption, 93–94
Masseria, Joe (The Boss), 168
Materiality, 123
Mayhem, 65
McVeigh, Timothy, 173
Megan's Law, 103
Menacing, 66
*Mens rea,* 12–13, 14, 14*f. See also specific offenses and defenses*
Mental defect, 151, 152
Mental disease, 151, 152
Mental retardation, 26, 154–155
Mercy killings, 77
Merger doctrine, 139
*Michael M. v. Superior Court of Sonoma County* (1981), 95 (exh.)
Milk, Harvey, 156
*Miller v. California* (1973), 111
*Miranda v. Arizona* (1966), 22
*Miranda* warnings, 22
Misadventure, 75, 75*f*
Misappropriation, 42
Misbranding, 185
Misdemeanor-manslaughter rule, 85–86
Misdemeanors, crimes classified as, 9, 10*f*
Misfortunes, 75
Mislaid property, 39–40
Misprision of felony, 9, 125
Misprision of treason, 177
Mistake of age, 156
Mistake of fact, 156
Mistake of law, 156–158
Mitigating circumstances, 30
M'Naghten Rule, 150–152
Model Penal Code
accomplices, 142
blackmail, 127
conspiracy, 138, 139, 140
depraved-heart murder, 82
development of, 44
extortion, 127
incest, 101
indecent exposure, 110
insanity defense, 150, 151, 152
obscenity, 110
perjury, 123
prostitution, 109
renunciation, 135, 137
resisting arrest, 122
solicitation, 137
Money damages, 7
Money laundering, 170, 171
Money Laundering Control Act of 1986, 171

*Moore; United States v.* (1980), 157
Moral turpitude, crime classification as, 9, 10*f*
Moral values, crimes against, 107–119
alcohol and drugs, 112–116, 113*f*, 115*f*
gambling, 112
hate crimes, 116–117
morality legislation in America, 107
overview, 107
sex offenses, 107–111
*Moran v. Ohio* (1984), 155
*Morrissey v. Brewer* (1972), 31
Moscone, George, 156
Most wanted list (FBI), 176
Motor vehicles
DUI. *See* Driving under the influence
theft, 14, 35, 38, 45, 63
traffic offenses, 130
Murder
deadly weapon doctrine, 78
defined, 77, 78–82, 79*f*
depraved-heart murder, 82, 83 (exh.)
federal law, 86–87
felony murder, 80–82, 81*f*
in general, 77–78
intent-to-cause-serious-bodily-injury murder, 80
intent-to-kill murder, 79–80
malice aforethought, 77–78
statistics, 71
transferred intent, doctrine of, 80
as violent crime, 14
Murrah Federal Building bombing, 173
Mutual affray, 63–64

## N

Narco terrorism, 174
National Stolen Property Act of 1934, 189
Native Americans and gambling, 112
Natural law, 3, 108, 112
Necessity, 158–160
Necrophilia, 98
Negligence, 13
Ness, Elliot, 182
*New Jersey v. T.L.O.* (1985), 20
*New York v. Ferber* (1982), 111
Nichols, Terry, 173
Nighttime defined, 56
No bill of indictment, 20
Noncriminal homicide, 74–75, 75*f*
Nontestimonial evidence, 22
Norman Conquest, 3
North Korea
counterfeiting by, 170
removal from state-sponsors of terrorism list, 173
Notice of charges, 24
*NOW v. Schedler* (1994), 171 (exh.)
Nude dancing, 110
Nuisance, 128
*Nulla poena sine lege,* 2
Nystagmus gaze, 116

## O

Obstruction of justice, 121–122
"Of another," defined, 38, 52
Oklahoma City bombing, 173
Old Testament, 3
Omissions and commissions, 8–9
Omnibus Crime Control and Safe Streets Act of 1968, 169
Online gambling, 112

Ordinance, 9, 11
Ordinary misdemeanor, 9, 10f
Ordinary negligence, 75, 82, 87 (exh.)
Organized crime, 167–172
    defined, 168
    emerging issues in, 172
    gangs, 172, 172 (exh.)
    history of, 168
    legal issues in, 169–170
    misconceptions about, 167
    nature of, 168–169
    offenses associated with, 170–172
    overview, 167
Organized Crime Control Act of 1970, 169
Origins of law, 2–4
    ancient, 3
    common law, 3–4
    natural law, 3, 108, 112
Others, defense of. *See* Self-defense, defense of others, and defense of property
Overbreadth, doctrine of, 22
Overt act, 139
Ownership, 36, 36f

## P

*Papachristou v. City of Jacksonville* (1972), 22, 129
*Parens patriae*, 19
Parental Kidnapping Prevention Act of 1980, 67–68
Parents, rights of, 22
*Parsons v. State* (1887), 150
Party liability, 141–142
PATRIOT Act. *See* USA PATRIOT Act of 2001
Pederasty, 97
Peeping Toms, 102–103
Penalties. *See also* Punishment; Sentencing
    accessory, 141
    armed robbery, 59
    assault, 65
    bad checks, 189
    bank robbery, 63
    blackmail, 127
    bribery, 127
    child molestation, 101
    child pornography, 111
    driving under the influence (DUI), 130
    environmental offenses, 186, 187
    escape and rescue, 126
    espionage, 177
    ethical violations, 127
    extortion, 63
    false advertising, 184
    forcible rape, 94
    harmful products, 185
    hate crimes, 117
    identity theft, 191–192
    indecent exposure, 110
    insider trading, 187
    mail fraud, 188
    malicious mischief, 129
    misprision of treason, 177
    money laundering, 171
    online gambling, 112
    pornography, 111–112
    possession of burglar tools, 58
    prostitution, 109
    RICO, 170
    sedition, 177
    sexual battery, 98
    solicitation, 141
    statutory rape, 96
    tax evasion, 183
    terrorism, 174
    traffic offenses, 130
    uttering, 48
    wire fraud, 188
    worthless checks, 189, 190
Penetration, 92
Peremptory challenges, use of, 24
Perfect self-defense, 149
Perjury and subornation of perjury, 122–123
Personal property, 38
Petit (petty) larceny, 38
Petty misdemeanor, 9, 10f
Physical proximity test, 134
Pinto Papers, 186 (exh.)
Policeman at the elbow test, 150
Police officers and correction officers, use of deadly force by, 146, 148–149
Police powers, 18
Political-dissident terrorism, 173
Polygamy, 108
*Pope v. Illinois* (1987), 111
Pornography
    child, 101, 102 (exh.), 111, 160, 161 (exh.)
    Internet, 111–112
    moral values, crimes against, 111–112
Positive law, 3
Possession
    of burglar tools, 58
    drug, 113–114
    with intent to distribute, 114
    in theft offenses, 36, 36f
Posttraumatic stress disorder (PTSD), 153, 155
Precedent, 4, 6
Precursors, 114
Predicate crimes, 29, 171 (exh.)
Predisposition, 160
Premeditation, 79
Premenstrual syndrome (PMS), 155–156
Preponderance of the evidence, 11
President's Commission on Law Enforcement and Administration of Justice report, 169
President, threats against, 63
Primary sources of criminal law, 4–6, 7f
    administrative law, 6
    case law, 6, 7f
    common law, 4
    constitutional law, 4, 6
    statutory law, 4–5, 5 (exh.)
Principal at the fact, 141
Prison breaks, 125
Private wrongs, 6–8, 8f
Privilege against self-incrimination, 21–22
Probable cause, 20
Probable desistance test, 134
Probation, 30–31
Procedural due process, 22
Procedural law, 6, 8
Prohibition (1920–1933), 115, 168, 169
Property crime, 14
Property, defense of. *See* Self-defense, defense of others, and defense of property
Proportionality of punishment, 26
Prostitution, 108–109
Protect America Act of 2007, 174
Protection of Children from Sexual Exploitation Act of 1977, 101, 111
Proximate cause, 11–12, 14t, 76
Prurient interest, 111
PTSD. *See* Posttraumatic stress disorder

Public intoxication, 115, 115f
Public officials, corruption of judicial process by, 126–127
Public order and safety, crimes against, 127–130
    disturbing the peace and disorderly conduct, 128
    fighting, 128
    nuisance, 128
    traffic offenses, 130
    trespass, 128
    unlawful assembly, rout, and riot, 127–128
    vagrancy and loitering, 129
    vandalism and malicious mischief, 129
Public trials, 23
Public wrongs, 8, 8f
Punishment. *See also* Penalties; Sentencing
    alternatives to incarceration, 30–31
    capital punishment, 26, 29–30, 94, 148–149, 174, 178 (exh.)
    cruel and unusual punishment, 25, 26, 94
    determinate sentences, 28
    as deterrence, 26–27
    fines, 27
    forfeiture, 28
    habitual offender statute and "third strike" laws, 29
    as incapacitation, 27
    incarceration, 28–30
    indeterminate sentences, 28
    mandatory sentences, 29
    as rehabilitation or reformation, 27
    restitution, 31
    restorative justice and, 27
    as retribution, 26
    sentencing disparity, 28
    sentencing guidelines, 28–29
    types of, 27–31
Punitive damage, 7, 8f
Purported marriage, 108

## Q

Quasi-political terrorism, 174
Quick fetus, 73

## R

Race
    equal protection, 19
    juries and peremptory challenge use, 19
Racketeer Influenced and Corrupt Organizations Act of 1961 (RICO), 169–170, 171 (exh.), 172
Racketeering, 170, 171 (exh.)
Raleigh, Walter, 24
Rape. *See* Forcible rape
Rape by instrumentation, 92
Rape shield laws, 94
*R.A.V. v. City of St. Paul* (1992), 117
Reagan, Ronald, 152 (exh.)
Real property, 38
Reasonable-person standard, 123
Receiving stolen property, 40–41
Recklessness, 13, 75, 82, 86
Reformation, 27
Registration of sex offenders, 103
Rehabilitation, 27
Religion and origin of law, 3, 108, 109
Religious-extremist terrorism, 173–174
*Reno v. American Civil Liberties Union* (1997), 112
Renunciation, 135–136, 137, 142

Republic, 2
Rescue, 125–126
Resisting arrest, 122
Restitution, 31
Restorative justice, 27
Retardation, 26, 154–155
Retribution, punishment as, 26
RICO. See Racketeer Influenced and Corrupt Organizations Act of 1961
Right of locomotion, 67
Right to attorney, 25, 31
Riot, 127–128
Robbery, 58–61, 59f
   common law, 58–59
   defined, 58–59
   federal law, 62–63
   felonious taking through use or threat of force, 59–60
   money or property of another person or presence, 60–61
   as violent crime, 14
*Roe v. Wade* (1972), 6, 7 (exh.), 22, 73, 173
Roman civilization and origin of law, 3
*Roper v. Simmons* (2005), 26, 29–30
Rosenberg, Julius and Ethel, 178 (exh.)
*Rosenberg v. United States* (1953), 178 (exh.)
*Roth v. United States* (1957), 110–111
Rothstein, Arnold "The Brain," 168
Rout, 127
Roving wiretaps, 174, 175 (exh.)
Ruby Ridge incident, 173
Rule of Consistency, 139
"Rules of the Road," 130

## S

Sabotage, 178
Safety, crimes against, 127–130
Scheduled drugs, 113
Schultz, Dutch, 168
Scuggs, Dickie, 125
Scuggs, Zach, 125
Search and seizure, 19–20
SEC. See Securities and Exchange Commission
Secondary traumatization, 91
Second-degree murder, 79
Secret Service, U.S., 170–171, 175
Securities and Exchange Commission (SEC), 182, 187
Securities fraud and insider trading, 187, 187f
Sedition, 177
Seduction, 102, 103
Selective incorporation, 19
Self-defense, defense of others, and defense of property, 85, 129, 146–148
Self-incrimination, privilege against, 21–22
Sentencing. See also Penalties; Punishment
   alternatives to incarceration, 30–31
   capital punishment, 26, 29–30, 148–149, 174, 178 (exh.)
   determinate sentences, 28
   disparity, 28
   guidelines, 28–29
   habitual offender statute, 29
   indeterminate sentences, 28
   mandatory sentences, 29
   "three strikes" laws, 29
Sentencing Reform Act of 1984, 28–29
Separation of powers, 18
Serious bodily injury, 65
Service, theft of, 41–42
Sex offenders, registration of, 103

Sex offenses, 91–105
   bigamy and polygamy, 108
   child molestation, 99–101, 100f
   child pornography and exploitation, 101, 102 (exh.), 111, 160, 161 (exh.)
   federal law, 103–104
   forcible rape, 91–94, 92f
   fornication and adultery, 109
   incest, 101–102
   indecent exposure, 110, 110f
   Megan's law, 103
   moral values, crimes against, 107–111
   obscenity, 110–111
   overview, 91
   pornography, 111–112
   prostitution, 108–109
   registration of offenders, 103
   seduction, 102, 103 (exh.)
   sexual battery, 97–98, 97f
   sodomy, 98–99, 99f, 109–110
   statutory rape, 13, 93, 94–96, 95t, 97, 156
   victim, effect on, 91
   voyeurism, 102–103
Sexual assault, 91. See also Forcible rape
Sexual battery, 97–98, 97f
   of another, 97–98
   consent, 98
   penalties, 98
   penetration, 97
*Shannon v. United States* (1935), 158
Shoplifting, 40
Short-sale orders, 187
Siegel, Benjamin (Bugsy), 168
Simple assault, 65–66
Single legal entity theory, 93
Sixth Amendment, 23–25
   Bill of Rights, 18, 205
   compel witnesses, right to, 24–25
   confrontation of witnesses, right to, 24
   counsel, right to, 25, 31
   impartial jury, right to, 23–24, 123
   notice of charges, 24
   speedy and public trial, right to, 23, 162–163, 163 (exh.)
Skimming, till, 190
Sleep deprivation as defense, 155
Sleepwalking as defense, 155
Smith Act of 1940, 177
Sneak-and-peek search warrants, 174, 175 (exh.)
Social construction of law, 2
Social contract theory, 2, 142
Sodomy, 98–99, 99f, 109–110
*Solesbee v. Balkcom* (1950), 22
Solicitation, 133f, 136–137, 137f
   defenses, 137
   defined, 136
   federal law, 140–141
   intermediaries, 137
   of prostitution, 109
Somnambulism, 155
*Sorrells v. United States* (1932), 160
Soviet Union and espionage, 178 (exh.), 179 (exh.)
Specific deterrence, 27
Specific intent, 13, 14f
Speck, Richard, 155
Speech, freedom of, 22, 177
Speedy trial, right to, 23, 162–163, 163 (exh.)
Sports
   bribery, 126
   event injuries, consent of victim as defense, 158
   wagering, 112

*Stack v. Boyle* (1951), 25
Stalking, 66, 67
Stand your ground laws, 146, 148
*Stare decisis*, 4, 6
State lotteries, 112
States
   authority to enact laws by, 18
   dual sovereignty, 21
   equal protection, 19
   incorporation, 19
   terrorism laws, 176
State-sponsored terrorism, 173
Statute of limitations, 163, 164 (exh.)
Statute of Westminster, 4
Statutory law, 4–5, 5 (exh.)
Statutory rape, 94–96, 95t
   age, 96, 156, 158
   carnal knowledge and gender of participants, 95
   chaste character, 96
   consent, 93, 96
   example, 97 (exh.)
   intent, 96
   penalties, 96
   strict liability, 13
Stealth, 36
Stewart, Potter, 110
Stop and frisk, 20
Stop-loss orders, 187
*Strickland v. Washington* (1984), 25
Strict liability, 12–13
Strong-armed robbery, 60
Structural degradation, 52
Structured transactions, 171
Subornation of perjury, 123
Subpoena of witnesses, 24
Substantial capacity test, 150, 151, 152, 152 (exh.)
Substantial factor test, 11
Substantial product hazard, 185, 185f
Substantial step test, 134
Substantive criminal law, 1–16
   classification of crime, 9–10, 10f
   crime and deviance distinguished, 10
   crime defined, 8–9, 9f
   crime statistics, 14
   elements of crime and liability, 10–13, 14f
   legal wrongs, types of, 6–8, 8f
   origins of law, 2–4
   overview, 1–2
   primary sources of criminal law, 4–6, 5 (exh.), 7f
   republic and democratic form of government, 2
   social construction of law, 2
Substantive due process, 22
Substantive law, defined, 8
Sudan and state-sponsored terrorism, 173
Sumeria and origin of law, 3
Superior right of possession, 38
Surety, 42
Surveillance, 174, 175 (exh.)
Suspect classifications, 19
Syria and state-sponsored terrorism, 173

## T

Taking and carrying away as elements of larceny, 37–38, 37f
Taliban, 173
Tammany Hall, 168
Tangible property, 38
Tax evasion, 182–183

Technology, crimes using, 172, 181. *See also* White collar crime
Ten Commandments, 3
*Tennessee v. Garner* (1985), 148, 149 (exh.)
Terrorism, 172–178
   defined, 172–173
   identity theft, 192
   legal issues in, 174–176
   overview, 167
   related crimes, 176–178
   state laws regarding, 176, 176 (exh.)
   types of, 173–174
*Terry v. Ohio* (1968), 20
Theft offenses and fraudulent practices, 35–49
   custodial thefts, other, 40–42
   custody, possession, and ownership, differentiating, 36, 36f
   embezzlement, 42–43
   false pretenses, 36, 36f, 43–44, 44f
   federal theft law, 44–45
   forgery and uttering, 45–48, 45f, 47f, 189
   incidents of theft, 35–36
   larceny at common law, 36, 36–37f, 37–40
   lost, mislaid, and abandoned property, 39–40
   modern consolidation of theft statutes, 44
   motor vehicles, 14, 35, 38, 45, 63
   theft in general, 36
Theft of service, 41–42
Third-party exclusion rule, 81–82
Threatened battery, 63
Threats against the President, 63
"Three strikes" laws, 29
Till skimming, 190–191
"Tokyo Rose," 176
Tolling, 163
Tools, burglar's, 58
Tortfeasors, 6, 7
Torts, 6, 128
Toxic Substances Control Act of 1976, 187
Traffic offenses, 130
Transferred intent, 13, 14f, 71, 80
Transportation Security Administration (TSA), 175
Treason, 176–177
Treble damages, 43
Trespass
   burglary and criminal trespass distinguished, 57–58
   civil, 128
   continuing trespass, 39
   ownership, custody, and possession, 36, 36f
   public order and safety, crime against, 128
*Trop v. Dulles* (1958), 26
True bill of indictment, 20
TSA (Transportation Security Administration), 175
Tumultuous conduct, 127

Tweed, William March (Boss), 168
Twenty-First Amendment, 115, 169
Twinkie defense, 156
Two-witness rule, 122

## U

UCR (Uniform Crime Reports), 14
Unabomber, 173
Unauthorized use, 41
Unborn child, death of, 72–73
Uniform Child Custody Jurisdiction Act of 1968, 68
Uniform Controlled Substances Act of 1970, 113
Uniform Crime Reports (UCR), 14
Uniform Determination of Death Act of 1980, 74, 74f
Uniform Vehicle Code, 130
Unilateral theory of conspiracy, 138, 139
United States government, form of, 2
Unity in marriage theory, 93
Unlawful act manslaughter, 85–86
Unlawful assembly, 127
Unlawful fleeing, 122
"Untouchables," 182
USA PATRIOT Act of 2001, 174–175, 175 (exh.)
   Additional Reauthorizing Amendments Act of 2006, 175
Usury, 170, 171–172
Uttering, 45, 46–48, 47f, 189

## V

Vagrancy, curfew, and loitering statutes, 22, 129
Vagueness, laws void for, 18, 22, 111
Vandalism, 129
Veterans, diminished capacity defense for, 153
Viable fetus, 73
Vice Lords, 172 (exh.)
Victim and Witness Protection Act of 1982, 124
Vigorish (vig), 172
Violations, crime classification as, 9, 10f
Violent crime, 14
Violent Crime Control and Law Enforcement Act of 1994, 63, 117
Void for vagueness, 18, 22, 111
*Voir dire*, 24, 30
Volition, 151, 152
Volstead Act of 1919, 115, 169
Voluntary intoxication, 153–154
Voluntary manslaughter, 82–85, 83f
   adequate provocation, 84
   causal connection, 84–85
   heat of passion, 84

   imperfect self-defense, 85
   no cooling period, 84
Voyeurism, 102–103

## W

Waco, Texas, Branch Davidian compound raid, 173
Waiver, 153
Walker, L.E., 155
Warrants
   FISA, 174
   Fourth Amendment, 19–20
   sneak-and-peek, 174, 175 (exh.)
*Washington v. Texas* (1967), 24–25
*Waucaush v. United States* (2004), 171 (exh.)
Web-based organized crime, 172
*Weeks v. United States* (1914), 20
*Weems v. United States* (1910), 25, 26
Wergild, 27
Wharton's Rule, 139
White collar crime, 181–194
   bad checks, 189–190
   corporate crime and liability, 182
   credit card theft/fraud, 190–191
   defined, 181
   environmental offenses, 185–187
   false advertising, 183–184
   and Food and Drug Administration (FDA), 185
   harmful products, 184–185, 185f, 186 (exh.)
   identity theft, 191–192
   mail fraud, 187–188
   overview, 181–182
   securities fraud and insider trading, 187, 187f
   tax evasion, 182–183
   wire fraud, 172, 188
White, Dan, 156
William the Conqueror, 3
Wire fraud, 172, 188
Wiretaps, 175 (exh.)
*Wisconsin v. Mitchell* (1993), 117
Witnesses
   compel, right to, 24
   confrontation, right to, 24
   tampering, 124
Wobblers, 9
Worthless checks, 189, 190
Writing samples as evidence, 22

## X

XYY chromosome abnormality, 155

## Y

*Yates v. United States* (1957), 177
Year-and-a-day rule, 76–77